Castings Practice

The 10 Rules of Castings

Dedication

To Merton C. Flemings of MIT for inspirational teaching and research

Castings Practice

The 10 Rules of Castings

John Campbell

BUTTERWORTH
HEINEMANN

Amsterdam • Boston • Heidelberg • London • New York • Oxford
Paris • San Diego • San Francisco • Singapore • Sydney • Tokyo

Elsevier Butterworth-Heinemann
Linacre House, Jordan Hill, Oxford OX2 8DP
200 Wheeler Road, Burlington, MA 01803

First published 2004

British Library Cataloguing in Publication Data
A catalogue record for this book is available from the British Library

Library of Congress Cataloguing in Publication Data
A catalogue record for this book is available from the Library of Congress

ISBN 07506 4791 4

For information on all Elsevier Butterworth-Heinemann publications visit our website at http://books.elsevier.com

Typeset by Newgen Imaging Systems (P) Ltd., Chennai, India
Printed and bound in Great Britain by Biddles Ltd., King's Lynn.

Contents

Preface

Castings can be difficult to get right. Creating things never is easy. But sense the excitement of this new arrival:

The first moments of creation of the new casting are an explosion of interacting events; the release of quantities of thermal and chemical energy trigger a sequence of cataclysms.

The liquid metal attacks and is attacked by its environment, exchanging alloys, impurities, and gas. The surging and tumbling flow of the melt through the running system can introduce clouds of bubbles and Sargasso seas of oxide film. The mould shocks with the vicious blast of heat, buckling and distending, fizzing with the volcanic release of vapours that flood through the liquid metal by diffusion, or reach pressures to burst the liquid surface as bubbles.

During freezing, liquid surges through the dendrite forest to feed the volume contraction on solidification, washing off branches, cutting flow paths, and polluting regions with excess solute, forming segregates. In those regions cut off from the flow, continuing contraction causes the pressure in the residual liquid to fall, possibly becoming negative (as a tensile stress in the liquid) and sucking in the solid surface of the casting. This will continue until the casting is solid, or unless the increasing stress is suddenly dispelled by an explosive expansion of a gas or vapour giving birth to a shrinkage cavity.

The surface sinks are halted, but the internal defects now start.

The subsequent cooling to room temperature is no less dramatic. The solidified casting strives to contract while being resisted by the mould. The mould suffers, and may crush and crack. The casting also suffers, being stretched as on a rack. Silent, creeping strain and stress change and distort the casting, and may intensify to the point of catastrophic failure, tearing it apart, or causing insidious thin cracks. Most treacherous of all, the strain may *not quite* crack the casting, leaving it apparently perfect, but loaded to the brink of failure by internal residual stress.

These events are rapidly changing dynamic interactions. It is this rapidity, this dynamism, that characterizes the first seconds and minutes of the casting's life. An understanding of them is crucial to success.

This new work is an attempt to provide a framework of guidelines together with the background knowledge to ensure understanding; to avoid the all too frequent disasters; to cultivate the targeting of success; to encourage a professional approach to the design and manufacture of castings.

The reader who learns to guide the production methods through this minefield will find the rare reward of a truly creative profession. The student who has designed the casting method, and who is present when the mould is opened for the first time will experience the excitement and anxiety, and find himself asking the question asked by all foundry workers on such occasions: 'Is it all there?' The casting design rules in this text are intended to provide, so far as present knowledge will allow, enough predictive capability to know that the casting will be not only all there, but all right!

The clean lines of the finished engineering casting, sound, accurate, and strong, are a pleasure to behold. The knowledge that the casting contains neither defects nor residual stress is an additional powerful reassurance. It represents a miraculous transformation from the original two-dimensional form on paper or the screen to a three-dimensional shape, from a mobile liquid to a permanently shaped, strong solid. It is an achievement worthy of pride.

The reader will need some background knowledge. The book is intended for final year students in metallurgy or engineering, for those researching in castings, and for casting engineers and all associated with foundries that have to make a living creating castings.

Good luck!

This new book is the second of three books dealing with castings. The three books are (i) *Principles* (the new metallurgy of cast metals; the metallurgist's book) (ii) *Practice* (the practical founder's book) and (iii) *Processes* (an appraisal of the various methods of making castings; perhaps a casting buyer's book). The three are intended as a sequence, dealing with the theory and practice of the casting of metals. At the rate at which new understanding is emerging, an additional text may also be required; *(iv) Properties* (a book for everyone).

The second in the series is devoted to the Ten Rules. These are my own checklist to ensure that no key aspect of the design of the manufacturing route for the casting is forgotten.

The Ten Rules listed here are proposed as necessary, but not, of course, sufficient, for the manufacture of reliable castings. It is proposed that they are used in addition to existing necessary technical specifications such as alloy type, strength, and traceability via international standard quality systems, and other well-known and well-understood foundry controls such as casting temperature etc.

Although not yet tested on all cast materials, there are fundamental reasons for believing that the Rules have general validity. They have been applied to many different alloy systems including aluminium, zinc, magnesium, cast irons, steels, air- and vacuum-cast nickel and cobalt, and even those based on the highly reactive metals titanium and zirconium. Nevertheless, of course, although all materials will probably benefit from the application of the Rules, some will benefit almost out of recognition, whereas others will be less affected.

The Ten Rules are first listed in summary form. They are then addressed in more detail in the following ten chapters with one chapter per Rule.

The Rules originated when emerging from a foundry on a memorable sunny day. The author was discussing with indefatigable Boeing enthusiasts for castings, Fred Feiertag and Dale McLellan, that the casting industry had specifications for alloys, casting properties, and casting quality checking systems, but what did not exist but was most needed was a *process specification*. Dale threw out a challenge: 'Write one!'. The Rules and this book are the outcome. It was not perhaps the outcome that either Dale or I originally imagined. A Process Specification has proved elusive, proving so difficult that I have concluded that it will need a more accomplished author.

The Rules as they stand therefore constitute a first draft of a Process Specification; more like a checklist of casting guidelines. A buyer of castings would demand that the list were fulfilled if he wished to be assured that he was buying the best possible casting quality. If he were to specify the adherence to these Rules by the casting producer, he would ensure that the quality and reliability of the castings was higher than could be achieved by any amount of expensive checking of the quality of the finished product.

Conversely, of course, the Rules are intended to assist the casting manufacturer. It will speed up the process of producing the casting right first time, and should contribute in a major way to the reduction of scrap when the casting goes into production. In this way the caster will be able to raise standards, without any significant increase in costs. Quality will be raised to the point at which castings of quality equal to that of forgings can be offered with confidence. Only in this way will castings be accepted by the engineering profession as reliable, engineered products, and assure the future prosperity of both the casting industry and its customers.

It is recognized that many users of this book will be students of casting technology. For completeness therefore, the strict description of the Rules as intended as the caster's checklist has been relaxed a little. A small addition has been made to paragraph 10, extending the section describing the requirement for location points. This extension includes related aspects not included elsewhere, such as the accuracy of the whole mould assembly, and the many-sided problems of mould design.

A further feature of the work that emerged as the book was being written was the dominance of Chapter 2, the design of the filling systems of castings. It posed the obvious question 'why not devote the book completely to filling systems?'. I decided against this option on the grounds that both caster and customer require products that are good in every respect. The failure of any one aspect may endanger the casting. Therefore, despite the enormous disparity in length of chapters, none could be eliminated; they were all needed.

Finally, it is worth making some general points about the whole philosophy of making castings.

For a successful casting operation, one of the revered commercial goals is the attainment of product sales being at least equal to manufacturing costs. There are numerous other requirements for the successful business, like management, plant and equipment, maintenance, accounting, marketing, negotiating etc. All have to be adequate, otherwise the business can suffer, and even fail.

This text deals only with the technical issues of the quest for good castings. Without good castings it is not easy to see what future a casting operation can have. The production of good castings can be highly economical and rewarding. The production of bad castings is usually expensive and damaging.

The 'good casting' in this text is defined as one that meets or exceeds the customer's specification.

It is also worth noting at this early stage, that we hope that meeting the customer's specification will be equivalent to meeting or exceeding service requirements. However, occasionally it is necessary to live with the irony that the aims of the customer and the requirements for service are sometimes not in the harmony one would like to see.

These problems illustrate that there are easier ways of earning a living than in the casting industry. But few are as exciting.

J.C.
West Malvern
3 September 2003

The 10 Rules: Summary

1. Start with a good quality melt

Immediately prior to casting, the melt shall be prepared, checked, and treated, if necessary, to bring it into conformance with an acceptable minimum standard. However, preferably, prepare and use only near-defect-free melt.

2. Avoid turbulent entrainment of the surface film on the liquid

This is the requirement that the liquid metal front (the meniscus) should not go too fast. Maximum meniscus velocity is approximately $0.5\,ms^{-1}$ for most liquid metals. This maximum velocity may be raised in constrained running systems or thin section castings. This requirement also implies that the liquid metal must not be allowed to fall more than the critical height corresponding to the height of a sessile drop of the liquid metal.

3. Avoid laminar entrainment of the surface film on the liquid

This is the requirement that no part of the liquid metal front should come to a stop prior to the complete filling of the mould cavity. The advancing liquid metal meniscus must be kept 'alive' (i.e. moving) and therefore free from thickened surface film that may be incorporated into the casting. This is achieved by the liquid front being designed to expand continuously. In practice this means progress only uphill in a continuous uninterrupted upward advance; i.e. (in the case of gravity poured casting processes, from the base of the sprue onwards). This implies

- Only bottom gating is permissible.
- No falling or sliding downhill of liquid metal is allowed.
- No horizontal flow of significant extent.
- No stopping of the advance of the front due to arrest of pouring or waterfall effects etc.

4. Avoid bubble entrainment

No bubbles of air entrained by the filling system should pass through the liquid metal in the mould cavity. This may be achieved by:

- Properly designed offset step pouring basin; fast back-fill of properly designed sprue; preferred use of stopper; avoidance of the use of wells or other volume-increasing features of filling systems; small volume runner and/or use of ceramic filter close to sprue/runner junction; possible use of bubble traps.
- No interruptions to pouring.

5. Avoid core blows

- No bubbles from the outgassing of cores or moulds should pass through the liquid metal in the mould cavity. Cores to be demonstrated to be of sufficiently low gas content and/or adequately vented to prevent bubbles from core blows.
- No use of clay-based core or mould repair paste unless demonstrated to be fully dried out.

6. Avoid shrinkage

- No feeding uphill in larger section thickness castings because of (i) unreliable pressure gradient and (ii) complications introduced by convection.
- Demonstrate good feeding design by following all Feeding Rules, by an approved computer solidification model, and by test castings.
- Control (i) the level of flash at mould and core joints; (ii) mould coat thickness (if any); and (iii) temperatures of metal and mould.

7. Avoid convection

Assess the freezing time in relation to the time for convection to cause damage. Thin and thick section castings automatically avoid convection problems. For intermediate sections either (i) reduce the problem by avoiding convective loops in the geometry of the casting and rigging, (ii) avoid feeding uphill, or (iii) eliminate convection by roll-over after filling.

8. Reduce segregation

Predict segregation to be within limits of the specification, or agree out-of-specification compositional regions with customer. Avoid channel segregation formation if possible.

9. Reduce residual stress

No quenching into water (cold or hot) following solution treatment of light alloys. (Polymer quenchant or forced air quench may be acceptable if casting stress is shown to be negligible.)

10. Provide location points

All castings to be provided with agreed location points for pickup for dimensional checking and machining.

Rule 1

Achieve a good quality melt

1.1 Background

It is a requirement that either the process for the production and treatment of the melt shall have been shown to produce good quality liquid, or the melt should be demonstrated to be of good quality, or, preferably, both. A good quality liquid is one that is defined as

(i) Substantially free from suspensions of non-metallic inclusions in general, and bifilms in particular.
(ii) Relative freedom from bifilm-opening agents. These include gas in solution and certain alloy impurities (such as Fe in Al alloys) in solution.

It should be noted that such melts are not to be assumed, and, without proper treatment, are probably rare. (Additional requirements, not part of this specification, may also be placed on the melt. For instance, low values of particular solute impurities that have no effect on bifilms.)

Unfortunately, many melts start life with poor, sometimes grossly poor, quality in terms of content of suspended bifilms. Figure 1.1 gives several examples of different poor qualities of liquid aluminium alloy. The figures show results from reduced pressure test (RPT) samples observed by X-ray radiography. Since the samples are solidified under only one tenth of an atmosphere (76 mm compared to 760 mm of full atmospheric pressure) any gas-containing defects, such as bubbles, or bifilms with air occluded in the centres of their sandwich structures, will be expanded by ten times. Thus rather small defects can become visible for the first time.

We can assume [following the conclusions of *Castings* (2003)] that bifilms always initiate pores, and that the formation of rounded pores simply occurs as a result of the bifilm being opened by excess precipitation of gas, finally achieving a diameter greater than its original length. Thus the RPT is an admirably simple device for assessing (i) the number of bifilms, but (ii) gas content is assessed by the degree of opening of the bifilms from thin crack-like forms to fairly spherical pores.

If the melt contained no gas-containing defects the radiographs would be clear.

However, as we can see immediately, and without any benefit of complex or expensive equipment, the melts recorded in Figure 1.1 are far from this desirable condition. Figure (a) shows a melt with small rounded pores indicating that the bifilms that initiated these defects were particularly small, of the order of 0.1 mm or less. The density of these defects, however, was high, between 10 and 100 defects per cm^3. Sample (b) has a similar defect distribution, but with slightly higher hydrogen content. Sample (c) illustrates a melt that displayed a deep shrinkage pipe, normally interpreted to mean good quality, but showing that it contained a scattering of larger pores, probably as a result of fewer bifilms, so that the available gas was concentrated on the fewer available sites. Melt (d) has considerably larger bifilms, of size in the region of 5 mm in length, and in a concentration of approximately 1 per cm^3. Samples (e) and (f) show similar samples but with increasing gas contents that have inflated these larger bifilms to reasonably equiaxed pores.

Naturally, it would be of little use for the casting engineer to go to great lengths to adopt the best designs of filling and feeding systems if the original melt was so poor that a good casting could not be made from it.

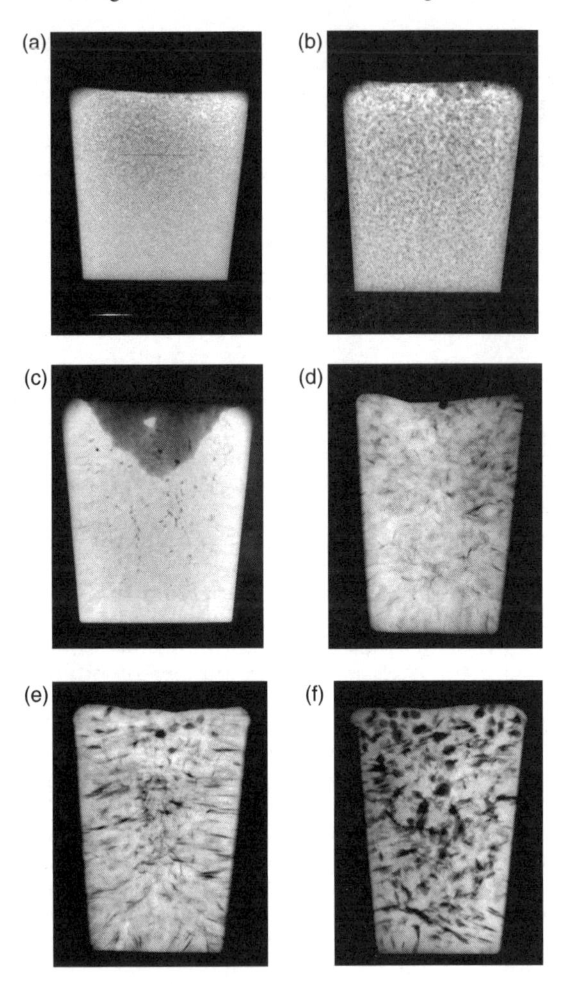

Figure 1.1 *Radiographs of RPT samples of Al–7Si – 0.4Mg alloy illustrating different bifilm populations (courtesy S. Fox)*

Thus this section deals with some of the aspects of obtaining a good quality melt. It should be noted that many of these aspects have been already been touched upon in *Castings* (2003).

In some circumstances it may not be necessary to reduce both bifilms and bifilm-opening agents. An interesting possibility for future specifications for aluminium alloy castings (where residual gas in supersaturated solution does not appear to be harmful) is that a double requirement may be made for the content of dissolved gas in the melt to be high, but the percentage of gas porosity to be low. The meeting of this double requirement will ensure to the customer that bifilms are not present. Thus these damaging but undetectable defects will, if present, be effectively labelled and made visible on X-ray radiographs and polished sections by the precipitation of dissolved gas. It is appreciated that such a stringent specification might be viewed with dismay by present suppliers. However, at the present time we have mainly only rather poor technology, making such quality levels out of reach.

(For steels, the content of hydrogen is a more serious matter, especially if the section thickness of the casting is large. In some steel castings of section thickness above about 100 mm or so, the hydrogen cannot escape by diffusion during the time available for cooling or during the time of any subsequent heat treatments. Thus the high hydrogen content retained in these heavy sections can lead to hydrogen embrittlement, and catastrophic failure of the section by cracking.)

The possible future production of Al alloys for aerospace, with high hydrogen content but low porosity, is a fascinating challenge. As our technology improves such castings may be found not only to be manufacturable, but offer guaranteed reliability of fatigue life, and therefore command a premium price.

The prospect of producing ultra-clean Al alloys that can be demonstrated in this way to be actually extremely clean, raises the issue of contamination of the liquid alloy from the normal metallurgical additions such as the various master alloys, and grain refiners, modifiers, etc. It may be that for clean material, normal metallurgical additions to achieve refinement of various kinds will be found unnecessary, and possibly even counter-productive.

For ductile iron production the massive amounts of turbulence that accompany the addition of magnesium in some form, such as magnesium ferro-silicon, are almost certainly highly damaging to the liquid metal. It is expected that immediately after such nodularization treatment that the melt will be massively dirty. It will be useful therefore to ensure that the melt can dwell for sufficient time for the entrained magnesium-oxide-rich films to float out. The situation is analogous to the treatment of cast iron with CaSi to effect inoculation (i.e. to achieve a uniform distribution of graphite of desirable form). In this case the volume of calcium-oxide-rich films is well known, so that the CaSi treatment is known as a 'dirty' treatment compared to FeSi inoculation. The author is unsure about in-mould treatments therefore with such oxidizable elements as Ca and Mg. Do they give results as good as external treatments? If so, how is this possible? Some work to clarify this situation would be valuable.

For nickel-based superalloys melted and cast in vacuum, it is with regret that the material is, despite its apparently clean melting environment, found to be sometimes as crammed with oxides (and/or nitrides) as an aluminium alloy (Rashid and Campbell 2004). This is because the main alloying element in such alloys is aluminium, and the high temperature favours rapid formation and thickening of the surface film on the liquid. This occurs even in vacuum, because the vacuum is only, of course, dilute air. Although thermodynamics indicates that aluminium oxide (and perhaps also the nitride) is unstable in vacuum, there is no doubt that a film forms rapidly, and can become entrained, thus damaging the liquid and any subsequent casting. The melting and pouring processes of these alloys also leaves much to be desired. The alloy is melted by induction, and is poured under gravity via a series of sloping launders, falling several times, and finally falling one or two metres or more into steel tubes that act as moulds. This awfully turbulent primary production process for the alloy impairs all the downstream products.

A recent move towards the production of Ni-base alloy bar by horizontal continuous casting is to be welcomed as the first step towards a more appropriate production technique for these key ingredients of our modern aircraft turbines. Production of improved material is currently limited, but should be the subject of demand from customers. Poor casting technology has been the accepted norm within the aircraft industry for too long. (However, as we all know, the aircraft industry is not alone.)

1.2 Melting

The quality of melts can be significantly affected by the type of melting furnace.

For instance the melting of the charge in a single pot is usual. Furnaces of this type automatically ensure that all the oxides on the surface of the charge material will be incorporated into the melt. Thus crucible furnaces, whether bale-out or tilting, or whether heated by gas or electrical resistance, are all of this type. Also included are induction furnaces and reverbatory (meaning reflected heat from a roof) melting units.

The remelting of aluminium alloy sand castings probably represents one of the worst cases, since the oxide skin on a sand casting is particularly thick, having cooled from high temperature in a reactive environment. When the skin is submerged in the melt it can float about, substantially complete. When remelting scrapped cylinder heads by adding them in to one end of a combined melting and holding furnace, the author has fished out the skin of a complete cylinder head from the opposite end.

The remelting of aluminium alloy gravity die (permanent moulded) castings have an oxide skin that is much thinner, and seems to give less problems. Whether this is a real or imagined advantage in view of the damage that can be caused by any entrained oxide skin, irrespective of its thickness, is not clear at this time.

Induction furnaces enjoy the great advantage of extremely rapid melting. However, they have long been regarded with some reserve by aluminium melters because of the electromagnetic stirring, with the suspicion that oxides may therefore be entrained. With normal inductive coil geometry there is a high-pressure region near the centre of the wall that drives a double torroid (a torroid is a ring shape like a doughnut) in directions away from this point. However, there is no evidence known to the author that the stirring is sufficiently rapid to entrain oxide, although such a problem cannot be ruled out. What is certain is that any oxide will have no opportunity to settle out, but this is also true of most of the above crucible furnaces because of the presence of natural convection. Because of the heat input via the walls of the crucible, and heat loss from the top surface, the convective stirring will be expected to take the form of a simple torroid, the flow direction being upwards at the wall, and downwards in the centre. The only significant difference, if any, between these two stirring modes is the rate of stirring. It is possible that the higher energy in the induction furnace may shred films, whereas the natural convective regimes in other furnaces would be expected to conserve the original film size distribution.

The dry hearth type of furnace is quite different. The charge to be melted is heaped onto a dry, sloping refractory floor, called a hearth. As the charge melts, the liquid alloy flows out of its oxide skin and down the sloping hearth into the main melt. The oxide skins present on the surface of the charge materials remain behind, accumulating on the hearth. The pile of dross is raked off the hearth at intervals via a side door. Such melting units are useful in aluminium and copper alloy production.

1.3 Holding

Holding furnaces can also have a significant effect on melt quality.

Holding furnaces were originally selected for their utility in smoothing the supply of molten metal between batch melters and a fluctuating demand from the casting requirements. An additional advantage was the smoothing of temperature and chemical analysis that was unavoidably variable from batch to batch.

The Cosworth Process was perhaps the first to acknowledge that for liquid aluminium alloys the oxide inclusions in a holding furnace could be encouraged to separate simply by a sink and float principle; the metal for casting being taken by a pump from a point at about midway depth where the best quality metal was to be expected.

In contrast, the holding of melts in closed vessels for the low-pressure die casting of aluminium, or the dosing of the liquid metal, are usually impaired by the initial turbulent pour of the melt to fill the furnace. The total pour height is often of the order of a metre. Not only are new oxides folded into the melt in this pouring action, but those oxides that have settled to the floor of the furnace since the last filling operation are stirred into the melt once again. Finally, these enclosed units suffer from the inaccessibility of the melt. This usually restricts any thorough action to improve the melt by any kind of degassing technique.

The Alotech approach to the design of a holding furnace is patented and not available for publication at the time of writing. It is hoped to rectify this in future editions of this work as details of the process are published. Why even mention it at this stage? The purpose of mentioning it here is to illustrate that even apparently simple equipment such as a holding furnace is capable of considerable sophistication, leading to the production of greatly improved processing and products. It takes the concept of melt cleaning and degassing to an ultimate level that probably represents a limit to what can be achieved. In addition, the technique is simple, low capital cost, low running cost, has no moving parts and is operator-free. At this time the technique is being applied only to aluminium alloys.

1.4 Pouring

Most foundries handle their metal from one point to another by ladle. The metal is, of course, transferred out of the ladle by pouring. In most foundries multiple pours are needed to transfer the liquid metal from the melting unit to the mould.

At every pouring operation, it is likely that large areas of oxide film will be entrained in the melt because pour heights are usually not controlled. It is known that pour heights less than the height of the sessile drop cannot entrain the surface oxide. However, such heights are very low; 16 mm for Mg, 13 mm for Al, and only 8 mm for dense metals such as copper-base, and iron and steel alloys.

However, this theoretical limit, while absolutely safe, may be exceeded for some metals with minimum risk. As long ago as 1928 Beck described how liquid magnesium could be transferred from a ladle into a mould by arranging the pouring lip of the ladle to be as close as possible to the pouring cup of the mould, and relatively fixed in position. In this way the semi-rigid oxide tube that formed automatically around the jet remained unbroken, and so protected the falling stream.

Experiments by Din and Campbell (2002) on Al–7Si–0.4Mg alloy have demonstrated that in practice the damage caused by falls up to 100 mm appears controlled and reproducible. This is in close agreement with early observations by Turner (1965) who noted that air was taken into the melt, reappearing as bubbles on the surface when the pouring height exceeded about 90 mm.

Above 200 mm, Din and Campbell (2003) found that random damage was certain. At these high energies of the plunging jet, bubbles are entrained, with the consequence that bubble trails add to the total damage in terms of area of bifilms.

In general, it seems that the lower the pour height the less damage is suffered by the melt. In addition, of course, less metal is oxidized, thus directly saving the costs of unnecessary melt losses. Ultimately, however, it is, of course, best to avoid pouring altogether. In this way losses are reduced to a minimum and the melt is maintained free from damage.

Until recent years, such concepts have been regarded as pipe dreams. However, the development of the Cosworth Process has demonstrated that it is possible for aluminium alloy castings to be made without the melt suffering any pouring action at any point of the process. Once melted, the liquid metal travels along horizontal heated channels, retaining its constant level through the holding furnace, and finally to the pump, where it is pressurized to fill the mould in a counter-gravity mode. Such technology would also appear to be relatively easily applied to magnesium alloys.

The potential for extension of this technology to other alloy base systems such as copper-based or iron-based alloys is less clear. This is because many of these other alloy systems either do not suffer the same problems from bifilms, or do not have the production requirements of some of the high volume aluminium foundries. Thus in normal circumstances, many irons and steels are relatively free from bifilms because of the large density difference between the inclusions and the parent melt, encouraging rapid flotation. Alternatively, many copper-based and steel foundries are more like jobbing shops, where the volume requirement does not justify a counter-gravity system, and high technology pumps, if they were available, would have problems surviving the oxidation and thermal shock of a stop-go production requirement.

Despite these reservations, the counter-gravity Griffin Process has been impressively successful for the volume production of steel wheels for rail rolling stock. The process produces wheels that require no machining (apart from the centre hub). The outer cast rim runs directly on the steel rail. The products outperform forged steel wheels in terms of reliability in service, earning the process 80 per cent of the market in the USA. What a demonstration of the soundness of the counter-gravity concept, contrasting dramatically with steel castings produced world-wide by gravity pouring, in which defects and expensive upgrading of the casting are the norm.

1.5 Melt treatments

1.5.1 Degassing

Gases dissolved in melts are disadvantageous because they precipitate in bifilms and cause them to unfurl, and even open further as cracks or voids, thereby progressively reducing the mechanical properties of the cast product.

Treatments to reduce the gas content of melts include vacuum degassing. Such a technique has only been widely adopted by the steel industry. If a melt were simply to be placed under vacuum the rate of degassing would be low because the dissolved gas would have to diffuse to the free surface to escape. The slow convection of the melt will gradually bring most of the volume near to the top surface, given time. The process is greatly speeded by the introduction of millions of small bubbles of inert gas via a porous plug or other technique. In this way the process is accomplished in a fraction of the time.

Traditionally the steel industry has used the technique of a carbon boil, in which the creation and floating out of carbon monoxide bubbles from the melt carries away unwanted gases such as oxygen (actively by chemical reaction with carbon) and hydrogen (passively by simple flushing action). The nitrogen in the melt may go up or down depending on its starting value and the nitrogen content of the environment above the melt, since it will tend to equilibrate with the environment. The more recent oxygen steelmaking processes in which oxygen is injected into the melt certainly reduce hydrogen and nitrogen, but require the oxygen to be reduced either chemically by reaction with C or other deoxidizers such as Si, Mn or Al etc., or use even more modern techniques such as AOD (argon-oxygen-decarburization). We shall not dwell further on these sophistications, since these specialized techniques, so well understood and well developed for steel, are almost unused elsewhere in the casting industry.

By comparison, the approaches to the degassing of aluminium alloys, containing only hydrogen, has for many years been primitive. Only recently have effective techniques been introduced.

For instance until approximately 1980, aluminium was commonly degassed using immersed tablets of hexachlorethane that thermally degraded in the melt to release large bubbles of chlorine and carbon (the latter as smoke). Alternatively, a primitive tube lance was used to introduce a gas such as nitrogen or argon. Again large bubbles of the gas were formed. These techniques involving the generation of large bubbles were so inefficient that little dissolved gas could be removed, but the creation of large areas of fresh melt to the atmosphere each time a bubble burst at the surface of the melt provided an excellent opportunity for the melt to equilibrate with the environment. Thus on a dry day the degassing effect might be acceptable. On a damp day, or when the flue gases from the gas-fired furnace were suffering poor extraction, the melt could gain hydrogen faster than it could lose it. This poor rate of degassing, combined with the high rate of regassing from interaction with the environment, led to variable and unsatisfactory results.

Rotary degassing came to the rescue. The use of a rapidly rotating rotor to chop bubbles of inert gas into fine clouds raised the rate of degassing and lowered the rate of regassing. Thus effective degassing could be reliably achieved in times that were acceptable in a production environment.

It is to be hoped that this is not the end of the story for the degassing of aluminium alloys. For instance the use of an ‘inert’ gas is a convenient untruth. Even sources of high purity inert gas contain sufficient oxygen as an impurity to create an oxide film on the inside of the surface of the bubbles. Thus millions of very thin bifilms must be created during this degassing process. For a new rotor, or after a weekend, the rotor and its shaft will have absorbed considerable quantities of water vapour (most refractories can commonly absorb water up to 10 per cent of their weight). Furthermore, the gas lines are usually not clean, or contain long rubber or plastic tubing. These materials absorb and therefore leak large quantities of volatiles into the degassing line. Thus by the time the gas arrives in the melt it is usually not particularly inert. At this time no-one knows whether the oxides created as by-products of degassing are negligible or whether they seriously reduce properties. What is known is that aluminium alloys are usually greatly improved by rotary degassing. Whether further improvements can be secured is not clear.

Part of the considerable benefit of rotary degassing is not merely the degassing action. It seems that the millions of tiny bubbles attach to oxides in the melt and float them to the top, where they can be skimmed off. After treatment, after the rotor has been raised out of the melt and the surface skimmed, a test of the efficiency of the cleaning action is simply to look at the surface of the melt. If the cleaning action is not complete, during the next few minutes small particles will be seen to arrive at the surface, under the surface oxide film, which is seen to be pushed upwards by the particle. The treatment can be repeated until no further arrival of debris can be seen. The melt surface then retains its pristine mirror smoothness. The melt can then be pronounced as clean as the treatment can achieve.

The action of rotary degassing on oxide films merits further examination. Much has been written on the benefits of small bubbles on the rate of degassing (although the rate of regassing from a damp rotor, or from the environment has been neglected). The action of the bubbles to eliminate films has in general been overlooked. However, the size of bubbles in relation to the efficiency of removal of films is probably critical. For instance, large bubbles will displace large volumes of melt during their rise to the surface, thus displacing films sideways, so that the film and bubble never make contact. On the other hand, small bubbles will displace relatively little liquid, and so be able to impact on relatively large films in their path. Thus such contacted films will be buoyed up to the surface. The mutual contact would be expected to become important when bubbles and films were approximately similar in size. Thus small bubbles will take out correspondingly smaller oxide films. This predicted effect deserves to be demonstrated experimentally at some future date.

An experience by the author illustrates some of the misconceptions surrounding the dual role of hydrogen and oxide bifilms in aluminium melts. An operator used his rotary degasser for 5 minutes to degas 200 kg of Al alloy. The melt was tested with a reduced pressure test (RPT; see below) sample that was found to contain no bubbles. The melt was therefore deemed to be degassed. The melt was immediately poured into a transfer ladle on a fork lift truck and conveyed to a low-pressure die casting furnace, into which it was poured. The melt in the low-pressure furnace was then tested again by RPT, and the sample found to contain many bubbles. The operator was baffled. He could not understand how so much gas could have re-entered the melt in only the few minutes required for the transfer.

The truth is, of course, that the melt was insufficiently degassed with only 5 minutes of treatment. In fact with a damp rotor the gas level is likely to rise initially (getting worse before it gets better!). The RPT showed no bubbles *not* because the hydrogen was low, but because the short treatment had clearly been sufficiently successful to remove a large proportion of the bifilms that were the nuclei for the bubbles. This high hydrogen metal was then poured twice, each time from a considerable height, re-introducing copious quantities of oxide bifilms that act as excellent nuclei, so that the RPT could now reveal its high hydrogen content.

1.5.2 Additions

Additions to melts are made for a variety of reasons. These can include additives for chemical degassing (as the addition of Al to steel to fix oxygen and nitrogen) or grain refinement (as in the addition of titanium and/or boron or carbon to Al alloys). Sometimes it seems certain that the poor quality of such materials (perhaps melted poorly and cast turbulently, and so containing a high level of oxides) can contaminate the melt directly.

Indirectly, however, the person in charge of making the addition will normally be under instructions to stir the melt to ensure the dissolution of the addition and its distribution throughout the melt. Such stirring actions can disturb the sediment at the bottom of melts, efficiently re-introducing and re-distributing those inclusions that had spent much time in settling out. The author has vivid memories of wrecking the quality of early Cosworth melts in this way: the addition of grain refiners gave wonderfully grain refined castings, and should have improved feeding, and therefore the soundness of the castings. In fact, all the castings were scrapped because of a rash of severe microporosity, initiated almost certainly on the stirred-up oxides that constituted the sediment in the holding furnace.

Additions of Sr to aluminium melts have often been accused of also adding hydrogen because of the porosity that has often been noted to follow such additions. Here again, it seems unlikely that sufficient hydrogen could be introduced by such a small addition. Perhaps the problem is with stirred sediment, or other more complex reasons as are discussed at some length in *Castings 2003*.

1.6 Filtration

Filtration is perhaps the most obvious way to remove suspended solids from liquid metals. However, it is not without its problems, as we shall see.

The action of a filter in the supply of liquid metal in the launder of a melt distribution system in a foundry or cast house (i.e. a foundry for continuous casting of long products) is rather different from its action in the running system of a shaped casting.

The filters used in castings act for only the few seconds or minutes that the casting is being poured. The velocities through them are high, usually several metres per second, compared to the more usual 0.1 m/s rates in launders. The filtration effect is further not helped by the concentration of flow through an area often a factor of up to 100 smaller than that used in launder systems. The total volume per second rates at which casting filters are used are therefore higher by a factor of around 1000. It is hardly surprising therefore that the filtration action is reduced nearly to zero. However, the effect on the flow is profound. The use of filters in running systems is dealt with in detail later in the book. We concentrate in this section on filters in launder systems.

The treatment of tonnage quantities of metal has been developed only for the aluminium casting industry. The central problem for most workers in this field is to understand how the filters work, since the filters commonly have pore sizes of around 1 mm, whereas, puzzlingly, they seem to be effective in removing a high percentage of inclusions of only 0.1 mm diameter. Most researchers expand at length, listing the mechanisms that might be successful to explain the trapping of such small solid particles. Unfortunately, these conjectures are probably not helpful, and are not repeated here.

The fact is that the important solids being filtered in aluminium alloys are not particles resembling small solid spheres, as has generally been assumed. The important particles are films (actually always double films that we have called bifilms). Once this is appreciated the filtration mechanism becomes much easier to understand. The films are often of size 1 to 10 mm, and so are, in principle, easily trapped by pores of 1 mm diameter. Such bifilms are not easily seen in their entirety in an optical microscope, the visible portions appearing to be much smaller, explaining why filters appear to arrest particles smaller than their pore diameters.

We need to take care. This explanation, while probably having some truth, may oversimplify the real situation. In *Castings* (2003), the life of the bifilm was described as starting as the folding in of a planar crack-like defect as a result of surface turbulence. However, internal turbulence wrapped the defect into a compact form, reducing its size by a factor of 10 or so. In this form it could pass through a filter, and finally open once again in the casting as the liquid metal finally came to rest, and bifilm-opening (i.e. unfurling) processes started to come into action.

In more detail now, the trapping of compact forms of films is explained by their irregular and changing form. During the compacted stage of their life, they will be constantly in a state of flux, ravelling and unravelling as they travel along in the severely turbulent flow. Two-dimensional images of such defects seen on polished sections always show loose trailing fragments. Thus in their progress through the filter, one such end could become attached, possibly wrapping over a web or wall of filter material. The rest of the defect would then roll out, unravelling in the flow, and be flattened against the internal surfaces of the filter, where it would remain fixed in the tranquil boundary layer.

1.6.1 Packed beds

In practice, in DC (direct chill) continuous casting plants, filters have been used for many years, sited in the launder system between the melting furnaces and casting units. Commonly, the filtration system is a large and expensive installation, comprising a crucible furnace that contains a divided crucible. One half is filled with refractory material such as alumina balls, or tabular alumina. The flow of the melt down one half of the crucible, through a connecting port, and up through the deep packed bed of such systems has been shown to be effective in greatly reducing the inclusion count (the number of inclusions per unit area). However, it is known that the accidental disturbance of such filters releases large quantities of inclusions into the melt stream. This has also been reported when enthusiastic operators see the filter becoming blocked by the increasing upstream level of the melt, and stir the bed with iron rods to ease the flow of metal. It is not easy to imagine actions that could be more counter-productive.

Some work has been carried out on filtering liquid aluminium through packed beds of tabular or ball alumina (Mutharasan *et al.* 1981) and through bauxite, alumino-silicates, magnesia, chrome-magnesite, limestone, silicon carbide, carbon, and steel wool (Hedjazi *et al.* 1975). This latter piece of work demonstrated that all of the materials were effective in reducing macro-inclusions. This is perhaps to be expected as a simple sieve effect. However, only the alumino-silicates were really effective in removing any micro-inclusions and films, whereas the carbon and chrome-magnesite removed only a small percentage of films and appeared to actually increase the number of micro-inclusions. The authors suggest that the wettability of the inclusions and the filter material is essential for effective filtration.

An interesting application of a chemically active packed bed is that by Geskin *et al.* (1986), in which liquid copper is passed through charcoal to provide oxygen-free copper castings. It is certain, however, that the charcoal will have been difficult to dry thoroughly, so that the final casting may be somewhat high in hydrogen. Because of the low oxygen content it is likely that hydrogen pores will not be nucleated (as discussed in *Castings* (2003)). The hydrogen is expected therefore to stay in solution and remain harmless.

1.6.2 Alternative varieties of filters

Other large and expensive filtration systems include the use of a pack of porous tubes, sealed in a large heated box, through which the aluminium is forced. The pores in this case are of the order of 0.25 mm, with the result that the filter takes a high head of metal to prime it. However, the technique is not subject to failure because of disturbance, and guarantees high quality of liquid metal.

Other smaller and somewhat cheaper systems that have been used include a ceramic foam filter, usually designed to be housed in a box, sited permanently below the surface of the melt. The velocities through the filter are usually low, encouraged by the large area of the filter, usually at least 300 × 300 mm. The filter is only brought out into the air to be changed when the metal level either side of the filter box shows a large difference, indicating that the filter is becoming blocked.

The efficiency of many filtration devices can be understood when it is assumed that the important filtration action is the removal of films (not particles). Thus glass cloth is widely used to good effect in many different forms in the Al casting industry.

1.6.3 Practical aspects

It is typical of most filtration systems that the high quality of metal that they produce (often at considerable expense) is destroyed by thoughtless handling of the melt downstream.

Even the filter itself can give difficulties in this way. For instance if the melt exits the filter downwards, or even horizontally, causing fine jets of metal to form in the air, and plunge into the melt, additional oxide defects are necessarily created downstream. The avoidance of this problem is an important aspect of the designing of effective filters into the running systems of shaped castings, as will be discussed later.

Many have wondered whether the filter itself causes oxides because the flow necessarily emerges in a divided state, and therefore must create double films in the hundreds of confluence events. Video observations by the author on a stream of aluminium alloy emerging from a ceramic foam filter with a pore diameter close to 1 mm have helped to clarify the situation. It seems true that the flow emerges divided as separate jets. However, within a few millimetres (apparently depending on the flow rate of the metal) the separate jets merge. Thus the oxide tubes formed around the jets appear to be up to 10 mm long when the melt travelled at around 500 mm s^{-1} but remained attached to the filter. The oxide tubes did not extend further because after the streams merged oxygen was necessarily excluded. The forest of tubes was seen to wave about in the flow like underwater grass. It is possible that more rapid flows might cause the grass to detach as a result of its greater length and the higher speed of the metal. This seems much more likely in the conditions of the running system of a casting.

Rule 2

Avoid turbulent entrainment (the critical velocity requirement)

The avoidance of surface turbulence is probably the most complex and difficult Rule to fulfil when dealing with gravity pouring systems.

The requirement is all the more difficult to appreciate by many in the industry, since everyone working in this field has always emphasized the importance of working with 'turbulence-free' filling systems for castings. Unfortunately, despite all the worthy intentions, all the textbooks, all the systems, and all the talk, so far as the author can discover, it seems that no-one appears to have achieved this target so far. In fact, in travelling around the casting industry, it is quite clear that the majority (at least 80 per cent) of all defects are directly caused by turbulence. Thus the problem is massive; far more serious than suspected by most of us in the industry.

To understand the fundamental root of the problem, it will become clear in Section 2.1 that any fall greater than the height of the sessile drop (of the order of 10 mm) causes the metal to exceed its critical velocity, and so introduces the danger of defects in the casting. As most falls are in fact in the range 10 to 100 times greater than this, and as the damage is likely to be proportional to the energy involved (i.e. proportional to the square of the velocity) the damage so created will usually be expected to be in the range 100 to 10 000 times greater. Thus in the great majority of castings that are poured simply under the influence of gravity, there is a major problem to ensure its integrity. In fact, the situation is so bad that the best outcome of many of the solutions proposed in this book is damage limitation. Effectively, it has to be admitted that at this time it seem impossible to guarantee the avoidance of some damage when pouring liquid metals.

This somewhat depressing conclusion needs to be tempered by a number of factors.

First; the world has come to accept castings as they are. Thus any improvement will be welcome. This book described techniques that will create very encouraging improvements.

Second; this book is merely a summary of what has been discovered so far in the development of filling system design. Better designs are to be expected now that the design parameters (such as critical velocity, critical fall height, etc.) are defined.

Third; there *are* filling systems that can yield, in principle, perfect results.

Of necessity, such perfection is achieved by fulfilling Rule 2 by avoiding the transfer of the melt into the mould by pouring downhill under gravity. Thus, considering the three directions of filling a mould:

(i) downhill pouring under gravity;
(ii) horizontal transfer into the mould (achieved by tilt casting in which the tilt conditions are accurately controlled);
(iii) uphill (counter-gravity) casting in which the melt is caused to fill the mould in only an uphill mode;

only the last two processes have the potential to deliver castings of near perfect quality. In my experience, I have found that in practice it is often difficult to make a good casting by gravity, whereas by a good counter-gravity process (i.e. a process observing all the 10 Rules) it has been difficult to make a bad casting. The jury is still out on horizontal transfer by tilting. This approach has great potential, but requires a dedicated effort to achieve the correct conditions.

Thus in summary, filling of moulds can be carried out down, along, or up. Only the 'along' and 'up' modes totally fulfil the non-surface turbulence condition.

However, despite all its problems, it seems more than likely that the downward mode, gravity casting, will continue to be with us for the foreseeable future. Thus we shall devote some considerable time to the damage limitation exercises that can offer considerably improved products, even if, unfortunately, those products cannot be ultimately claimed as perfect. Most will shed no tears over this conclusion. Although potential perfection in the along and up modes is attractive, the casting business is all about making adequate products; products that meet a specification and at a price a buyer can afford.

The question of cost is interesting; perhaps the most interesting. Of course the costs have to be right, and often gravity casting is acceptable and sufficiently economical. However, more often than might be expected, high quality and low cost can go together. An improved gravity system, or even one of the better counter-gravity systems, can be surprisingly economical and effective. Such opportunities are often overlooked. It is useful to watch out for such benefits.

2.1 Maximum velocity requirement

One day, I was seated in the X-ray radiographic room using an illuminated screen to study a series of radiographic films of cylinder head castings made by our recently developed casting system at Cosworth. Each radiograph in turn was beautiful, having a clear, 'wine glass' perfection that every founder dreams of. I was at peace with the world. However, suddenly, a radiograph appeared on the screen that was a total disaster. It had gas bubbles, shrinkage porosity, hot tears, and sand inclusions. I was shocked, but sensed immediately what had happened. I shot out to query Trevor, our man on the casting station. 'What happened to this casting?' He admitted instantly 'Sorry. I put the metal in too fast.'

This was a lesson that remained with me for years. This chance experiment by counter-gravity, using an electromagnetic pump allowing independent control of the ingate velocity, had kept constant all the other casting variables

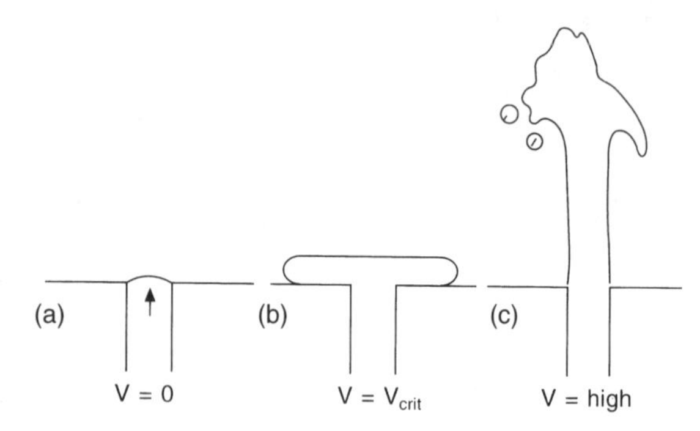

Figure 2.1 *The extremes of velocity entering the mould from the gate; zero, critical and high.*

(temperature, metal quality, alloy content, mould geometry, aggregate type, binder type, etc.), showing them to be of negligible importance. Clearly, the ingate velocity was dominant. By only changing the speed of entry of metal we could move from complete success to complete failure.

Thinking further about this, common sense tells us all that there is an optimum velocity at which a liquid metal should enter a mould. The concept is outlined in Figure 2.1. At a velocity of zero the melt is particularly safe (Figure 2.1a), being free from any danger of damage. Regrettably, this condition is not helpful for the filling of moulds. Conversely, at extremely high velocities the melt will enter like a jet of water from a fire-fighter's hosepipe (Figure 2.1c), and is clearly damaging to both metal and mould. At a certain intermediate velocity the melt rises to just that height that can be supported by surface tension around the periphery of the spreading drop (Figure 2.1b). The theoretical background to these concepts is dealt with at length in the first book in this series *Castings (Principles)* (2003). For nearly all liquid metals this critical velocity is close to $0.5\,\mathrm{m\,s^{-1}}$. This value is of central importance in the casting of liquid metals, and will be referred to repeatedly in this section.

The liquid drop, emerging close to its critical velocity, and spreading slowly from the ingate is closely in equilibrium, its surface tension holding the drop in its compact shape, just balancing the head of pressure tending to spread it because of its density. This slowly expanding drop is closely similar to a sessile (Latin 'sitting') drop. (The word contrasts with glissile drop, meaning a gliding or sliding drop.) A sessile drop of Al sitting on a non-wetted substrate is approximately 12.5 mm high. Corresponding values for other liquids are Fe 10 mm, Cu 8 mm, Zn 7 mm, Pb 4, water about 5 mm.

Recent research has demonstrated that if the liquid velocity exceeds the critical velocity there is a danger that the surface of the liquid metal may be folded over by surface turbulence. If a perturbation to the surface exceeds the height of the sessile drop the liquid can no longer be supported by its surface tension. It will therefore fall back under gravity and may thus entrain its own surface in an enfolding action. At risk of overly repeating this important phenomenon, this entrainment of the surface can occur if there is sufficient energy in the form of velocity in the bulk liquid to perturb the surface against the smoothing action of surface tension. In addition, notice that damage is not necessarily created by the falling back of the metal. The falling is likely to be chaotic, so that any folding action may or may not occur. The significance of the critical velocity is clear therefore: above the critical velocity there is the danger of surface entrainment leading to defect creation. Below the critical velocity the melt is safe from entrainment problems.

(It is perhaps useful to remind the reader that entrainment may still occur as a result of surface contraction. Thus the steady forward advance of the meniscus, Rule 3, remains another central axiom of good filling systems.)

Japanese workers optimizing the filling of their design of vertical stroke pressure die casting machine using both experiment and computer simulation (Itamura 1995) confirm the critical velocity of $0.5\,m\,s^{-1}$, finding that air bubbles have a chance to become entrained above this value. (These workers go further to define the amount of liquid that needs to be in the mould cavity to suppress entrainment at higher ingate velocities.)

Normally, the surface is covered with an oxide film, although many other types of films are possible in different circumstances. A common alternative is a graphitic film. There is a chance therefore that if the speed of the liquid exceeds this critical velocity its surface film may be folded into the bulk of the liquid. This folding action is an entrainment event. It leads to a variety of problems in the liquid that we can collectively call entrainment defects. The major entrainment defects are air bubbles and doubled-over oxide films. The author has named these folded-in films 'bifilms' to emphasize their double, folded-over nature. Because the films are necessarily folded dry side to dry side, there is little or no bonding between these dry interfaces, so that the double films act as cracks. The cracks (alias bifilms) become frozen into the casting, lowering the strength and fatigue resistance. Bifilms may also create leak paths, causing leakage failures.

The folding-in of the oxide is a random process, leading to scatter and unreliability in the properties and performance of the product on a casting to casting, day to day, and month to month basis during a production run.

The different qualities of metals arriving in the foundry from batch to batch will also be expected to contain different quantities and different forms of bifilms. Thus the performance of the foundry will suffer further variation. This is the reason for Rule 1. The foundry needs to have procedures in place to smooth variations of its incoming raw material.

Looking a little more closely at the detail of critical velocities for different liquids, it is close to $0.4\,m\,s^{-1}$ for dense alloys such as irons, steels and bronzes and about $0.5\,m\,s^{-1}$ for liquid aluminium alloys. The value is 0.55 to $0.6\,m\,s^{-1}$ for Mg and its alloys. Taking an average of about $0.5\,m\,s^{-1}$ for all liquid metals is usually good enough for most purposes related to the design of filling systems for castings, and will be generally used in this book.

The maximum velocity condition effectively forbids top gating of castings (i.e. the planting of a gate in the top of the mould cavity, causing the metal to fall freely inside the mould cavity). This is because liquid aluminium reaches its critical velocity of about $0.5\,m\,s^{-1}$ after falling only 12.5 mm under gravity. The critical velocity of liquid iron or steel is exceeded after a fall of only about 10 mm (these are, of course, the heights of the sessile drops). Naturally, such short fall distances are always exceeded in practice in top gated castings, leading to the danger of the incorporation of the surface films, and consequent leakage and crack defects.

Castings that are made in which velocities everywhere in the mould never exceed the critical velocity are consistently strong, with high fatigue resistance, and are leak-tight (if properly fed, of course, so as to be free from shrinkage porosity).

Experiments on the casting of aluminium have demonstrated that the strength of castings may be reduced by as much as 90 per cent or more if the critical velocity is exceeded. The corresponding defects in the castings are not always detected by conventional non-destructive testing such as X-ray radiography or dye penetrant, since, despite their large area, the folded oxide films are thin, and do not necessarily give rise to any significant surface indications.

The speed requirement automatically excludes conventional pressure die-castings as having significant potential for reliability, since the filling speeds are usually 10 to 100 times greater than the critical velocity.

Over recent years there have been welcome moves, introducing some special developments of high-pressure technology that are capable of meeting this requirement. These include the vertical injection squeeze casting machine, and the shot control techniques. Such techniques can, in principle, be operated to fill the cavity through large gates at low speeds, and without ingress of air into the liquid metal. Such castings require to be sawn, rather than broken, from their filling systems of course. Unfortunately, the castings remain somewhat impaired by the action of pouring into the shot sleeve. Even here, these problems are now being addressed by some manufacturers, with consequent benefits to the integrity of the castings.

Other uphill filling techniques such as low-pressure filling systems are capable of meeting Rule 2. Even so, it is regrettable that the critical velocity is practically always exceeded during the filling of the low-pressure furnace itself because of the severe fall of the metal as it is transferred into the pressure vessel, so that the metal is damaged even prior to casting. In addition, many low-pressure die casting machines are in fact so poorly controlled on flow rate, that the speed of entry into the die greatly exceeds the critical velocity, thus negating one of the most important potential benefits of the low-pressure system. Processes such as the Cosworth Process avoid these problems by never allowing the melt to fall at any stage of processing, and control its upward speed into the mould by electromagnetic pump.

Metals that can also suffer from entrained surface films are suggested to be the ZA (zinc–aluminium) alloys and ductile irons. Carbon and stainless steels are thought to be similar, although in some of these systems the entrained bifilms agglomerate as a result of being partially molten and therefore somewhat sticky. They therefore remain more compact, and float out more easily to form surface imperfections in the form of slag macroinclusions on the surface. For a few materials, particularly alloys based on the Cu–10Al types (aluminium- and manganese-bronzes) the critical velocities were originally thought to be much lower, in the region of only $0.075\,\mathrm{m\,s^{-1}}$. However, from recent work at Birmingham this low velocity seems to have been a mistake, probably resulting from the confusion caused by bubbles entrained in the early part of the filling system. With well-designed filling systems, the aluminium-bronzes accurately fulfil the theoretically predicted $0.4\,\mathrm{m\,s^{-1}}$ value for a critical ingate velocity (Halvaee and Campbell 1997).

Because of the central importance of the concept of critical velocity, the reader will forgive a re-statement of some aspects in this summary.

(i) Even if the melt does jump higher than the height of a sessile drop, when it falls back into the surface there is no certainty that it will enfold its surface film. These tumbling motions in the liquid can be chaotic, random events. Sometimes the surface will fold badly, and sometimes not at all. This is the character of surface turbulence; it is not predictable in detail. The key aspect of the critical velocity is that at velocities less than the critical velocity the surface is safe. Above the critical velocity there is the danger of entrainment damage. The criterion is a necessary but not sufficient condition for entrainment damage.

(ii) If the whole, extensive surface of a liquid were moving upwards at a uniform speed, but exceeding the critical velocity, clearly no entrainment would occur. Thus the surface disturbance that can lead to entrainment is more accurately described not merely as a velocity but in reality a velocity difference. It might therefore be defined more accurately as a critical velocity gradient measured across the liquid surface. For those of a theoretical bent, the critical gradient might be defined as the velocity difference achieving the critical velocity along a distance in the surface of the order of the sessile drop radius (approximately half its height) in the liquid surface. To achieve reasonable accuracy, this approach requires one to allow for the reduction in drop height with velocity. Hirt (2003) solves this problem with a delightful and novel approach, modelling the surface disturbances as arrays of turbulent eddies, and achieves convincing solutions for the simulation of entrainment at hydraulic jumps and plunging jets. Such niceties are neglected here. The problem does not arise when considering the velocity of the melt when emerging from a vertical ingate into a mould cavity. In that situation, the ingate velocity and its relation to the critical velocity is clear.

(iii) If the melt is travelling at a high speed, but is constrained between narrowly enclosing walls, it does not have the room to fold-over its advancing meniscus. Thus no damage is suffered by the liquid despite its high speed, and despite the high risk involved. This is one of the basic reasons underlying the design of extremely narrow channels for filling systems that are proposed in this book.

2.2 The 'no fall' requirement

It is quickly shown that if liquid aluminium is allowed to fall more than 12.5 mm then it exceeds the critical $0.5\,m\,s^{-1}$. The critical fall height can be seen to be a kind of re-statement of the critical velocity condition. Similar critical velocities and critical fall heights can be defined for other liquid metals. The critical fall heights for all liquid metals are in the range 3 to 15 mm.

It follows immediately that top gating of castings almost without exception will lead to a violation of the critical velocity requirement. Many forms of gating that enter the mould cavity at the mould joint, if any significant part of the cavity is below the joint, will also violate this requirement.

In fact, for conventional sand and gravity die casting, it has to be accepted that some fall of the metal is necessary. Thus it has been accepted that the best option is for a single fall, concentrating the loss of height of the liquid at the very beginning of the filling system. The fall takes place down a conduit known as a sprue, or down runner. This conduit brings the melt to the lowest point of the mould. The distribution system from that point, consisting of runners and gates, should progress only uphill.

Considering the mould cavity itself, the requirement effectively rules that all gates into the mould cavity enter at the bottom level, known as bottom gating. The siting of gates into the mould cavity at the top (top gating) or at the joint (gating at the joint line) are not options if safety from surface turbulence is required.

Also excluded are any filling methods that cause waterfall effects in the mould cavity. This requirement dictates the siting of a separate ingate at every isolated low point on the casting.

Even so, the concept of the critical fall distance does require some qualification. If the critical limit is exceeded it does not mean that defects will necessarily occur. It simply means that there is a risk that they may occur. This is because the energy of the liquid is now sufficiently high that the melt is potentially able to enfold in its own surface. Whether a defect occurs or not is now a matter of chance. (This contrasts, of course, with falls of less than the critical height. In this case there is no chance that a defect can occur, the regime being completely safe.)

There is, however, further qualification that needs to be applied to the critical fall distance. This is because the critical value quoted above has been worked out for a liquid, neglecting the presence of any oxide film. In practice, it seems that for some liquid alloys, the surface oxide has a certain amount of strength and rigidity, so that the falling stream is contained in its oxide tube and so is enabled to better resist the conditions that might enfold its surface. This behaviour has been investigated for aluminium alloys (Din, Kendrick and Campbell 2003). It seems that although the original fall distance limit of 12.5 mm continues to be the safest option, fall heights of up to about 100 mm might be allowable in some instances, possibly depending somewhat on the precise alloy composition. However, falls greater than 200 mm definitely entrain defects; the velocity of the melt in this case is about $2\,m\,s^{-1}$ so that entrainment seems unavoidable. Also, of course, other alloys may not enjoy the benefits of the support of a tube of oxide around the falling jet. This benefit requires to be investigated in other alloy systems to test what values beyond the theoretical limits may be used in practice.

The initial fall down the sprue in gravity-filled systems does necessarily introduce some oxide damage into the metal. For this reason it seems reasonable to conclude that gravity-poured castings will never attain the degree of reliability that can be provided by counter-gravity and other systems that can avoid surface turbulence.

Of necessity therefore, it has to be accepted that the no-fall requirement applies to the design of the filling system downstream of the base of the sprue. The damage encountered in the fall down the sprue has to be accepted; although with a good sprue and pouring basin design this initial fall damage can be reduced to a minimum as we shall see.

It is a matter of good luck that it seems that for some alloys much of the oxide introduced in this way does not appear to find its way through and into the mould cavity. It seems that much of it remains attached to the walls of the sprue. This surprising effect is clearly seen in many top-gated castings, where most of the oxide damage (and particularly any random leakage problem) is confined to the area of the casting under the point of pouring, where the metal is falling. Extensive damage does not seem to extend into those regions of the casting where the speed of the metal front decreases, and where the front travels uphill, but there does appear to be some carry-over of defects. Thus the provision of a filter immediately after the completion of the fall is valuable. It is to be noted, however, that significant damage will still be expected to pass through the filter.

The requirement that the filling system should cause the melt to progress only uphill after the base of the sprue forces the decision that the runner must be in the drag and the gates must be in the cope for a horizontal single jointed mould (if the runner is in the cope, then the gates fill prematurely, before the runner itself is filled, thus air bubbles are likely to enter the gates). The 'no fall' requirement may also exclude some of those filling methods in which the metal slides down a face inside the mould cavity, such as some tilt casting type operations. This undesirable effect is discussed in more detail in the section devoted to tilt casting.

It is noteworthy that these precautions to avoid the entrainment of oxide films also apply to casting in inert gas or even in vacuum. This is because the oxides of Al and Mg (as in Al alloys, ductile irons, or high temperature Ni-base alloys for instance) form so readily that they effectively 'getter' the residual oxygen in any conventional industrial vacuum, and form strong films on the surface of the liquid.

Rule 2 applies to 'normal' castings with walls of thickness over 3 or 4 mm.

For channels that are sufficiently narrow, having dimensions of only a few millimetres, the curvature of the meniscus at the liquid front can keep the liquid front from disintegration. Thus narrow filling system geometries are valuable in their action to conserve the liquid as a coherent mass, and so acting to push the air out of the system ahead of the liquid. The filling systems therefore fill in one pass.

A good filling action, pushing the air ahead of the liquid front as a piston in a cylinder, is a critically valuable action. Such systems deserve a special name such as perhaps 'one pass filling (OPF) designs'. Although I do not usually care for such jargon, the special name emphasizes the special action. It contrasts with the turbulent and scattered filling often observed in systems that are over-generously designed, in which the melt can be travelling in two directions at once along a single channel. A fast jet travels under the return wave that rolls over its top, rolling in air and oxides.

For a wide, narrow, horizontal channel, any effect of surface tension is clearly limited to channels that have dimensions smaller than the sessile drop height for that alloy. Thus for Al alloys the maximum channel height would be 12.5 mm, although even this height would exert little influence on the melt, since the roof would just touch the liquid, exerting no pressure on it. Similarly, taking account that the effect of surface tension is doubled if the curvature of the liquid front is doubled by a second component of the curvature at right angles, a channel of square section could be 25 mm square, and be contained just by surface tension. In practice, however, for any useful restraint from the walls of channels, these dimensions require to be at least halved, effectively compressing the liquid into the channel.

For very thin walled castings, of section thickness less than 2 mm, the effect of surface tension in controlling filling becomes predominant. The walls are so much closer than the natural curvature of a sessile drop that the meniscus is effectively compressed, and requires the application of pressure to force it into such narrow gaps. The liquid surface is now so constrained that it is not easy to break the surface, i.e. once again there is no room for splashing or droplet formation. Thus the critical velocity is higher, and metal speeds can be raised by approximately a factor of 2 without danger.

In very thin walled castings, with walls less than 2 mm thickness, the tight curvature of the meniscus becomes so important that filling can sometimes be without regard to gravity (i.e. can be uphill or downhill) since the effect of gravity is swamped by the effect of surface tension. This makes even the uphill filling of such thin sections problematical, because the effective surface tension exceeds the effect of gravity. Instabilities therefore occur, whereby the moving parts of the meniscus continue to move ahead in spite of gravity because of the reduced thickness of the oxide skin at that point. Conversely, other parts of the meniscus that drag back are further suppressed in their advance by the thickening oxide, so that a run-away instability condition occurs. This dendritic advance of the liquid front is no longer controlled by gravity in very thin castings, making the filling of extensive sections, whether horizontal or vertical, a major filling problem.

The problem of the filling of thin walls occurs because the flow happens, by chance, to avoid filling some areas because of random meandering. Such chance avoidance, if prolonged, leads to the development of strong oxide films, or even freezing of the liquid front. Thus the final advance of the liquid to fill such regions is hindered or prevented altogether.

The dangers of a random filling pattern problem are relieved by the presence of regularly spaced ribs or other geometrical features that assist organizing the distribution of liquid. Random meandering is thereby discouraged and replaced by regular and frequent penetration of the area, so that the liquid front has a better chance to remain 'live', i.e. it keeps moving so that a thick restraining oxide is given less chance to form.

The further complicating effect of the microscopic break-up of the front known as micro-jetting (*Castings 2003*) observed in sections of 2 mm and less in sand and plaster moulds is not yet understood. The effect has not yet been investigated, and may not occur at all in the dry mould conditions such as are found in gravity die casting.

2.3 Filling system design

Getting the liquid metal out of the crucible or melting furnace and into the mould is a critical step when making a casting: it is likely that most casting scrap arises during this few seconds of pouring of the casting.

Recent work observing the liquid metal as it travels through the filling system indicates that most of the damage is done to castings by poor filling system design. It is also worth reflecting on the fact that every gram of metal in the casting has, of necessity, travelled through the filling system. Leaving its design to chance, or even to the patternmaker (with all due respect to all our invaluable and highly skilful patternmakers), is a risk not to be recommended.

The early part of this section presents the design background, outlining the general thinking and some of the detailed logic behind the design of filling systems. The detailed calculations that are required to determine the precise dimensions of the various parts of the system are presented later (Section 2.3.7).

2.3.1 Gravity pouring of open-top moulds

Most castings require a mould to be formed in two parts: the bottom part (the drag) forms the base of the casting, and the top half (the cope) forms the top of the casting. However, some castings require no shaping of the top surface. In this case only a drag is required. The absence of a cope means that the mould cavity is open, so that metal can be poured directly in. The foundryman can therefore direct the flow of metal around the mould using his skill during pouring (Figure 2.2). Such open-top moulds represent a successful and economical technique

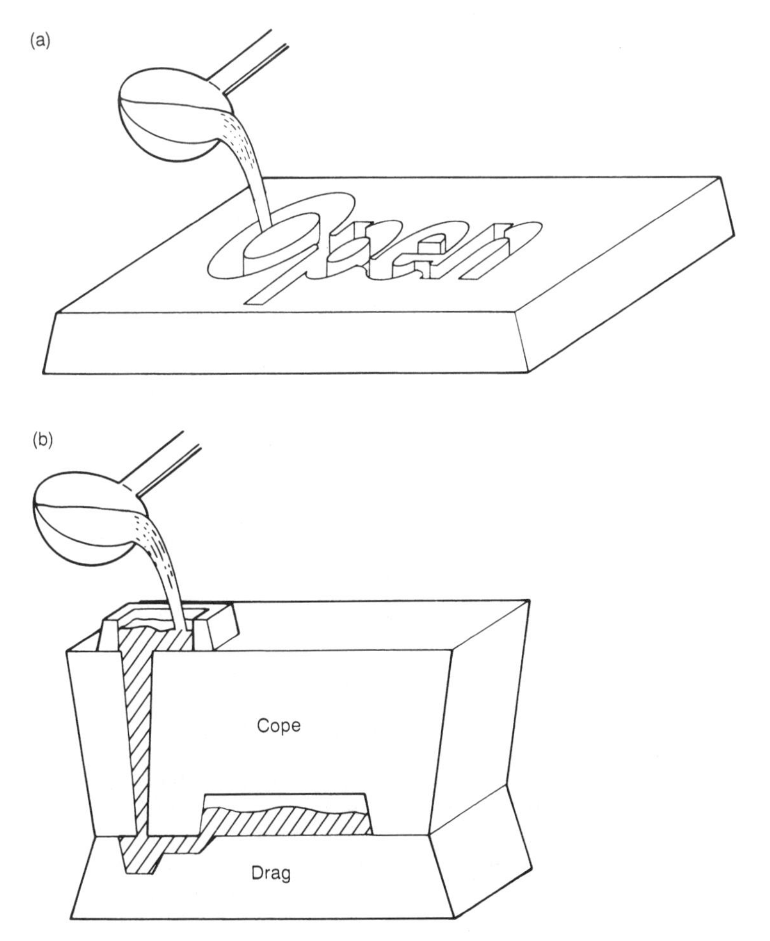

Figure 2.2 *(a) An open and (b) closed mould partially sectioned.*

for the production of aluminium or bronze wall plaques and plates in cast iron, which do not require a well-formed back surface. The first great engineering structure, the Iron Bridge built across the River Severn by the great English ironmaster Abraham Darby in 1779, had all its main spars cast in this way. This spectacular feat, with its main structural members over 23 m long cast in open-top sand moulds, heralded the dawn of the modern concept of a structural engineering casting.

Other viscous and poorly fluid materials are cast similarly, such as hydraulic cements, concretes, and organic resins and resin/aggregate mixtures that constitute resin concretes. Molten ceramics such as liquid basalt are poured in the same way, as witnessed by the cast basalt curb stones outside the house where I once lived, that have lined the edge of the road for the last hundred years or so and whose maker's name is still as sharply defined as the day it was cast.

The remainder of this section concentrates on the complex problem of designing filling systems for castings in which all the surfaces are moulded, i.e. the mould is closed. In all such circumstances, a bottom-gated system is adopted (i.e. the melt enters the mould cavity from one or more gates located at the lowest point, or if more than one low point, at each lowest point).

2.3.2 Gravity pouring of closed moulds

The series of funnels, pipes and channels to guide the metal from the ladle into the mould constitutes our liquid metal plumbing, and is known as the filling system, or running system. Its design is crucial; so crucial, that this is without doubt the most important chapter in the book.

However, the reader needs to keep in mind that the elimination of a running system by simply pouring into the top of the mould (down an open feeder, for instance) may be a reasonable solution in certain cases. Although apparently counter to much of the teaching in this book, there is no doubt that a top-poured option has often been demonstrated to be preferable to some poorly designed running systems, especially poorly designed bottom-gated systems. There are fundamental reasons for this that are worth examining right away.

In top gating the plunge of a jet into a liquid is accompanied by relatively low shear forces in the liquid, since the liquid surrounding the jet will move with the jet, reducing the shearing action. Thus although some damage is always done by top pouring, in some circumstances it may not be too bad, and may be preferable to a costly, difficult, or poor bottom-gated system.

In poor filling system designs, velocities in the channels can be significantly higher than the free-fall velocities. What is worse, the walls of the channels are stationary, and so maximize the shearing action, encouraging surface turbulence and the consequential damage from the shredding and entraining of bubbles and bifilms.

Ultimately, however, a bottom-gated system, if designed well, has the greatest potential for success.

Most castings are made by pouring the liquid metal into the opening of the running system, using the action of gravity to effect the filling action of the mould. This is a simple and quick way to make a casting. Thus gravity sand casting and gravity die-casting (permanent mould casting in the USA) are important casting processes at the present time. Gravity castings have, however, gained a poor reputation for reliability and quality, simply because their running systems have in general been badly designed. Surface turbulence has led to porosity and cracks, and unreliability in leak-tightness and mechanical properties.

Nevertheless, there are rules for the design of gravity-running systems that, although admittedly far from perfect, are much better than nothing. Such rules were originally empirical, based on transparent-model work and some confirmatory tests on real castings. We are now a little better informed by access to real-time video radiography of moulds during filling, and sophisticated computer simulation, so that liquid aluminium or liquid steel can be observed as it tumbles through the mould. Despite this, many uncertainties still remain. The rules for the design of filling systems are still not the mature science that we all might wish for. Even so, some rules are now evident, and their intelligent use allows castings of the highest quality to be made. They are therefore described in this section, and constitute essential reading!

It is hoped to answer the questions 'Why is the running system so complicated?' and 'Why are there so many different features?' It is a salutary fact that the apparent complexity has led to much confused thinking.

An invaluable general rule that I recommend to all those studying running and gating systems is 'If in doubt, visualize water'. Most of us have clear perceptions about the mobility and general flow behaviour of water in the gentle pouring of a cup of tea, the splat as it is spilled on the floor, the flow of a river over a weir, or the spray from a high-pressure hose pipe. A general feeling for this behaviour can sometimes allow us to cut through the mystique, and sometimes even the calculations! In addition, the application of this simple criterion can often result in the instant dismissal of many existing filling systems intended for the production of a reliable quality of casting as being quite clearly useless!

Closed moulds represent the greatest challenge to the casting engineer. There are numerous ways to get the metal into the mould, some disastrously bad, some tolerable, some good. To appreciate the good we shall have to devote some space to the bad (Figure 2.3). If this reads like a sermon, then so be it. *The Good Running System* is a *Good Cause* that deserves the passionate concern of the casting engineer. Too many castings with hastily rigged running systems have appeared to be satisfactory in limited prototype trials, but have proved to have disastrous levels of scrap when put into production. This is normally the result of surface turbulence during filling that produces non-reproducible castings, some apparently good, some definitely bad. This result confirms the nature of turbulence. Turbulence implies chaos; and chaos implies unpredictability. When using a running system that generates surface turbulence a typical scrap rate for a commercial vehicle casting might be 15 per cent, whereas a turbine blade subjected to much more stringent inspection can easily reach 75 per cent rejections.

In general, however, experience shows that foundries that use exclusively turbulent filling methods such as most investment foundries, experience on average about 20–25 per cent scrap, of which 5–10 per cent is the total of miscellaneous minor processing problems such as broken moulds, castings damaged during cut-off, etc. The remaining 15 per cent is composed of random inclusions, random porosity, and misruns—the standard legacy of turbulent running systems: the inclusions are created by the folding of the surface, as are the random pockets of porosity; and the misruns by the unpredictable ebb and flow in different parts of the casting during filling. In sand casting foundries, most of the so-called mould problems leading to sand inclusion are actually the result of the poor filling system designs. With good filling systems sand problems such as mould erosion and sand inclusions usually disappear.

In a foundry making a variety of castings, the 15 per cent running system scrap is made up of difficult castings which might run at 85–95 per cent scrap (almost never 100 per cent!) and easy castings which run at 5 per cent scrap (almost never zero!). The non-repeatable results continuously raise the characteristic false hope that the problems are solved, only to have the hopes dashed again by the next few castings. The variability is baffling, because the foundry engineer will often go to extreme lengths to ensure that all the variables believed to be under control are held constant.

Only a carefully worked out running system will give filling that is characterized by low surface turbulence, and which is therefore reproducible every time. Interestingly, this can mean 100 per cent scrap. However, this is not such a bad result in practice because the defect will be reproducibly repeated in every casting. It is therefore easy to identify and correct, and when corrected, stays corrected. After the first trials, the good running system should yield reliable, repeatable castings, and be characterized by a scrap rate close to zero.

A good running system, perhaps something like that shown in Figure 2.3b, will also be tolerant of wide variations in foundry practice, in contrast with the normal experience accompanying turbulent filling, in which pouring conditions are critical. Many foundries will know the problem that certain castings can only be poured successfully by certain operators. The good running system will ensure that pouring speed will now be under the control of the running system, not the pourer, and casting temperature will no longer be dictated by the avoidance of misruns, but can be set independently to control grain size without the addition of grain refiners. It is clear, therefore, that a good running system is a good ally in the creation of economical products of high quality.

The elements of a good system are:

1. Economy of size. A lightweight system will increase yield (the ratio of finished casting weight to total cast weight), allowing the foundry to make more castings from the existing melt supply. It may also help to get more castings into a given mould size. This has a big effect on productivity and economy.
2. The filling of the mould at the required speed. In the method proposed in this book, the whole running system is designed so that the velocity of the metal in the gates is below the critical value. This value varies from one alloy to another, but is generally close to $0.5\,\mathrm{m\,s^{-1}}$. There is now much experimental and theoretical data to support this value (Runyoro 1992). Data on the density of castings produced by gating uphill have shown that air entrapment can occur above approximately $0.5\,\mathrm{m\,s^{-1}}$ (Suzuki 1989). In computer simulations of flow, Lin and Hwang (1988) show that when liquid aluminium enters the mould horizontally at $1.1\,\mathrm{m\,s^{-1}}$ it hits the far wall with such force that the reflected wave breaks, causing

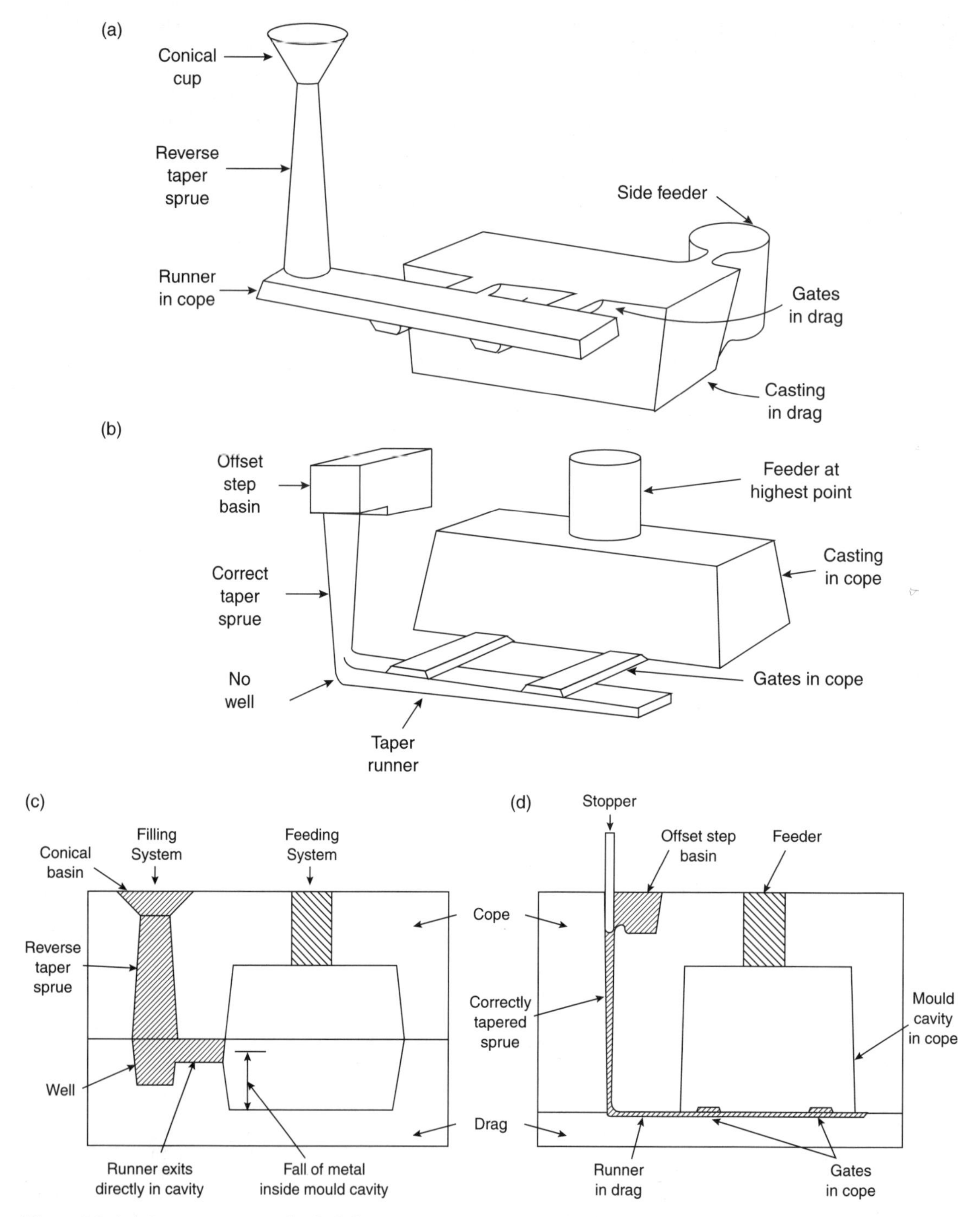

Figure 2.3 *(a) Poor top gates and side-fed running system, compared with (b) a more satisfactory bottom-gated and top-fed system (c) poor system gated at joint and (d) recommended economical and effective system.*

surface turbulence. These figures confirm the safety of $0.5\,m\,s^{-1}$, and the danger of exceeding $1\,m\,s^{-1}$.

3. The delivery of only liquid metal into the mould cavity, i.e. not other phases such as slag, oxide, and sand. However, in most cases the overwhelmingly common and unwelcome phase is air (probably contaminated with other mould gases of course). The design of filling systems to achieve the exclusion of air will constitute a major preoccupation in this book.
4. The elimination of surface turbulence, preferably at an early stage in the runner system, but certainly by the time that the metal arrives in the mould cavity. The problem here is that by the time the metal has fallen the length of the sprue to reach the lowest level of the casting, its velocity is well above the critical velocity for surface turbulence. Despite this danger, the running system should, so far as possible, prevent the resulting fragmentation of the stream. Any fragmentation will result in permanent damage to the casting in most alloys. However, if fragmentation occurs, the best that can now happen is that it should be followed by an action to gather the stream together again. In this way the melt enters the mould as a coherent, compact spreading front, preferably at a velocity sufficiently low that the danger of any further break-up of the front is eliminated.
5. Ease of removal. Preferably the system should break off. As a next best option, it should be removable with a single stroke of a clipping press, or a straight cut. Curved cuts take more time and are more difficult to dress to finished size by grinding or linishing. Internal or shielded gates may need to be machined off, in which case the expense of setting up the casting for machining might be avoidable by carrying out this task later, during the general machining of the casting.

(Note that in general practice it is usually best to assume that there is no requirement for the filling system to act as a feeder, i.e. to compensate for the contraction on solidification. We should ensure that the feeding function if necessary at all, is carried out by a separate feeder placed elsewhere, preferably high up, on the casting (Figure 2.3b). In some cases it is possible to use a running system that can also act as a feeder. These special systems should be used whenever possible. They are considered in Chapter 6. It is worth noting that in investment casting the almost universal confusion between filling and feeding systems is deeply regrettable. In this book the two functions are treated totally separately.)

Because the above list of criteria have been so difficult to meet in practice, there has been a move away from gravity casting as a result of what have been believed to be insoluble barriers to the attainment of high quality and reliability. Uphill filling, against gravity, known as counter-gravity casting (and, more colloquially and less helpfully, as low-pressure casting), has provided a solution to the elimination of surface turbulence. It has seemed to be the ultimate development of bottom gating (Figure 2.4). This development has therefore provided the impetus for the growth of low-pressure die casting, low-pressure sand casting, and various forms of counter-gravity filling of investment castings. A form of high-pressure die-casting has also been developed to take advantage of the quality benefits associated with counter-gravity filling followed by high-pressure consolidation. These different techniques of getting the metal into the mould will all be discussed later.

However, although counter-gravity filling fulfils all the above requirements, our main aim

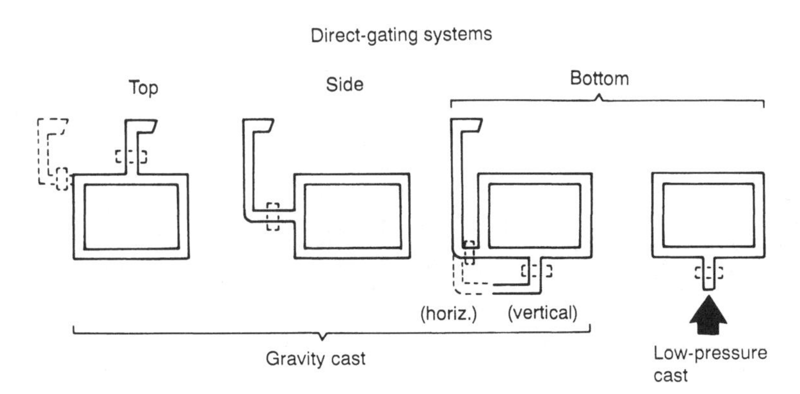

Figure 2.4 *Various direct gating systems applied to a box shaped casting. Possible filter locations are shown as dashed outlines. Note that all of the gravity systems shown here are poor: the sprue base connects directly with the ingate into the casting. All need mechanisms (not shown for clarity) to reduce the velocity of the melt.*

in this section is to evaluate gravity filling, to see how far it can meet this difficult set of criteria.

Requirement 3 for good gating is important: only liquid metal should enter the casting. Thus all bubbles entrained by the surface turbulence characterizing the early part of the running system should have been eliminated by this stage. If the running system is poor, and bubbles are still present, their rise and bursting at the liquid surface in the mould violates Rule 4. This violation results in a number of problems, including bubble trails, splash defects, and the retention of the scattering of smaller bubbles that remain trapped under the oxide skin of the rising metal. These cause concentrations of medium-sized pores (0.5–5 mm diameter) at specific locations in the casting, usually at upper surfaces of the casting above the ingates.

The other point in Requirement 3, that dross or slag does not enter the mould cavity is interesting. In the production of iron castings it is normal for the runner to be placed in the cope and the gates in the drag, as is illustrated in Figure 2.3a. The thinking behind this design of system is that slag will float to the top of the runner, and thus will not enter the gates. Such thinking is at fault because it is clear that at least some of the first metal to enter the runner will fall down the first gate that it meets, taking with it not only the first slag but also air. This premature delivery of metal into the mould before the runner is full is clearly unsatisfactory. The metal has had insufficient time to settle down, to organize itself free from dross, oxide and bubbles. The fact that such systems are widely used, and are found in practice to reduce bubble defects in the casting actually reveals how poor the front end of the running system is. Clearly, bubbles are being generated throughout the pour, so the off-take of gates at the base of the runner is valuable in this case.

A more satisfactory system is illustrated in Figure 2.3b. Here the runner is in the drag and the gates in the cope. In this system the runner has to fill first before the gates are reached. Thus the metal has a short but valuable time to rid itself of bubbles and dross, most of which can be trapped in the dross trap or against the upper surface of the runner. Only a limited amount of slag or dross will be unfortunately placed to enter the gate. Provided the velocity of the metal in the gate is not too high, even this slag still has a good chance of being held against the ceiling of the gate, and thus not entering the casting. Figure 2.3d illustrates an optimum system (contrasting with 2.3c), designed to resist the entrainment of air at all stages of the system.

Statement 4 is deceptively simple. However, the requirement of no surface turbulence is so important, and so central to the quest for good castings, that we have to consider it at length.

Texts elsewhere often refer to turbulence-free filling as laminar filling. The implication here is that turbulence as defined by Reynold's number is involved, and that the desirable criterion is that of laminar flow of the bulk. As discussed in *Castings 2003*, it is not bulk turbulence that is relevant since turbulent flow in the bulk liquid can still be accompanied by the desirable smooth flow of the surface. Our attention requires to be concentrated on the behaviour of the liquid surface. Thus provided we ensure that by 'laminar fill' we mean 'surface laminar fill', then we shall have our concepts correct, and our thinking accurate.

Requirement 4 above is clearly violated by splashing during filling. It can be seen immediately that top gating will probably therefore always introduce some defects (the exception is very thin wall castings where surface tension takes over control of surface turbulence). Figure 2.4 illustrates a poor running system where the metal enters from top or side gates that allow the metal to suffer a free fall into the mould cavity. Bottom-gated systems are always required if surface turbulence is to be eliminated.

However, although bottom gating is necessary, it is not a sufficient criterion. It is easy to design a bad bottom-gating system! In fact, it is possible to state the case more forcefully: a bad bottom-gated system is usually worse than most top-gated systems.

For instance, it is common to see bottom-gated systems proudly displayed with the base of the runner turned so that metal directly enters the mould (Figure 2.5). Such systems are

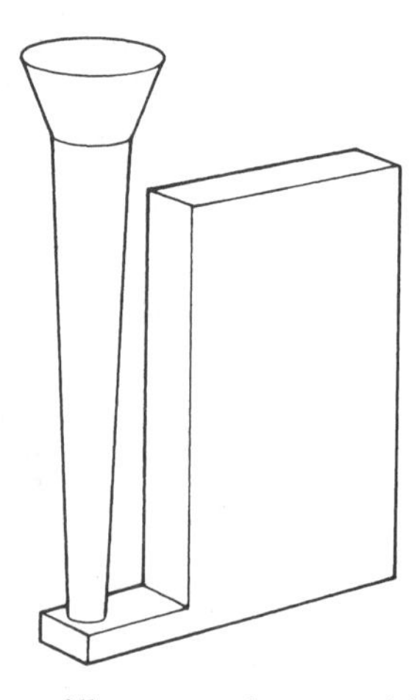

Figure 2.5 *A poor filling system because of direct entry of high velocity metal into the mould cavity.*

compact, and appear economical until the percentage scrap figures are inspected. The sequence of events is clear if we consider the fall of the first liquid down the length of the sprue. The high velocity of the metal on its impact at its base is not contained. The resulting splash may be likened to an explosion of high-velocity drops or jets fired like projectiles directly into the mould. The bulk of the metal follows in an untidy fashion, mixed with air and mould gases, and ricochets from the far wall, causing more surface turbulence as the rebounding wave breaks, rolling over and entraining yet more surface and more gas. The elimination of the entrained bubbles by bursting as they rise to the surface of the melt causes additional droplets to be created by splashing. It is important, therefore, to design the down-runner with care so that it will fill quickly, excluding air as quickly as possible, and to design the runner and gate to constrain the metal, avoiding any provision of room for splashing (Figure 2.6a). Further improvements might be allowable as in Figures 2.6b and 2.6c in which the fall heights down the sprue are progressively reduced, reducing velocities in the mould, by simply re-orienting the casting.

The base of the sprue should be the lowest point in the whole system: having reached here, all subsequent flow of the liquid should be uphill, displacing the air ahead in a controlled and progressive advance. So far as possible, the liquid should be slowed as it goes, experiencing as much opportunity as possible to become quiescent before entering the mould. It should finally enter the mould at a velocity less than its critical velocity for the entrainment of defects. In this way a good and reproducible casting is favoured.

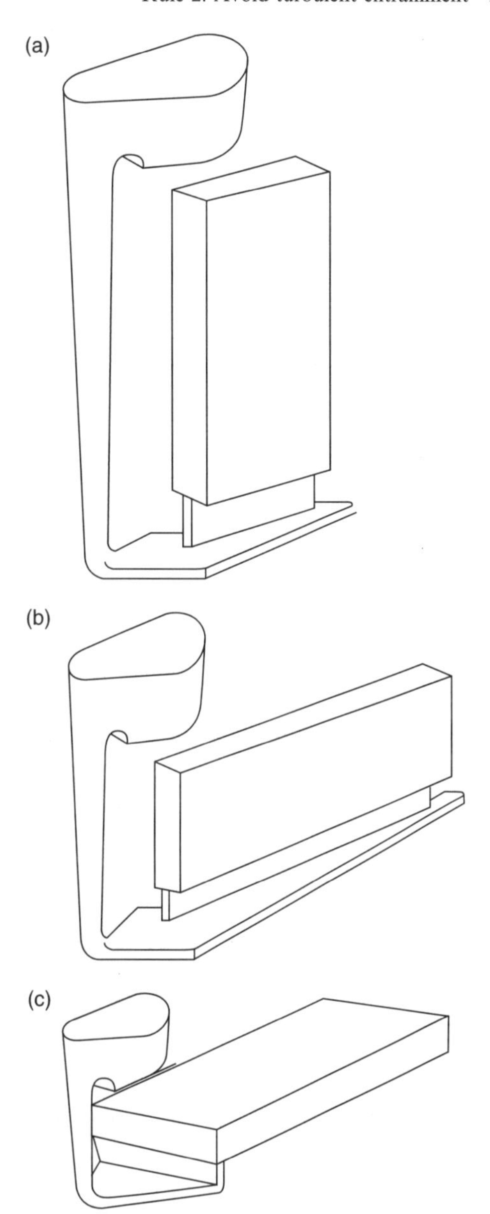

Figure 2.6 *(a) An improved bottom-gated system; (b) and (c) further improved by height reductions.*

2.3.2.1 Pressurized versus unpressurized

In the book *Castings 1991* the author recommended the achievement of velocity reduction by the progressive enlargement of the area of the flow channels at each stage, with the aim of progressively reducing the rate of flow. This is known as an unpressurized running system. The aim was to ensure that the gate was of a sufficient area to make a final reduction to the speed of the melt, so that it entered the mould at a speed no greater than its critical velocity. More recent research, however, has demonstrated that the enlargement of the system, by, for instance, a factor of two as the flow emerges from the exit of the sprue and enters the runner, usually fails to fill the runner. Thus the unpressurized systems unfortunately behaved poorly, entraining bubbles and oxides, because much of the system runs only partly full. The other standard criticism (but incidentally of much less importance) was that unpressurized systems are heavy, thus reducing metallic yield, and thus costly.

In fact, video radiography reveals that at the abrupt increase in cross-section at the base of the sprue on the entry to the runner, the entrainment of air occurs with dramatic effectiveness. This is because the melt jets along the base of the runner (not filling the additional area provided) and hits the end of the runner.

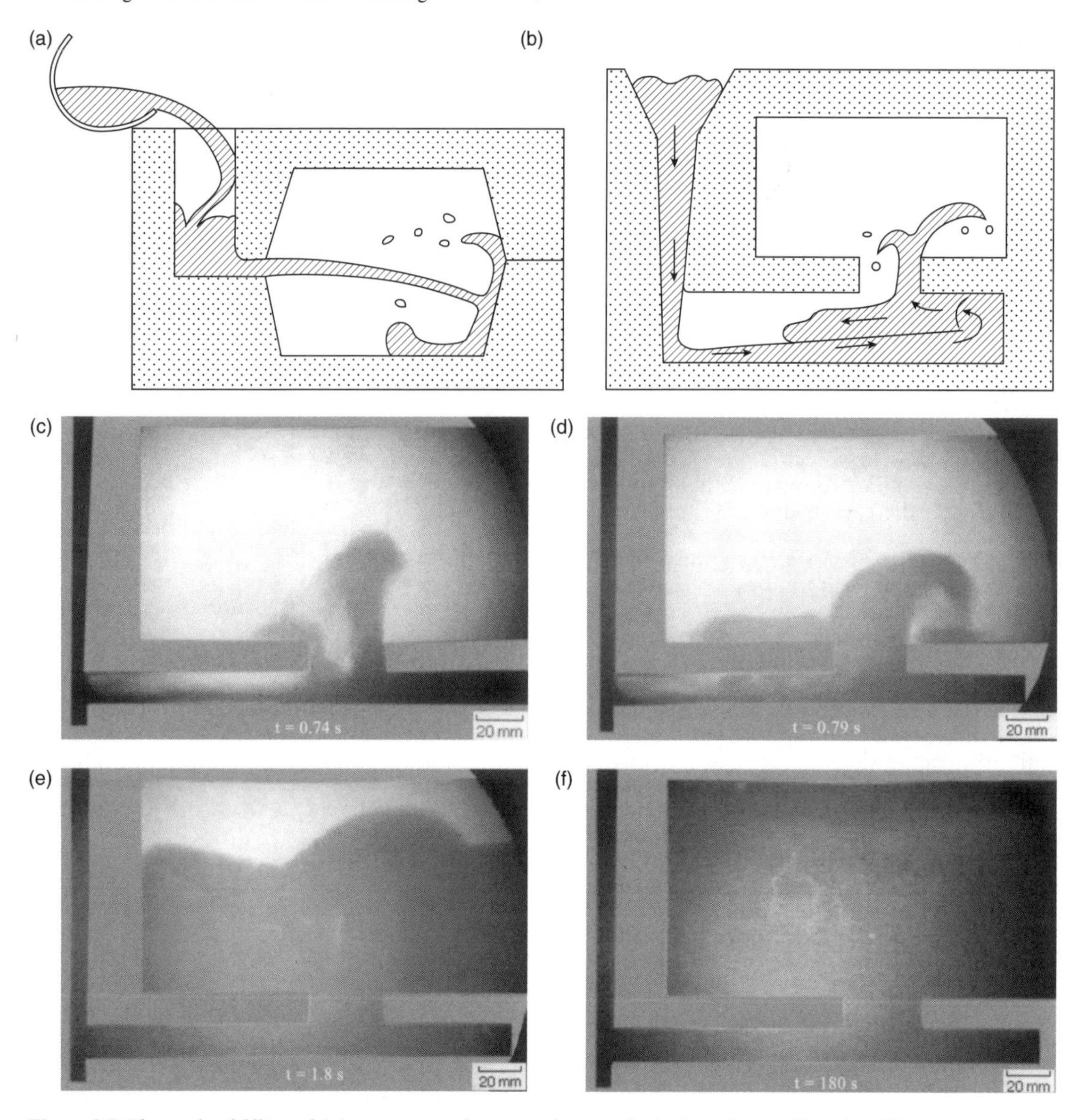

Figure 2.7 *The mode of filling of (a) a pressurized system, showing the jet into the mould cavity; (b) an unpressurized system, showing the fast underjet, and the rolling back wave in the oversized runner (c, d, e) X-ray video frames of an Al alloy filling a mould 100 mm high × 200 mm wide × 20 mm deep illustrating the unpressurized system; (f) the final casting showing subsurface bubbles and internal cracks.*

The reflected back wave rolls over the underlying fast jet, rolling in oxides and bubbles at the interface between the two (Figure 2.7a). The effect can be long-lived, developing into a stable *hydraulic jump*. The bubbles travel along the interface between the two opposing streams (probably because of the presence of two non-wetting oxide films separating the two flowing streams) and progress to the ingate, usually collecting in a low pressure zone on one side of the ingate, before proceeding to swim up through the metal in the mould cavity (2.7e). Naturally, these bubbles and oxides bequeath serious permanent damage to the casting (2.7f).

The cast iron foundryman had some justification therefore to champion his own favourite *pressurized* systems. For the benefit of the reader, the so-called *pressurized running system* is one in which the metal flow is choked (i.e. limited by constriction) at the gate; i.e. its rate of flow into the mould is controlled by the area of the gate, the last point in the running system (Figure 2.7b). This causes the running system to back-fill from this point, and become pressurized with liquid, forcing the system to fill and exclude air. Thus the system entrains fewer bubbles and oxides. However, it also forced the metal into the mould as a jet. Clearly this system violates one of our principal rules, since the metal is now entering the mould above its critical speed. The resulting splashing and other forms of surface turbulence inside the mould introduces its own spectrum of problems, different from those of the unpressurized system, but usually harming both the quality of the mould and the casting.

Thus neither the unpressurized nor the pressurized traditional systems are seen to work satisfactorily. This is a regrettable appraisal of present casting technology.

Because for many years the pressurized systems were mainly used for cast iron, there were special reasons why the systems appeared to be adequate:

1. In the days of pouring grey iron into greensand moulds the problems of surface turbulence were minimized by the tolerance of the metal–mould system. The oxidizing environment in the greensand mould produced a liquid silicate film on the surface of the liquid iron. Thus when this was turbulently entrained it did not lead to a permanent defect (*Castings* (2003)). In fact, many good castings were produced by tipping the metal into the top of the mould, using no running system at all! Nowadays, with the use of certain core binders and mould additives that cause solid graphitic surface films on the metal, and consequently reduce its tolerance to surface turbulence, the pressurized systems are producing defects where once they were working satisfactorily. This problem has become more acute as it has become increasingly common for irons to have alloy additions such as magnesium (to make ductile iron) and chromium (for many alloyed irons).
2. Over recent years the standards required of castings have risen to an extent that the traditional foundryman is shocked and dazed. Whereas the pressurized system was at one time satisfactory, it now needs to be reviewed. The achievement of quality is now being seen to be not by inspection, but by process control. Turbulence during filling introduces a factor that will never be predictable or controllable. This ultimately will be seen as unacceptable. Reproducibility of the casting process will be guaranteed only by systems that fill the mould cavity with laminar surface flow. At one time this was achievable only with counter-gravity filling systems. Nowadays, as we shall see, we can achieve some success with gravity systems, provided they are designed correctly.

The conclusion given by the author in *Castings* (1991) was 'Unpressurized systems are recommended therefore. Pressurized are not.' This bold statement now requires revision in the light of recent research since we now find that neither system is really satisfactory.

In summary, the unpressurized system had the praiseworthy aim to reduce the gate velocity to below the critical velocity. Unfortunately such systems usually run only part-full, causing damage to the castings because of entrained air bubbles and oxides. The pressurized system probably benefited greatly from its ability to fill quickly and to run full, greatly reducing the damage from bubbles. However, the high velocity of the melt as it jetted into the mould created its own contribution to havoc.

Turning now to another sacred cow of running system design that requires to be addressed. This is the concept of a choke. The choke is a local constriction designed to limit flow. In the non-pressurized system the choke was generally at the base of the sprue, whereas the pressurized system was choked at the ingates into the mould. Unfortunately, a choke is an undesirable feature. Flow rates are usually sufficiently high that the melt will be speeded up through a constriction and emerge as a jet, entraining air once again downstream, with much consequential damage.

All these systems were devised before the benefits of computer simulation and video X-ray radiography. They also pre-dated the development of the concepts of surface turbulence, critical velocity, critical fall height and bifilms. It is not surprising therefore that all these traditional approaches to the design of filling systems gave less than satisfactory results.

In the history of the development of filling systems most of the early work was of limited value because the emphasis was on steady state flow through fully filled pipework, following the principles of hydraulics. This does, of course, sometimes occur late during the filling process. However, the real problems of filling are associated with the priming of the filling system, i.e. its behaviour before the filling system is filled. Thus these early studies give us relatively little useful background on which to base effective designs for real castings.

A completely new approach is described in this book that attempts to address these issues. We shall abandon the concept of a localized choke. The whole of the length of the filling system should experience its walls in permanent contact and gently pressurized by the liquid metal. Thus, effectively, the whole length of the running system should be designed to act like a choke; a kind of continuous choke principle. In all probability, it seems that we really need uniformly pressurized systems. An alternative description might be 'naturally pressurized' systems, because the new design concept is based on designing the flow channels in the mould so as to follow the natural form that the flowing metal wishes to take.

For instance, at the base of the sprue we can define its area as unity. After the right angle bend into the runner, if the stream loses energy so that its velocity falls by 20 per cent, we can expand the channel by this amount. The runner can remain at this area of 1.2 along its length. After turning through a further right angle bend into the gate this gives a series of permissible area ratios of 1 : 1.2 : 1.4, although it will be noticed that the ingate velocity has only fallen by approximately 40 per cent from that at the sprue exit.

If the 20 per cent expansion of area after each bend is not entirely allowed (for instance if only 10 per cent expansion were provided) the stream will experience a gentle pressurization. This modest pressure against the walls of the running system will be valuable to counter any effect of bubble formation and will act to support the walls of the running system against collapse (a special problem in large running systems for large castings). Thus to be more sure of maintaining the system completely full, and slightly pressurized, a ratio of 1 : 1.1 : 1.2 or even 1 : 1 : 1 might be used.

Examples of area ratios are shown in Table 2.1.

Table 2.1 Examples of area ratios (sprue exit area : runner area : gate area)

	Examples of area ratios
Pressurized	1:0.8:0.6
	1:1:0.8
Unpressurized	1:2:4
	1:4:4
Natural	1:1.2:1.4
Slightly pressurized	1:1:1
	1:1.1:1.2
With foam filter in gate	1:1:4
With speed reduction or by-pass designs	1:1:10

From the ratios it is clear that the naturally (or slightly) pressurized system is part-way between the pressurized and unpressurized systems.

However, there is a major problem with the use of these new systems that the reader may already have noticed. The naturally pressurized system has no built-in mechanism for any significant reduction in velocity of the stream. Thus the high velocity at the base of the sprue is maintained (with only minor reduction) into the mould. Thus the benefits of complete priming of the filling system to exclude air are lost once again on entering the mould cavity.

This fundamental problem alerts us to the fact that the naturally pressurized approach requires completely separate mechanisms to reduce the velocity of the melt through the ingates. The options include

(i) the use of filters;
(ii) the provision of specially designed runner extension systems such as flow-offs;
(iii) a surge control system;
(iv) the use of a vertical fan gate at the end of the runner. Additional mechanisms might be possible in the future, when properly researched, such as
(v) the use of vortices to absorb energy while avoiding significant surface turbulence.

We shall consider all these options in detail in due course, but the reader needs to be aware that, unfortunately, at this time the use of naturally pressurized systems is in its infancy. In particular, the rules for such designs are not yet known for some features such as joining round or square section sprues to rectangular runners. Filters are not easily incorporated, nor are vortex systems fully understood.

This creates a familiar problem for the foundry person: in the real world, the casting engineer has to take decisions on how to make things, whether or not the information is available at the time to help make the best choice. Thus insofar as the rules are presently understood for the majority of castings, they are set out below, for good or for bad. I hope they assist the caster to achieve a good result. One day I hope we in the industry will all have the better answers that we need.

In the meantime, computers are starting to simulate successfully the flow of metal in filling systems. At the present time such simulations are highly computationally intensive, and therefore slow and/or not particularly accurate. It is necessary to be aware that some simulation packages are still highly inaccurate. However, time will improve this situation, to the great benefit of casting quality.

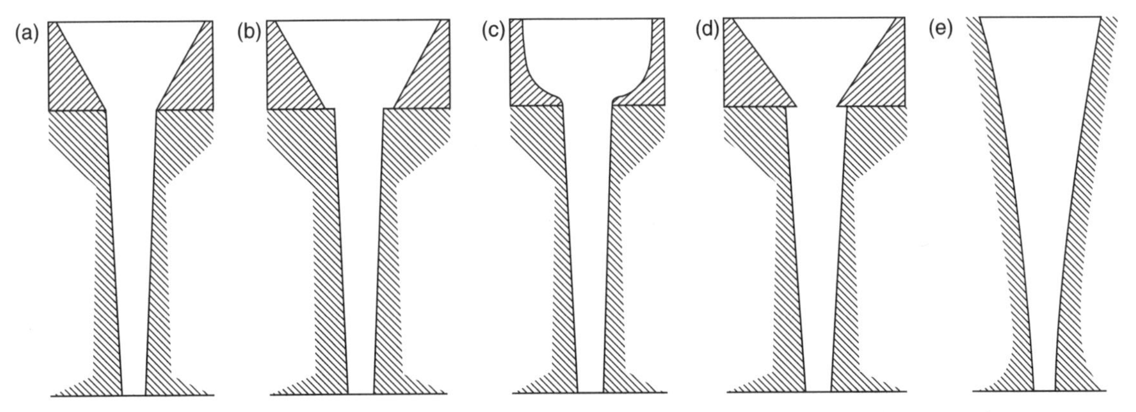

Figure 2.8 *A rogues gallery of non-recommended scrap generating systems. Conical basin and sprue combinations showing (a) perhaps least damaging; (b) basin too large; (c) cup form; (d) basin too small; (e) enlarged sprue to act as a combined basin and sprue.*

2.3.2.2 Design of pouring basin

The conical basin

The in-line conical basin (Figure 2.8a), used almost everywhere in the casting industry, appears to be about as bad as could be envisaged for most casting operations. It is probably responsible for the production of more casting scrap than any other single feature of the filling system. It is not recommended.

The problems with the use of the conical basin arise as a result of a number of factors:

1. The metal enters at an unknown velocity, making the estimation of the design of the remainder of the running system problematical.
2. The metal enters at high, unchecked velocity. Since the main problem with running systems is to reduce the velocity, this adds to the difficulty of reducing surface turbulence.
3. Any contaminants such as dross or slag that enter with the melt are necessarily taken directly down the sprue.
4. The device works as an air pump, concentrating air into the flow (the action is analogous to other funnel-shaped pumps in which a fast stream of fluid directed down the centre of the funnel is designed to entrain a second surrounding fluid. Good examples are steam ejectors and the vacuum suction device that can be driven from a compressed air supply). Because air is probably the single most important contaminant in running systems, this is probably the most severe disadvantage, yet is not widely appreciated to be a problem.

   An example that the author has witnessed many times can be quoted. Bottom-teemed steel ingots were produced by a conventional arrangement that consisted of pouring the steel into a central conical cup, affixed to the top of a spider distribution system of ceramic tubes connected to the centre of the base of a group of four or six surrounding ingot moulds. Because the top of the ingot mould remained open during filling, the upwelling cascade of air bubbles in the centre of the rising metal was clear for all to see. (The bottom fill technique was designed to deliver an improved surface condition of the ingot as a result of the gentle rolling action of the liquid meniscus against the wall of the ingot mould as the metal ascended. However, the overall cleanness of the ingot would have been significantly impaired by the passage of so much air. It would have been useful to retain the benefits of the bottom-teemed ladles and yet achieve improved castings by reducing the entrainment of air into the system.)
5. The small volume of the basin makes it difficult for the pourer to keep full (its *response time* is too short, as explained later), so that air is automatically entrained as the basin becomes partially empty from time to time during pouring. The pourer is usually unaware of this, since the aspiration of air usually takes place under the surface at the basin/sprue junction.
6. The mould cavity fills differently depending on precisely where in the basin the pourer directs the pouring stream, whether at the far side of the cone, the centre, or the near side. Thus the castings are intrinsically not reproducible.
7. This type of basin is most susceptible to the formation of a vortex, because any slight off-axis direction will tend to start a rotation of the pool. There has been much written

about the dire dangers of a vortex, and some basins are provided with a flat side to discourage its formation. In fact, however, this so-called disadvantage would only have substance if the vortex continued down the length of the sprue, along the runner and into the mould cavity. This is unlikely. Usually, a vortex will 'bottom out,' giving an air-free flow into the remaining runner system as will be discussed later. This imagined problem is almost certainly the least of the difficulties introduced by the conical basin.

If this long list of faults was not already damning enough, it is made even worse for a variety of reasons. A basin that is too large for the sprue entrance (Figure 2.8b) jets metal horizontally off the exposed ledge formed by the top of the mould, creating much turbulence and preventing the filling of the sprue. The problem is unseen by the caster, who, because he is keeping the basin full, imagines he is doing a good job. The cup shape of the basin (Figure 2.8c) is bad for the same reason. The basin that is too small (Figure 2.8d) has painful memories for the writer: a casting with an otherwise excellent running system was repeatedly wrecked by such a simple oversight! Again, the caster thought he was doing a good job. However, the aspirated air caused a staggering amount of bubble damage in an aluminium sump casting.

The expansion of the sprue entrance to act as a basin (Figure 2.8e) may hold the record for air entrainment (however the author has no plans to expend effort investigating this black claim). Worse still, the top of this awful device is usually not sufficiently wide that the pourer can fill it because it is too small to hit with the stream of metal without the danger of much metal splashed all over the top of the mould and surroundings. Thus this combined 'basin/sprue' necessarily runs partially empty for most of the time. Furthermore, the velocity of the melt is increased as the jet is compressed into the narrow exit from the sprue (this point is discussed in detail later). The elongated tapered basin system has been misguidedly chosen for its ease of moulding. There could hardly be a worse way to introduce metal to the mould.

For very small castings weighing only a few grams, and where the sprue is only a few millimetres diameter, there is a strong element of control of the filling of the sprue by surface tension. For such small castings the conical pouring cup probably works tolerably well. It is simple and economical, and, probably fills well enough. This is as much good as can be said about the conical basin. Probably even this is praising too highly.

Where the conical cup is filled with a hand ladle held just above the cone, the fall distance of about 50 mm above the entrance to the sprue results in a speed of entry into the sprue of approximately 1 m s^{-1}. At such speeds the basin is probably least harmful. On the other hand, where the conical cup is used to funnel metal into the running system when poured directly from a furnace, or from many automatic pouring systems, the distance of fall is usually much greater, often 200 to 500 mm. In such situations the rate of entry of the metal into the system is probably several metres per second. From the bottom-poured ladles in steel foundries the metal head is usually over 1 m giving an entry velocity of 5 m s^{-1}. This situation highlights one of the drawbacks of the conical pouring basin; it contains no mechanism to control the speed of entry of liquid.

The pouring cup needs to be kept full of metal during the whole duration of the pour. If it is allowed to empty at any stage then air and dross will enter the system. Many castings have been spoiled by a slow pour, where the pouring is carried out too slowly, allowing the stream to dribble down the sprue, or simply poured down the centre without touching the sides of the sprue, and without filling the basin at all (which is the trouble with the expanded sprue type). Alternatively, harm can be done by inattention, so that the pour is interrupted, allowing the bush to empty and air to enter the down-runner before pouring is restarted. Even so, because of the small volume of the basin, it is not easily kept full so that these dangers are a constant threat to the quality of the casting.

Unfortunately, even keeping the pouring cup full during the pour is no guarantee of good castings if the cup exit and the sprue entrance are not well matched, as we have seen above. This is the most important reason for moulding the cup and the filling system integral with the mould if possible.

Finally, even if the pour is carried out as well as possible, any witness of the filling of a conical basin will need no convincing that the high velocity of filling, aimed straight into the top of the sprue, will cause oxides and air to be carried directly into the running system, and so into the casting. For castings where quality is at a premium, or where castings are simply required to be adequate but repeatable, the conical basin is definitely not recommended.

Inert gas shroud

A shroud is the cloth draped as a traditional covering over a coffin. This sober meaning does convey the sense in which the word is used in the foundry.

The inert gas shroud has been adopted in some steel foundries. The device is a protective shield around the metal stream issuing from a bottom-poured ladle, rather like a collar, providing an inert gas environment, usually argon. Its purpose is to reduce re-oxidation of the steel during casting.

It is difficult to believe that a user would think that the short distance between the ladle and the conical basin was influential in any substantial reduction of the oxidation of the melt. Usually, the time involved in this short journey will be probably only a few hundred milliseconds. It is not easy therefore to escape the conclusion that users were in fact tacitly acknowledging the air pump action of the conical basin. The shroud therefore encourages argon to be sucked into the cone instead of air, assuming that the rate of delivery of argon is sufficient (since such pumps usually transfer roughly equal volumes of pumping fluid and entrained fluid).

The beneficial action of an argon shroud is that the reactive gas is simply replaced by an inactive gas. Thus although volumes of bubbles will continue to be entrained with the flow, they at least do not react to produce oxides or nitrides.

In fact of course, the shroud will never be completely protective for various reasons: the gas itself will be contaminated with oxygen, water vapour and other gases and volatiles in the plumbing system that delivers the gas. More important still, the seal of the shroud around the stream cannot be made proof against leakage of air; and finally the outgassing from the mould, especially in the case of an aggregate (sand) mould, will be massive.

Even so, when used appropriately, the shroud is useful. It greatly reduces re-oxidation problems of steels during casting as demonstrated by research carried out by the Steel Founders Society of America (2000). The result emphasizes the damage done by the emulsion of steel and air bubbles that characterizes the average poorly designed casting system.

The shroud has been taken to an extreme form as a long silica tube mounted directly to the underside of a bottom-pour ladle (Harrison Steel, USA, 1999). The tube acts as a re-usable sprue, and is inserted through the top of the mould and lowered carefully, so that its exit reaches the lowest point of the filling system. The stopper is then opened. If the seal between the ladle and tube is good, the filling rate of the mould is high. If leakage of air occurs at the seal the rate of mould filling is significantly reduced, implying the strong pumping action of the falling stream to create a vacuum in the upper part of the tube, drawing in air if it can, and thus diluting the falling stream with air. Several castings in succession can be poured from one tube. However, after the tube cools the silica fragments, and requires to be replaced. Although this solution to the protection of the metal stream from oxidation is to be admired for its ingenuity, it does appear to the author to be awkward in use. The leakage problem is always an attendant danger.

In general, the author has not opted for the shroud solution, but has preferred to put in place systems that avoid the ingestion of gases into the filling system. These various systems are described below.

Contact pouring

The attempt to exclude air during the pouring of castings is carried to its ultimate logical solution in the concept of contact pouring. In this system the metal delivery system and the mould are brought into contact so that air is effectively sealed out.

The direct contact system is of course necessary, and taken for granted in the case of counter-gravity systems, in which the mould is placed directly over a source of metal. The metal is then displaced upwards by pump or differential pressure.

In the case of gravity pouring, however, the author is only aware of one use in a foundry (VAW, now Hydro Aluminium Limited, Dilligem, Germany) casting aluminium alloy. The melt is brought to the casting station by launder (a horizontal channel). The mould is also brought up to the underside of the launder in the base of which is a nozzle closed by a stopper. When the mould is presented to and pressurized against the nozzle the stopper is opened. After the mould is filled the stopper is closed and the mould can be removed, in this particular case to be rolled immediately through 180 degrees to avoid convection and aid feeding. This system works reliably and well.

The thought of transferring the concept to steel castings, using the stopper in the base of the bottom-poured ladle to deliver directly into the mouth of a sprue is quite another matter. The engineering problems for steels are daunting at this time, but may be solved one day.

The offset basin

Another design of basin (sometimes called a bush) that has been recommended from time to time, is the offset basin (Figure 2.9a).

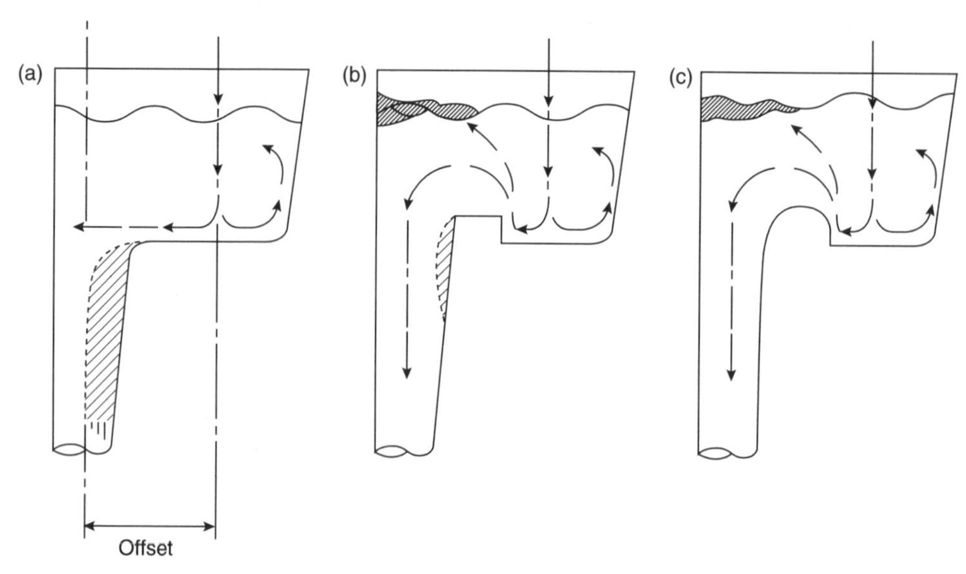

Figure 2.9 *Offset pouring basins (a) without step (definitely not recommended); (b) sharp step (not recommended); (c) radiused step (recommended).*

The floor of this basin is usually arranged to be horizontal (but sometimes sloping). The intention is that the falling stream is brought to rest prior to entering the sprue. This, unfortunately, is not true. The vertical component of flow is of course zero, but the horizontal component is practically unchecked. This sideways jet across the entrance to the sprue prevents approximately half of the sprue from filling properly, so that air is entrained once again. The horizontal component of velocity continues beneath the surface of the liquid throughout the pour, even though the basin may be filled.

There has been research using this type of basin over the years, in which the discharge coefficients from sprues have been measured and found to be in the region of 50 per cent or less. These low figures confirm that the sprue is only 50 per cent or less filled, so that the major fluid being discharged is air. The quality of any castings produced from such devices must have been lamentable.

This type of basin is definitely not recommended.

The offset step (weir) basin

The provision of a vertical step, or weir, in the basin (Figure 2.9b and c) brings the horizontal jet across the top of the sprue to a stop. It is an essential feature of a well-designed basin.

Interestingly, this basin has a long history. Sexton and Primrose described a closely similar design (but without a well-formed step) in their textbook on ironfounding published in 1911. If this basin is really valuable (as is recommended here) the reader will be curious as to why it has been known for so long, but has been extremely unpopular in foundries, whose experience of it has been discouraging. There are several reasons for this bad experience. Sometimes the basin has been made incorrectly, neglecting the important design features listed below. However, more serious than this, it has been usual to place this excellent design of basin on a filling system that completely undoes all the benefits provided by the basin. Thus the benefits of the basin are never realized, and the basin is unjustly blamed.

Despite the revered age of this basin design, the precise function and importance of each feature of the design had not been investigated until recent computer studies by Yang and Campbell (1998). These studies make it clear that

(i) The *offset* blind end of the basin is important in bringing the vertical downward velocity to a stop. The *offset* also avoids the direct *inline* type of basin, such as the conical basin, where the incoming liquid goes straight down the sprue, its velocity unchecked, and taking with it unwanted components such as air and dross, etc.

In older designs of this device the blind end of the basin was often moulded as a hemispherical cup. This was not helpful, since metal could easily be returned out of the basin by the sloping sides. The flat floor and near-vertical sides of the basin were therefore significant advantages. In fact the use of sharp corners to the offset side of the basin is positively helpful to avoid metal being ejected by the basin as discussed later.

(ii) The *step (or weir)* is essential to eliminate the fast horizontal component of flow over the top of the sprue, preventing it from filling properly. Basins without this feature commonly only approximately half fill the sprue, giving an effective so-called discharge coefficient of only approximately 0.5 (how could it be higher if the sprue is only half full?). The provision of the step yields a further bonus since it reverses the downward velocity to make an upward flow, giving some opportunity for lighter phases such as slag and bubbles to separate prior to entering the sprue. Floating debris that has separated in this way is shown schematically in Figure 2.9b, c). Again, early designs were less than ideal because the step was not vertical (Swift 1949) so that its effect was compromised. The step needs a vertical height at least equal to the height of the stream at that point to ensure that it brings the horizontal component of flow to a complete stop. Commonly, this height will be at least a few millimetres for a small casting, and might be 10 to 20 mm for a casting weighing several tonnes.

(iii) Finally, the provision of a generous *radius* over the top of the step (Figure 2.9c), smoothing the entrance into the sprue, further aids the smooth, laminar flow of metal. Swift and co-workers (1949) illustrated this effect clearly in their water models of various basins. The effect is also confirmed by the computer study by Yang and the author (1998).

The practice of placing a boom, or dam across the top of the basin (Figure 2.10) to hold back floating debris is probably counter-productive. It is seen to interfere with the natural circulation in the basin that will automatically favour the separation of buoyant phases. A dam is not recommended.

In practice, compared to the conical type, the offset step design of basin is so easy to keep full it becomes immediately popular with both caster and quality technologist alike. And, naturally, when teamed up with a well-designed filling system, the basin can demonstrate its full potential for quality improvement of the casting.

An understandable criticism is that the basins are so voluminous that they reduce yield and are

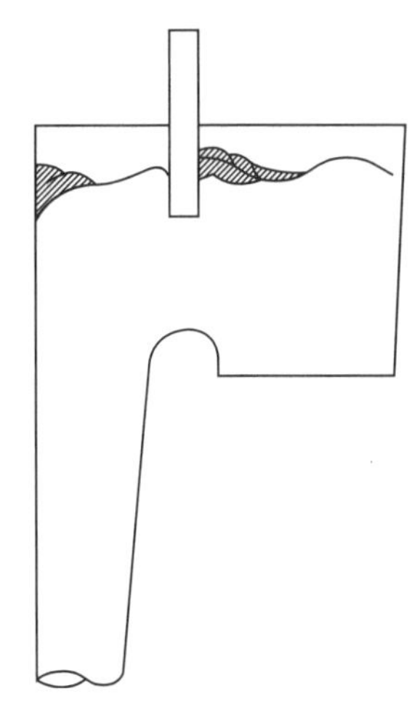

Figure 2.10 *Basin with dam (probably not helpful).*

thus costly. The usual design is shown in Figure 2.11a. Clearly the yield criticism can be completely met by ensuring that the basin drains as completely as possible by arranging it to be sufficiently higher than the casting. However, of those cases where the basin has to be placed lower and will not drain, the problem is to some extent addressed by the design variant shown in Figure 2.11b. In addition to saving money, this basin works even better because it constrains the melt more effectively. It encourages the funnelling of the melt into the sprue with excellent laminar directional guidance.

These offset step basins can be made as separate cores, stored, and planted on moulds, matching up with the sprue entrance when required. However, because they will be required for many different castings, and so will need to mate up with different sprue entrance diameters, there is concern about any mis-match of the basin exit and the sprue entrance. However, the problem is much less acute than mis-match of conical basins, because the speed of the falling stream at this point is considerably lower, in fact only at about its critical velocity. In these circumstances surface tension is able to bridge modest outstanding ledges without significant entrainment of the liquid surface. An overhanging ledge is probably more serious and to be avoided. Thus a selection of stored basins with excess exit diameter is to be preferred. In fact it may be preferable to arrange the bush to have its base completely removed on the sprue side. The bush will then fit practically any mould. Provided the entrance to the sprue on the top surface of the cope is nicely radiused, the metal will probably be adequately funnelled into the sprue (see Figure 2.23).

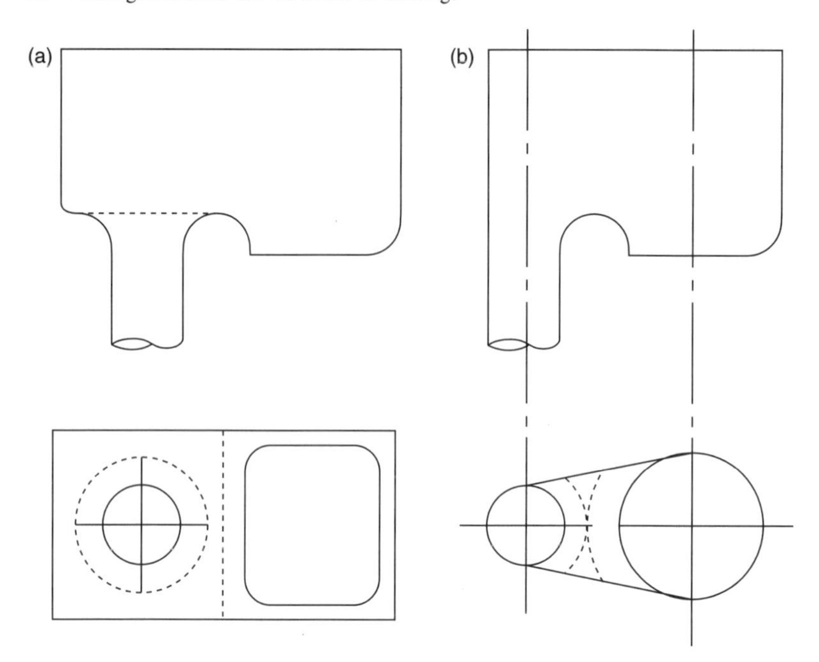

Figure 2.11 *Side and plan views of offset basins (a) conventional rectangular; (b) slimmed shape to streamline flow and improve metal yield.*

Ultimately, however, the author prefers to mould the basin integral with the sprue, and so avoiding the link-up and alignment problems. This is easily achieved with a vertical mould joint, but less easy, but still possible, with a horizontally jointed mould.

The basin is easier to use, and works more effectively, if its *response time* is approximately 1 second. To the author's knowledge there is no definition of response time. I therefore adopt a convenient measure as the time for the basin to empty completely if the pourer stops pouring. In practice, of course, the pourer does not usually stop pouring, so that the actual rate of change of level of the basin is usually at least double the response time as defined above. Such times are relatively leisurely, allowing the pourer to maintain a consistent level of melt in the basin. Different pourers or pouring systems may require times shorter or faster than this.

The volume of the basin V_b (m^3) to give a response time t_r (in seconds) at a pouring rate Q ($m^3 s^{-1}$) is given simply by

$$V_b = Q/t_r$$

Clearly, when $t_r = 1$ second, $V_b = Q$ when using the recommended SI units.

Offset stepped basin with a bottom-pour ladle

Ladles equipped with a nozzle in the base are common for the production of large steel castings. The benefits are generally described to be:

(i) the metal is delivered from beneath the surface of the melt, so avoiding the transfer of slag;
(ii) for large castings the tipping of a ladle to effect a lip pour becomes impractical;
(iii) the accuracy of the placing and the direction of the pour is valuable. Even so it is widely known in the trade that foundries using bottom pour ladles suffer dirtier castings than those steel foundries that use lip pour ladles. This follows as a natural consequence of the great difference in pouring speeds into the conical basin, with the consequent great difference in the rate of entrainment of air. (The use of bottom-pour ladles with an offset stepped basin at the entry to the mould has the potential to avoid this central problem. However, it is not without its own set of requirements that need to be studied carefully, as we shall see below.)

The common problem when using an offset stepped basin is that although a pourer using a lip pour ladle can continue to adjust the rate of pour to maintain the level of liquid at the required height in the basin, this is easier said than done if the melt is being supplied from a bottom-pour ladle whose rate of delivery often cannot be controlled, the stopper is either open or closed. Any attempt to adjust the rate of delivery results in sprays of steel in all directions.

In addition to this problem, as the bottom-teemed ladle gradually empties it reduces its rate of delivery. In the case of pouring a single casting from a ladle, it is fortunate that the filling system for the casting actually requires a falling rate of delivery as the net head (the level in the basin minus the level of metal in the mould) of metal driving the flow around the filling system gradually falls to zero. Even so, it is clear that the two rates are independently changing, and may be poorly matched at times. The match of speeds might be so bad that the basin runs empty, but even well before this moment, filling conditions are expected to be bad. At a filling level beneath the designed fill level in the basin the top of the liquid will appear to be covering the entrance to the sprue, but underneath, the sprue will not be completely filled, and so will be taking down air. It is essential therefore to ensure, somehow, that the level in the basin remains at least up to its designed level. At this time the problem of satisfactorily matching speeds can only be solved in detail by computer. Most software designed to simulate the filling of castings should be able to tackle this problem. However, it is perhaps more easily solved by simply having a basin with greatly increased depth, for instance perhaps up to four times the design depth. The ladle nozzle size is then chosen to deliver at a higher rate, causing the basin to overfill its design level, and so effectively running the casting at an increased speed. This increased speed is far preferable to the danger of underfilling the basin with the consequential ingestion of air into the melt.

In general therefore, a greatly increased depth to the basin is very much to be recommended. The problem of overfilling and increased speed of running may not be as serious as it might first appear. The reason is quickly appreciated. If the rate of delivery from the ladle is 40 per cent higher (a factor of $2^{1/2}$) than the designed rate of filling of the casting, the height of metal in the pouring basin will rise to a level twice as high (provided the basin has been provided with sufficient depth of course). A basin four times the minimum height will accommodate delivery from the ladle at up to twice as fast as the running system was designed for. The increase in pressure that this provides will drive the filling system to meet the higher rate. (Notice that the narrow sprue exit is not acting as a so-called choke, illustrating how wrong this concept is.) Thus the system is, within limits, automatically self-compensating if the basin has been provided with sufficient freeboard. It is important therefore to make sure that offset stepped basins in collaboration with a bottom poured ladle do have sufficient additional height.

The preferred option to overfill the basin in terms of height is valuable in the other common experience of using a large bottom-pour ladle to fill a succession of castings. Let us take as an example a 20 000 kg ladle that is required to pour nine castings each of 2000 kg. (The final 2000 kg in the ladle will probably be discarded because it will pour too slowly, contain too much slag and be too low in temperature; there are sometimes real problems when pouring successive castings from one ladle.) The first castings will be poured extremely rapidly because the head of metal in the ladle will be high. However, the most serious problem is that the final castings in the sequence will be poured slowly, perhaps too slowly, and so might suffer severe damage from air entrainment.

The important precaution therefore is to ensure that the final casting is still poured sufficiently quickly that the minimum height in the pouring basin is still met. This is a key requirement, and will ensure that the final casting is good. Thus all of the filling design should be based on the filling conditions for the last casting. Clearly, all the preceding castings will all be overpressurized by increased heights of metal in their pouring basins, and so will fill correspondingly faster, with correspondingly higher velocities entering the mould. This should be checked to ensure that the velocities are not so very high as to cause unacceptable damage. Usually, this approach can be made to work out well.

In some cases the first castings may have their pouring basins filled high, but the metal not yet arrived in the feeders to give a signal to the operator to stop pouring. In this case the only option is to monitor the progress of the pour by some other factor, such as precise timing, or better still, a direct read-out load cell on the overhead hoist carrying the ladle.

The matching of the speed of delivery from the ladle with the speed of flow out of the pouring basin is greatly assisted if the rate of delivery from the ladle is known. This is a complex problem dependent on the height of metal in the ladle, its diameter, and the diameter of the nozzle. The interaction of all these factors can be assessed using the nomogram provided in the Appendix.

The sharp-edged or undercut offset weir basin

In addition to the matching of the rate of flow between the ladle and the casting, there are

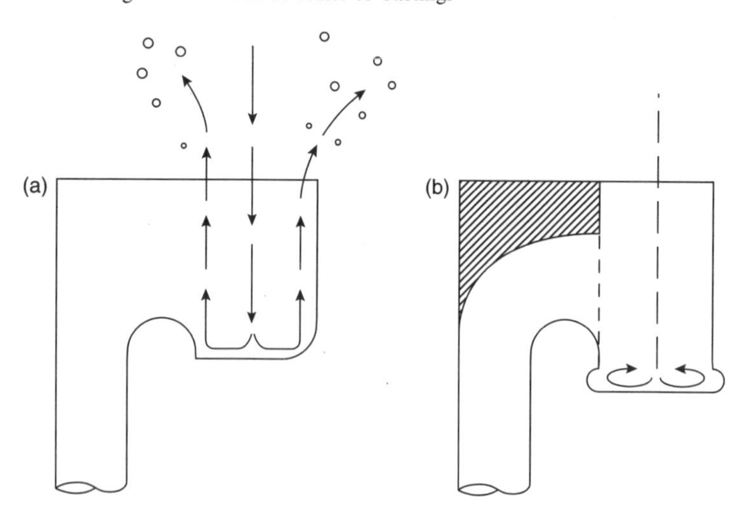

Figure 2.12 *Offset basins for high velocity input. (a) No undercut empties spectacularly upwards (not recommended). (b) The provision of an undercut gives a basin that does not splash a drop. The shaded area can be moulded in a vertical jointed mould to further improve flow and metal yield, and prevents any risk of pouring directly down the sprue.*

additional problems with the application of offset weir basins for use with bottom-poured ladles.

As we have discussed above, the velocity of the melt exiting the base of the bottom-poured ladle when the stopper is first opened is soberingly high. This is because the melt at the base of a full ladle is highly pressurized. Effectively it has fallen from the upper surface of the melt in the ladle; often as much as a metre or more. Thus the exit speed is often in the region of 4 or 5 $\mathrm{m\,s^{-1}}$. This is so high that if this powerful jet is directed into the blind end of a step basin, the liquid metal flashes outwards over the base, hits the radii in the corners of the vertical sides, where it is turned upwards to spray all over the foundry (Figure 2.12a). Such spectacular pyrotechnic displays are not recommended; little metal enters the mould.

The small radii around the four sides of the off-axis well of the basin are extremely effective in redirecting the flow upwards and out of the basin. One solution to this problem is therefore simply the removal of the radii. The provision of sharp corners to all four sides reduces the splashing tendency to a minimum (the top of the weir step leading over to the sprue entrance should still be nicely radiused of course).

The sharp cornered basin is a useful design. However, an ultimate solution to the splashing problem is provided by a simple re-entrant undercut at the base of the basin (Figure 2.12b). (The author demonstrated such a basin in a steel foundry while foundry personnel hid behind pillars and doors. On the opening of the ladle stopper the stream gushed into the basin, but not a drop emerged. The pouring process was quiet; its intense energy tamed for the first time. The foundry personnel emerged from their hiding places to gaze in wonder.)

The undercut is, of course, a problem for many greensand moulding operations making horizontally parted moulds. This is why the sharp-edged basin is so useful. Even so, where extreme incoming velocities are involved, an undercut edge to all four sides of the filling well of the basin may be the only solution.

The undercut may be difficult to mould, but it can be machined. The upgrading of a sprue cutter to 3-D machining unit equipped with a ball-ended high speed cutter would make short work of the basin, complete with its undercut and sprue entrance, and providing all this within the moulding cycle time. Such a unit would be an expensive sprue cutter, but would be a good investment.

The undercut is not a problem for vertically jointed moulds. Its use on machines such as Disamatics is popular and welcomed by the foundry operators. Its quiet filling is easily controlled, and there is complete absence of splashed metal (commonly seen as pools, sometimes nearly lakes, swimming around on the tops of moulds). The reduction of pouring overspill is a significant contribution to the raising of metal yield in the foundry.

The moulding of the sprue cover (Figure 2.12b) ensures that metal is never poured in error directly down the sprue, and saves a little metal, making a further small contribution to yield. (In some iron foundries, however, the design may be less good at holding back slag since there is now less volume provided for slag to accumulate.)

If the offset stepped basin is successfully maintained full, the head of metal provided by the height of the down-runner will be steady, and the rate of flow will be controlled by the sprue. The filling rate will be no longer at the mercy of the human operator on that day. The running system will the have the best chance to work in accord with the casting engineer's calculations.

Stopper

As a further sophistication of the use of the offset step basin, some foundries place a small sand core in the entrance to the sprue. The core floats only after the bush is full, and therefore ensures that only clean metal is allowed to enter the sprue. Alternatively, a wire attached to the core, or a long stopper rod lifted by hand accomplishes the same task. For a large casting the raising of the stopper will require a more ruggedly engineered solution, involving the benefit of the action of a long lever to add to the mechanical advantage and keep the operator well away from sparks and splashes. However it is achieved, the delayed opening of the down-runner is valuable in many foundry situations.

The early work on the development of filling systems at Birmingham concentrated on the use of the offset step basin. A stopper was not used because it was considered to be too much trouble. However, after about the first 12 months, as a gesture to scientific diligence, it was felt that the action of a stopper should be checked, if only once, by observing the filling of a sprue using the video X-ray radiographic unit, comparing filling conditions with and without a stopper. A stopper was placed in the sprue entrance, sealing the sprue. The metal was poured into the basin. When the basin was filled to the correct level the stopper was raised. The pouring action to keep the basin full was then continued until the mould was filled. The results were unequivocal. The use of a stopper greatly improved the filling of the sprue. It was with some resignation that the author affirmed this result. For all castings after that day, a stopper was always used.

Latimer and Read (1976) demonstrated that the use of a stopper reduced the fill time by 60 per cent. This is further proof that the system runs much fuller.

There seems little doubt therefore that, despite the inconvenience, when the best quality castings are required, a stopper is advisable. Thus the author always recommends its use for aerospace products.

In addition, the use of stoppers is particularly useful for very large castings where different levels of the filling system are activated by the progressive opening of stoppers as the melt level rises in the mould, so bringing into action new sources of metal to raise the filling speed.

2.3.2.3 Sprue (down-runner)

The sprue has the difficult job of getting the melt down to the lowest level of the mould while introducing a minimum of defects despite the high velocity of the stream.

The fundamental problem with the design of sprues is that the length of fall down the sprue greatly exceeds the critical fall height. The height at which the critical velocity is reached corresponds to the height of the sessile drop for that liquid metal. Thus for aluminium this is about 13 mm, whereas for iron and steel it is only about 8 mm. Since sprues are typically 100 to 1000 mm long, the critical velocity is greatly exceeded. How then is it possible to prevent damage to the liquid? This question is not easily answered and illustrates the central problem to the design of filling systems that work using gravity. (Conversely, of course, counter-gravity systems can solve the problem at a stroke, which is their massive technical advantage.)

For the sprue at least, the problem is soluble. It seems that the secret of designing a good sprue is to make it as narrow as possible, so that the metal has minimal opportunity to break and entrain its surface during the fall. The concept on protecting the liquid from damage is either (i) to prevent it from going over its critical velocity, or (ii) if the critical velocity has to be exceeded, to protect it by constraining its flow in channels as narrow as possible so that it is not able to jump and splash.

Theoretically a design of the sprue can be seen to be achieved by tailoring a funnel in the mould of exactly the right size to fit around a freely falling stream of metal, carrying just the right quantity of metal per second (Figure 2.13). We call the funnel the down-runner, or sprue for short. Many old hands call it the spue, or spew (which, incidentally, does not appear to be a joke).

Most sprues are oversized. This is bad for metallic yield, and thus bad for economy. However, it is much worse for the metal quality, which is damaged in two important ways:

(i) The sprue takes more time to fill. Air is therefore taken down with the metal, causing severe surface turbulence in the sprue. This, of course, leads to a build-up of oxide in the sprue itself, and much consequential damage downstream from oxide and entrained air. The amount of damage to the metal caused by a poor basin and sprue can

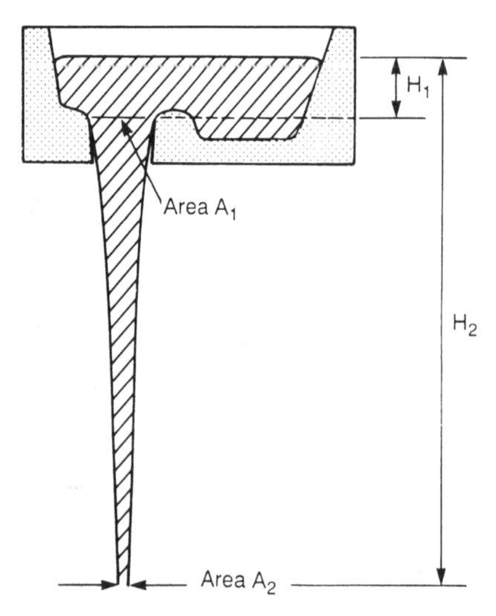

Figure 2.13 *The geometry of the stream falling freely from a basin.*

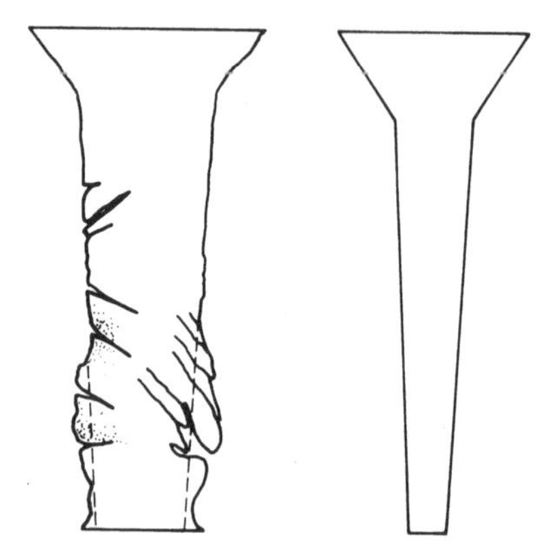

Figure 2.14 *An oversize sprue that has suffered severe erosion damage because of air entrainment during the pour. A correctly sized sprue shows a bright surface free from damage.*

be quickly appreciated from the common observation of the blockage of filters. Even with good quality liquid metal, a poor basin and sprue will create so much oxide that a filter is simply overloaded. Such poor front ends to filling systems are so common that filter manufacturers give standard recommendations of how much metal a filter can be expected to take before becoming choked. However, in contrast to what the manufacturers say, with a good basin and sprue (and providing, of course, the quality of the melt is not too bad) a filter seems capable of passing indefinite quantities of liquid metal without problem.

(ii) The free fall of the melt in an oversized sprue, together with air to oxidize away the binder in the sand, is a potent combined assault that is highly successful in destroying moulds. The hot liquid ricochets and sloshes about, its high speed and agitation punishing the mould surface with a hammering and scouring action. At the same time the pockets of air in this unsteady flow will be displaced through the sand like blasts from a blacksmith's bellows, causing the organic matter in the binder to glow, and, literally, to disappear in a puff of smoke! When the binder is burned away, reclaiming the sand back to clean, unbonded grains, the result is, of course, severe sand erosion. Figure 2.14 shows a typical result for an aluminium alloy casting in a urethane resin-bound mould. An oversize sprue is a liability.

Conversely, if the sprue is correctly sized the metal fills quickly, excluding air before any substantial oxidation of the binder has a chance to occur. The small amount of oxygen in the surface region of the mould is used up quickly by the burning of a small percentage of binder, but further oxidation has to proceed at the rate at which new supplies of air can arrive by diffusion or convection through the body of the mould. This is, of course, slow, and is therefore not important for those parts of the mould such as the sprue, that are required to survive for only the relatively short duration of the pour. Furthermore, since the liquid metal now fills the volume of the down-runner, the oxide film forming the metal–mould interface is stationary, protecting the mould material in contact with the sprue, and transmitting the gentle pressure of the steady head of metal to keep it intact. The result is a perfectly cast sprue (Figure 2.14), free from sand erosion and oxide laps. A correct-sized sprue for an aluminium alloy casting will shine like a new pin. (But beware, an undersized sprue will too!) Figure 2.3 illustrates some examples of good and bad systems. A test of a good filling system design in any metal is how well the *running system* has cast. It should be perfectly formed.

How then is it possible to be sure that the sprue is exactly the right size? The practical method of calculating the dimensions of the sprue is explained in Section 2.3.7 'Practical calculation of the filling system'. Basically, the sprue is designed to mimic the taper that the falling stream adopts naturally as a result of its acceleration due to gravity (Figure 2.15). The shape is a hyperbola (interestingly, not a parabola as widely stated). Because most sprues

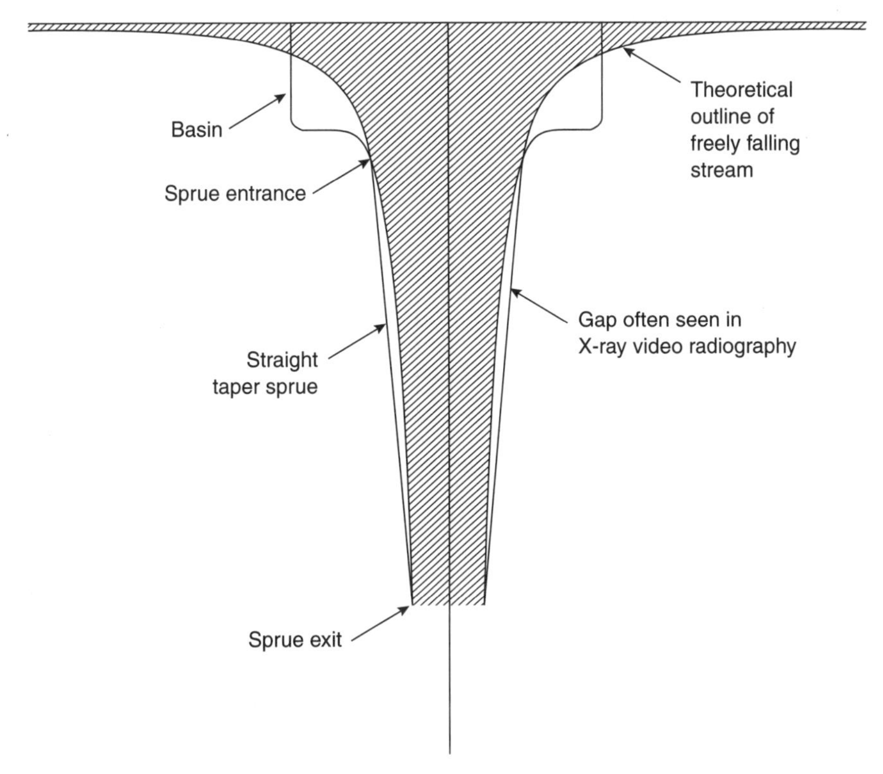

Figure 2.15 *The theoretical hyperbola shape of the falling stream, illustrating the complicating effects of the basin and sprue entrance.*

approximate the shape to a straight taper, the curved sides of the stream encourage the metal to become detached from the walls at about half-way down as shown in this figure. For modest-sized castings this (together with other errors, mainly due to the geometry and friction of the flow in the basin) is simply corrected by making the sprue entrance about 20 per cent larger in area (corresponding of course to about 10 per cent increase in diameter). Thus straight tapered sprues are commonly used, and appear to be satisfactory.

For very tall castings the straight tapered approximation to the sprue shape is definitely not satisfactory. In this case it is necessary to calculate the true diameter of the sprue at close intervals along its length. The correct form of the falling stream can then be followed with sufficient accuracy, and air entrainment during the fall can be avoided.

Using this detailed approach the author has successfully used sand sprues for very large castings (including a steel casting of about 50 000 kg and 7 m high. The sprue was assembled from a stack of tubular sand cores, accurately located by an annular stepped joint. Only one core box was required, but the central hole required a pile of separately turned loose pieces). The conventional use of ceramic tubes for the building of filling systems for steel castings was thereby avoided, with advantage to the quality of the casting. As an interesting aside, the appearance of this sprue after being broken from the mould was at first sight disappointing. It seemed that considerable sand erosion had occurred, causing the sprue to increase in diameter by over 10 mm (about 10 per cent). On closer examination however, it became clear that no erosion had occurred, but the chromite sand had softened and been compressed, losing its air spaces between the grains to become a solid mass. It had partially softened probably as a result of the use of a silicate binder system; the silicate had probably reacted with the chromite to form a lower melting point phase. Since such a growth in diameter would necessarily have occurred by a kind of creep process, in which pressure, temperature and time would be involved, it follows that much of this expansion would have happened after the casting had filled, since pressure was then highest, the sand fully up to temperature, and more time would be

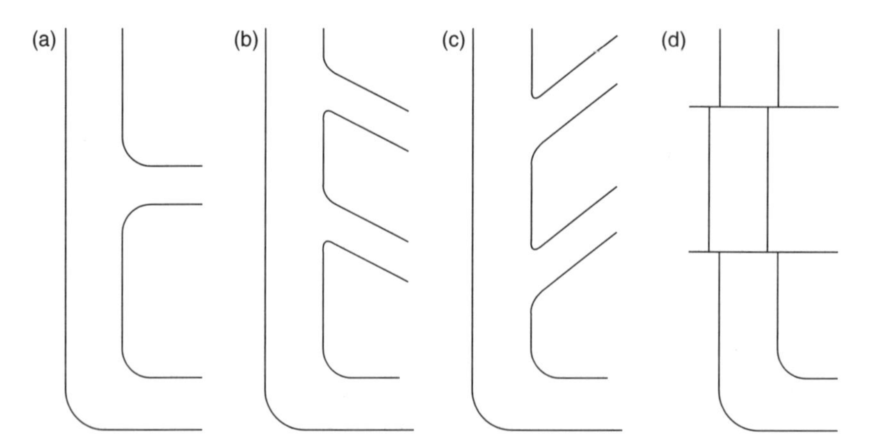

Figure 2.16 *An illustration of various kinds of common junctions or misalignments of the sprue. None are recommended.*

available because the time for the solidification of the sprue would be as much as ten times longer than the pouring time. Thus during the pour the sand-moulded sprue would almost certainly have retained a satisfactory shape, as corroborated by the predicted fill time being fulfilled, and the cleanness of the metal rising in the mould cavity was clearly seen to be satisfactory.

Sand-moulded filling systems for steel castings are, of course, prone to erosion if the system design is bad, and particularly if, as is usual, the system is oversized. In this case however, the sand-moulded sprue worked considerably better than the conventional ceramic tube system. However, there would be no doubt that the ceramic tubes would be excellent if they could be specifically designed and produced for the sprues for each individual steel casting. Naturally, at the present time this is not easily arranged. Even so, it may be found to be an economic option in view of the expensive sands and mould coatings required if sand is used alone. In addition, the ceramic tubes are extremely easy and quick to incorporate into a sand mould, often avoiding the problem of creating a new joint line in the mould.

The cross-section of the sprue can be round or square. Some authorities have strongly recommended square in the interests of reducing the tendency of the metal to rotate, forming a vortex, and so aspirating air. This probably was important in castings using conical pouring basins because any out-of-line pouring would induce rotation of the melt. However, the author has never seen any vortex formation with an offset step basin. The problem seems not to exist with good basin design.

In addition, of course, the vortex appears to be unjustifiably maligned. The central cone of air will only act to introduce air to the casting if the central cone extends into the mould cavity. This is unlikely, and in its use with the vortex sprue and other benign use of vortices, the design is specifically arranged to suppress this possibility. The vortex can be a powerful friend, as we shall see.

The attempt to provide gating or feeding off various parts of the sprue at various heights is almost always a mistake, and is to be avoided. Examples are shown in Figure 2.16. Overflow from such channels can introduce metal into the mould prematurely, where it can fall, splashing, and damaging the casting and mould before the general arrival of the melt via the intended bottom gate. Even if the channels are carefully angled backwards to avoid premature filling, they then act to aspirate air into the metal stream. Thus divided sprues usually either act to let out metal or let in air. They are not easily designed. Extreme caution is recommended. Perhaps one day we shall be able to design such features with complete safety as a result of high quality computer simulation. Those days are awaited patiently.

To summarize: for ease and safety of design at this time, the sprue should be a single, smooth, nearly vertical, tapering channel, containing no connections or interruptions of any kind. The rate of filling of the mould cavity should be under the absolute control of its cross-section area. If, therefore, the casting is found in practice to be filling a little too fast or too slow, then the rate can be modified without difficulty by slight adjustment of the size of the sprue.

Significantly, it is not simply the sprue exit that requires modification in this case. If correctly designed, the whole length of the sprue acts to control the rate of flow. This is what is meant by a naturally pressurized system. We can get the design absolutely correct for the sprue along its complete length. Although methoding engineers have been carrying out such calculations correctly for many years, somehow only the sprue exit has been considered to act as the choke. We need to take careful note of this widespread error, and perhaps take time to re-think our filling system concepts.

Turning now to a common problem with many automatic moulding units for the manufacture of horizontally parted moulds. It is regrettable that a reverse-taper sprue is usually the only practical option, flagging up a major problem with the design of nearly all of our modern automatic moulding machines. (What is worse, these units also cannot usually provide for a properly moulded basin. Despite such a basin being possible to be machined as mentioned above, production by cutting is usually never actioned.) The sprue pattern needs to be permanently fixed to the pattern plate, and therefore has to be mouldable (i.e. the mould has to be able to be withdrawn off the sprue when stripping the mould off the pattern) as seen in Figure 2.3a. In this case all is not yet lost. The top of the sprue should be designed to maintain its correct size, and the taper (now the wrong sign, remember) down the length of the sprue should be kept to a minimum. (A polished stainless steel sprue pattern can often work perfectly well with zero taper providing the stripping action is accurately square.)

Even though all precautions are taken in this way to reduce the surface turbulence to a minimum, the consequential damage to the melt by a reverse or zero tapered sprue is preferably reduced by the provision of a filter as soon as possible after the base of the sprue. The friction provided by the filter acts to hold back the flow, and thus assist the poorly shaped sprue to back-fill as completely and as quickly as possible, and so reduce the rate of damage. The filter will also act to filter out some of the damage, although it has to be realized that this filtering action is not particularly efficient. The use of filters is dealt with in detail later in Section 2.3.6.3.

We need to dwell a little longer on the importance of the use of the correct taper, so far as possible, for sprues.

The effect of too little, or even negative taper has been seen above to be detrimental to casting quality. Surely, one might expect that the opposite condition of too much taper would not be a problem, since it seems reasonable to assume that the velocity of the metal depends only on the distance of fall. However, this is not true. The head of metal in the pouring basin is the driving force experienced by the melt entering the sprue. If the sprue tapers to match the natural taper of the falling stream the only acceleration experienced by the melt *is* the acceleration due to gravity. If, however, the taper of the sprue is greater than this, the melt is correspondingly speeded up as the sprue constricts its area. This extra speed is unwelcome, since the task of the filling system designer is to reduce the speed. The effect of varying taper has been studied by video X-ray techniques. In experiments in which the sprue exit area was maintained constant, a doubling of the sprue entrance area was seen to nearly double the exit speed, with the generation of additional turbulence in the runner. Three times greater entrance area led to such increased velocities in the runner that severe bubble entrainment was created (Sirrell and Campbell 1997). This is one of the reasons why the elongated basin/sprue (Figure 2.8e) is so bad.

This effect is illustrated in Figure 2.17. For the negative tapers (a) and (b) the velocity at the sprue exit is merely that due to the fall of metal. The rate of arrival (kg s^{-1}) is of course controlled by the area of the sprue top. For the correctly sized sprue (c) the velocity and rate of delivery are substantially unchanged, although it will be noticed that the whole of the length of the sprue is now contacting and controlling the stream, to the benefit of the melt quality. Those sprues with too much taper (d) and (e) continue to deliver metal at nearly the same rate (in kg s^{-1} for instance), but at much higher speed (in m s^{-1} for instance) in proportion to the reduction in area of the exit. Far from acting as an effective restraint, the narrow sprue exit merely increases problems.

These effects were studied using real-time X-ray radiography (Sirrell *et al.* 1995) to optimize the taper, measuring the time for the sprue to back-fill, and the speed of the exiting melt (Figure 2.18). This work confirmed that the long-used 20 per cent increase of the area of the sprue entrance was a valuable correction. The consequential 20 per cent increase in velocity into the runner was an acceptable penalty to ensure that the sprue primed faster and more completely despite its straight-wall approximate shape.

Thus to summarize the effect of sprue taper; the taper has to be correct (within the 20 per cent outlined above). Too little or too much taper both lead to damage of the melt.

Multiple sprues

In magnesium alloy casting the widespread use of a parallel pair of rectangular slots to act as the sprue seems to be due to the desire for the

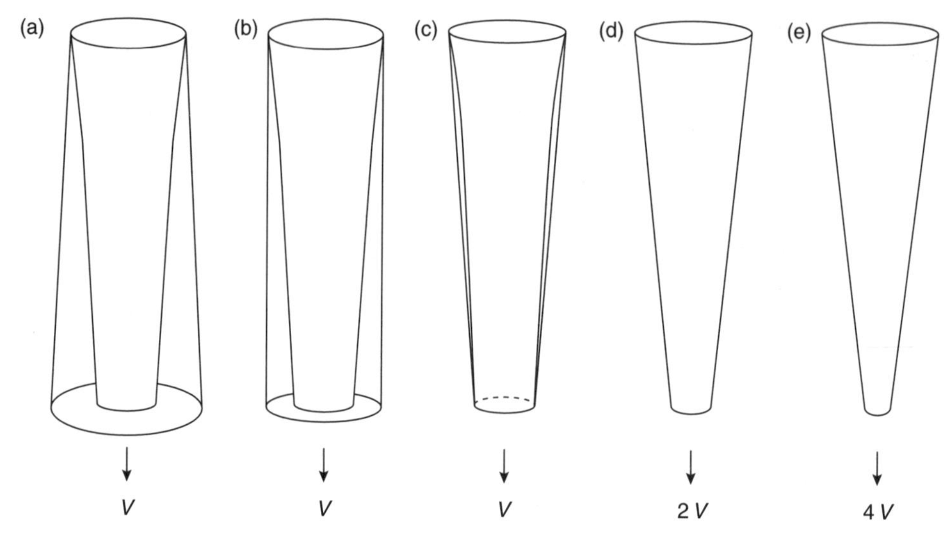

Figure 2.17 *A variety of straight tapered sprues. Too little or too much taper is bad. Only the centre taper to match the falling stream is recommended. Even this could be improved by 20 per cent additional entrance area, or better still, shaped to follow the shape of the stream.*

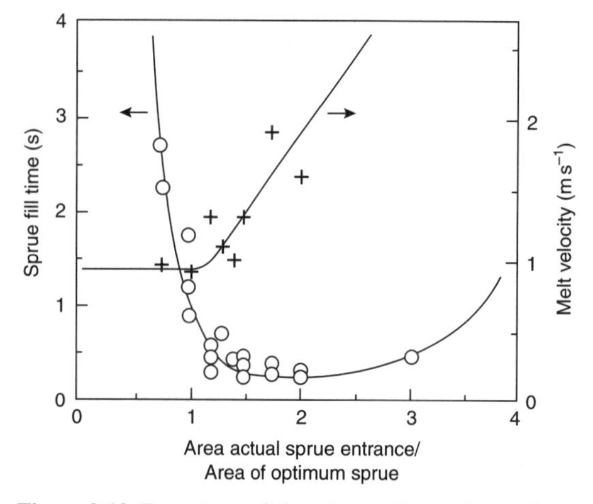

Figure 2.18 *Experimental data from video radiographic observations of sprue filling time and velocity of discharge. A taper of 1.2 is shown to be close to an optimum choice (Sirrell 1995).*

reduction in vortex formation (especially, as we have noted, if poor designs of pouring basin are employed). Swift *et al.* (1949) have used three parallel slots in their studies of the gating of aluminium alloys. However, the really useful benefit of a slot shape is probably associated with the reduction in stream velocity by the effect of friction on the increased surface area. The slots can be tapered to tailor their shape to that of the falling stream. However, to be strictly accurate, their area should be modified to make allowance for the additional frictional losses. Such sprues would probably benefit the wider casting industry. A study to confirm the extent of this expected benefit would be valuable.

A really important benefit from the use of a slot sprue appears to have been widely overlooked. This is the accuracy with which it can be attached to a slot runner to give an excellent filling pattern. This benefit is described in detail in the section below concerning the design problems of the sprue/runner junction. It seems that we should perhaps be making much more regular use of slot sprues.

When pouring a large casting whose volume is greater than can be provided from the ladle, it is common to use more than one ladle. The sequential pouring of one ladle after the other into a single basin has to be carried out smoothly because any interruption to the pour is almost certain to create defects in the casting. Simultaneous pouring is often carried out. Occasionally this can be accomplished with a single sprue, but using an enlarged pouring basin, often with a double end, either side of the sprue, allowing the ladles access from either side. Often, however, two or more sprues are used, sited at opposite ends of the mould, so as to give plenty of accessibility for ladles and cranes, and reduce the travel distance for the melt in the filling system. The correspondingly smaller area used when using more than one sprue is an advantage because they fill more easily and quickly, excluding their air more rapidly. Multiple sprues for larger castings are to be recommended and should be considered more often.

For very large castings, an interesting technique can be adopted. Several sprues can connect to runners that are arranged around the mould cavity at different heights. In the pour of a 3 m high iron casting weighing 37 000 kg described by Bromfield (1991), four sprues were arranged to exit from two pouring basins. The sprues were initially closed with graphite stoppers. The trough was first filled. The stoppers to the lowest level runner were then opened. The progress of the filling was signalled by the making of an electrical contact at a critical height of metal in the mould. In other instances witnessed by the author, the progress of filling could be observed by looking down risers or sighting holes placed on the runners. When the next level of runner was reached, announced by the bright glow of metal at the base of the sighting hole, the next level of sprues was brought into action to deliver the metal to this level of runner. In the case of the casting that was witnessed, three levels of runners were provisioned by six sprues. The technique had the great advantage that the rate of pouring did not start too fast, and then slowly decrease to zero during the course of the pour. The rate could be maintained at a more consistent level by the action of bringing in additional sprues as required. In addition, the temperature of the advancing front of the melt could also be maintained by the fresh supplies of hot metal arriving at the different levels, thus reducing the need for excessive casting temperatures to avoid misruns. Again, the significant advantages of multiple sprues are clear.

2.3.2.4 Sprue base

The point at which the falling liquid emerges from the exit of the sprue and executes a right-angle turn along the runner requires special attention. The design of this part of the liquid metal plumbing system has received much attention by researchers over the years, but with mixed results that the reader should note with caution.

The well

One of the widely used designs for a sprue base is a *well*. This is shown in Figure 2.19a. Its general size and shape has been researched in an effort to provide optimum efficiency in the reduction of air entrainment in the runner. The final optimization was a well of double the diameter of the sprue exit and double the depth of the runner. This optimization was confirmed in an elegant study by Isawa (1993) who found that the elimination of the hundreds and thousands of bubbles that were generated initially reduced exponentially with time. The exponential relationship gave a problem to define a finite time for the elimination of bubbles because the data could not be extrapolated to zero bubbles; clearly the extrapolation predicted an infinite time! He therefore cleverly extrapolated back to the time required to arrive at the last bubble, and used 'the time to the last bubble' to compare different well designs.

However, it should be noticed that both this and all the research into wells had been carried out on water models, and all had used runners of large cross-section that were not easy to fill. The result was a well design that, at best, cleared the liquid of bubbles after about 2 seconds.

For small castings that fill in only a few seconds we have to conclude that such well designs are counter-productive. In these cases it is clear that much of the filling time will be taken up conveying highly damaged metal into the mould

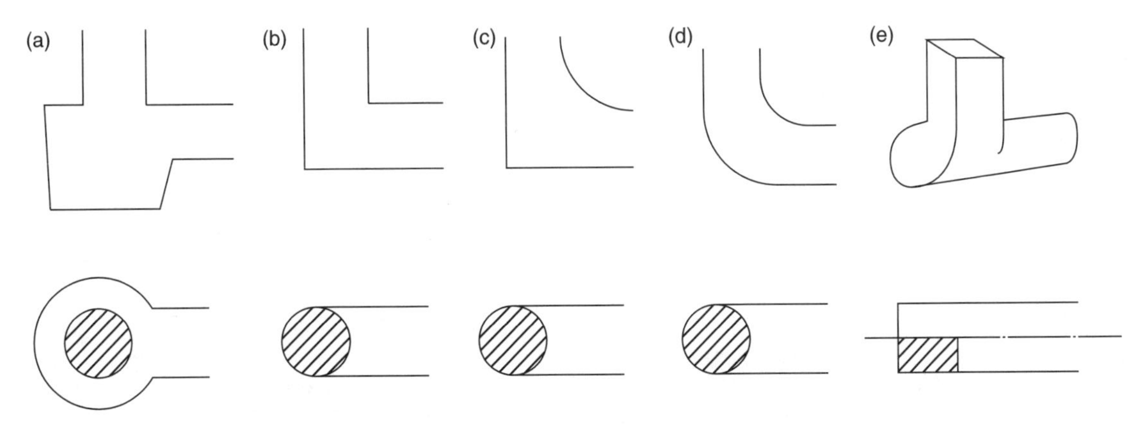

Figure 2.19 *A variety of sprue/runner junctions in side and plan views from poorest (a) to best (d). The offset junction at (e) forms a vortex flow along the cylindrical runner.*

cavity. Thus the comforting and widely held image of the well as being a 'cushion' to soften the fall of the melt is seen to be an illusion. In reality, the well was an opportunity for the melt to churn, entraining quantities of oxide and bubble defects.

Systematic X-ray radiographic studies started in 1992 have been revealing. They have shown that in a sufficiently narrow filling channel with a good radius at the sprue/runner junction, the high surface tension of the liquid metal assists in retaining the integrity of a compact liquid front, constraining the melt. These investigative studies on dramatically narrow channels in real moulds with real metals quickly confirmed that the sprue/runner junction was best designed as a simple turn (Figure 2.19c and d), provided that the channels were of minimum area.

The studies showed that if a well of any kind was provided, the additional volume created in this way was an opportunity for additional surface turbulence, so damaging the melt. Furthermore, after the well was filled, the rotation of the liquid in the well was seen to act as a kind of ball bearing, reducing the friction on the stream at the turn. In this way the velocity in the runner was increased. These higher speeds observed out of right-angle turns provided by a well were unhelpful. For a narrow turn without a well the velocity of the metal in the runner had the benefit of additional friction from the wall, giving a small (approximately 20 per cent) but useful reduction in metal speed. Thus the conclusion that the filling systems perform better without a well seems conclusive.

On a note of caution, it is perhaps necessary to bear in mind that all this research has been conducted on rather small castings. Even so, there seems no *a priori* reason why the principles should not also apply to large products.

It is unlikely that wells will disappear from the casting scene without strong defence from their supporters. It should be borne in mind that wells may once have been appropriate where large section runners were used.

In summary, despite what was recommended by the author in *Castings 1991*, more recent research confirms that wells are no longer recommended, particularly for narrow section filling systems.

The radius of the turn

It has been shown that for small castings, generally up to a few kilograms in weight, the melt can be turned through the right angle at the base of the sprue simply by putting a right-angle bend into the channel. However, if no radius is provided, the melt cannot follow the bend, so that a *vena contracta* is created (Figures 2.7a and 2.20). The trailing edge of this cavitated region is unstable, so that its fluttering and flapping action sheds bubbles into the stream.

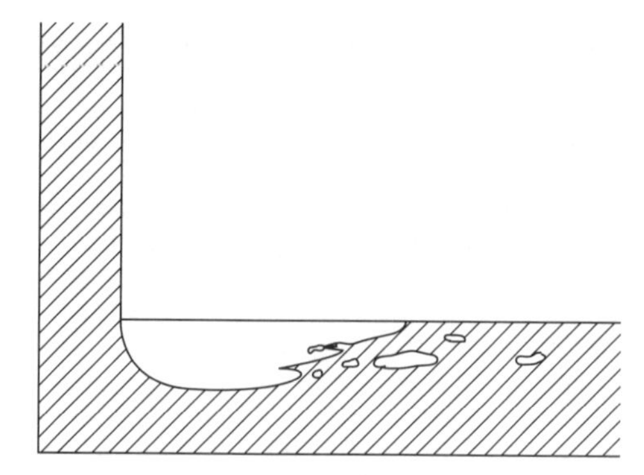

Figure 2.20 *The* vena contracta *problem at a right angle with inadequate radius.*

The *vena contracta* is a widely observed phenomenon in flowing liquids. It occurs wherever a rapid flow is caused to turn through a sharp change of direction. An important example has already been met in the offset pouring basin if no step is provided (Figure 2.9a). This creates a *vena contracta* that showers bubbles down the sprue. However, the base of the down-runner is probably an even more important example if, as is usually the case, speeds are much higher here. The loss of contact of the stream from the top of the runner immediately after the turn has been shown to be the source of much air in the metal. Experiments with water have modelled the low-pressure effect here, demonstrating the sucking of copious volumes of air into the liquid as streams and clouds of bubbles (Webster 1967). This is expected to be particularly severe for sand moulds, where the permeability will allow a good supply of air to the region of reduced pressure.

In fact, when pouring castings late at night, when the foundry is quiet, the sucking of air through into the liquid metal can be clearly heard, like bath water down the plug-hole! Such castings always reveal oxides, sand inclusions and porosity above the gates, which are the tell-tale signs of air bubbles aspirated into the running system.

In contrast, provided that the internal corner of the bend is given a sufficiently large radius, the melt will turn the corner without cavitation or turbulence (Figure 2.19c). In fact, the action of the advancing metal is like a piston in a cylinder: the air is simply pushed ahead of the advancing front, never becoming mixed. To be effective, the radius needs to be at least equal to the diameter of the sprue exit, and possibly twice this amount. The precise radius requires further research. The action of the internal radius is improved further if the outside of the bend is also provided with a radius (Figure 2.19d).

For larger casting where surface tension becomes progressively less important, the channels are filled only by the available volume of flow. Initially, during the first critical period as the filling system is priming, there is considerable danger of significant damage to the metal.

To limit such damage it is helpful to take all steps to prime the front end of the filling system quickly. This is assisted by the use of a stopper. However, in *Castings* (1991) the author considered the use of various kinds of choke at the entrance to the runner as a possible solution to these problems. Again, recent research has not upheld these recommendations. It seems that any such constriction merely results in the jetting of the flow into the more distant expanded part of the runner.

This finding emphasizes the value of the concept of the *naturally pressurized* system. It is clearly of no use to expand the running system to fulfil some arbitrary formula of ratios, in the hope that the additional area will persuade the flow velocity to reduce. The flow will obey its own rules, and we need to design our system to follow these rules.

The use of a vortex sprue, or even simply a vortex base or vortex runner (Figure 2.19e) to the conventional sprue represent exciting and potentially important new developments in running system design. These concepts are described more fully in Section 2.3.2.12.

2.3.2.5 Runner

The runner is that part of the filling system that acts to distribute the melt horizontally around the mould, reaching distant parts of the mould cavity quickly to reduce heat loss problems.

The runner is usually necessarily horizontal because it simply follows the normal mould joint in conventional horizontally parted moulds. In other types of moulds, particularly vertically jointed moulds, or investment moulds where there is little geometrical constraint, the runner would often benefit from being inclined uphill.

It is especially useful if the runner can be arranged under the casting, so that the runner is connected to the mould cavity by vertical gates. All the lowest parts of the mould cavity can then be reached easily this way. The technique is normally achieved only in a three-part mould in which the joint between the cope and the drag contains the mould cavity, and the joint between the lower mould parts (the base and the drag) contains the running channels (Figure 2.21a). The three-part mould is often an expensive option. Sometimes the three-level requirement can be achieved by use of a large core (Figure 2.21b), or the distribution system can be assembled from ceramic or sand sections, and built into the mould as the moulding box is filled with sand (Figure 2.21c). These options are often worth considering, and might prove an economic investment.

More usually, however, a two-part mould requires both casting and running system to be moulded in the same joint between cope and drag. To avoid any falls in the filling system the runner has to be moulded in the drag, and the gates and casting in the cope (Figure 2.3d).

The usual practice, especially in iron and steel foundries, of moulding the casting in the drag

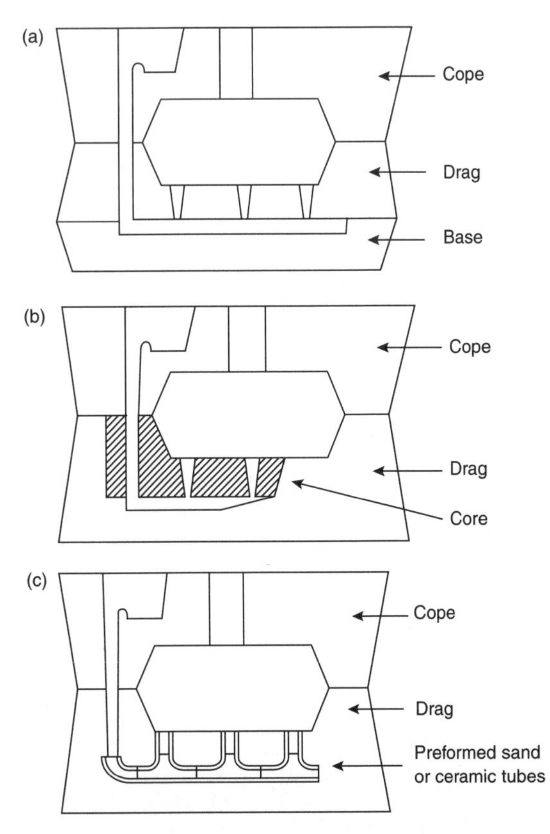

Figure 2.21 *Bottom-gated systems achieved by (a) a three-part mould with accurately moulded running system; (b) making use of a core; and (c) a two-part mould with preformed channel sections.*

(Figure 2.3a) is understandable from the point of vicw of minimizing the danger of run-outs. A leak at the joint, or a burst mould is a possible danger and a definite economic loss. This was an important consideration for hand-moulded greensand, where the moulds were rather weak (and was of course the reason for the use of the steel moulding box or flask). However, the placing of the mould cavity below the runner causes an uncontrolled fall into the mould cavity, creating the risk of imperfect castings. It is no longer such a danger for the dense, strong greensand moulds produced from modern automatic moulding machines, nor for the extremely rigid moulds created in chemically bonded sands. For products whose reliability needs to be guaranteed, the arrangement of the runner at the lowest level of the mould cavity, causing the metal to spread through the running system and the mould cavity only in an uphill direction is a challenge that needs to be met (Figure 2.22). Techniques to achieve this include the clever use of a core (Figure 2.21b) or for some hollow castings the use of central gating (Figures 2.23 and 2.24b).

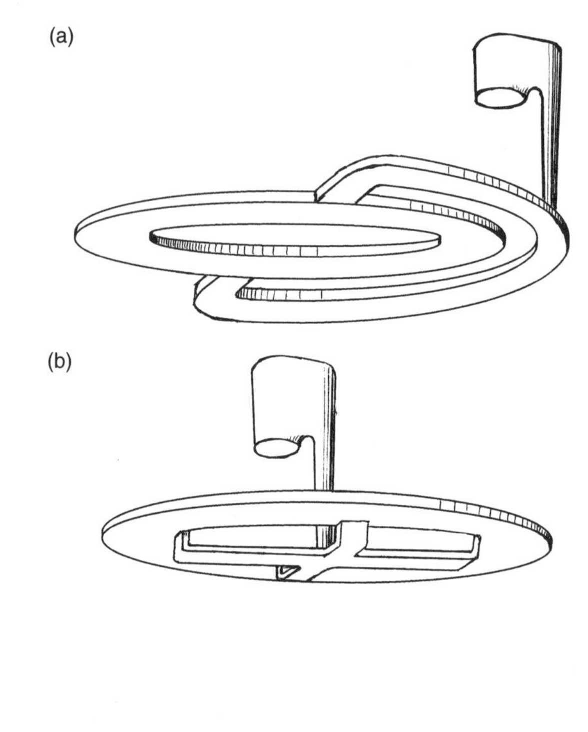

Figure 2.22 *An external running system arranged around an automotive sump (oil pan).*

Figure 2.24 *Ring casting produced using (a) an external and (b) an internal filling system.*

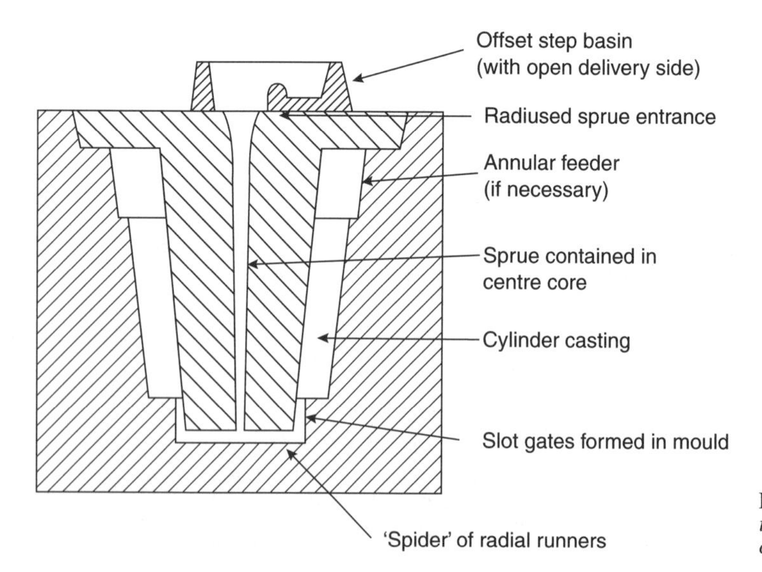

Figure 2.23 *Cross-section of an internal running system for the casting of a cylinder.*

Webster (1964) carried out some early exploratory experiments to determine optimum runner sizes. We can summarize his results in terms of the comparative areas of the runner/sprue exit. He found that a runner that has only the same area as the sprue exit (ratio 1) will have a metal velocity that is high. A ratio of 2 he claims is close to optimum since the runner fills rapidly and excludes air bubbles reasonably efficiently. A ratio of 3 starts to be difficult to fill; and a ratio of 4 is usually simply wasteful for most castings. Webster's work was a prophesy, foretelling the dangers of large runners that foundries have, despite all this good advice, continued to use.

For the best results, however, recent careful studies have made clear that even the expansion of the area of flow by a factor of 2 is not easy to achieve without a serious amount of surface turbulence. This is now known from video X-ray radiographic studies, and from detailed examination of the scatter of mechanical properties of castings using highly sensitive Weibull analysis.

The best that can easily be achieved without damage is merely the reduction of about 20 per cent in velocity by the friction of the sprue/runner bend, necessitating a 20 per cent increase in area of the runner as has been discussed above. Any greater expansion of the runner will cause the runner to be incompletely filled, and so permit conditions for damage.

Greater speed reductions, and thus greater opportunities for expansion of the runner occur if the number of right-angle bends is increased, since the factor of 0.8 reduction in speed is cumulative from one bend to the next. After three such bends the speed is reduced by half ($0.8 \times 0.8 \times 0.8 = 0.5$). Right-angle bends were anathema in filling system designs when large cross-sections were the norm. However, with very narrow systems, there is less room for surface turbulence. Even so, great care has to be taken. For instance video X-ray studies have confirmed that the bends operate best if their internal and external radii provide a parallel channel. The lack of an external radius can cause a reflected wave in larger channels.

One of the most effective devices to reduce the speed of flow in the runner is the use of a filter. The close spacing of the walls of its capillaries ensures a high degree of viscous drag. Flow rate can often be reduced by a factor of 4 or 5. This is a really valuable feature, and actually explains nearly all of the beneficial action of the filter (i.e. when using good quality metal in a well-designed filling system the filter does very little filtering. Its really important action in improving the quality of castings is its reduction of velocity). The use of filters is considered later (Section 2.3.6).

There has over the years been a considerable interest in the concept of the separation of second phases in the runner. Jeancolas *et al.* (1969) carried out experiments on ferrous metals to show that at Reynold's numbers below the range 7000–12 000, suspended particles of alumina could be deposited in the runner but at values in excess of 15 000, they could not precipitate. Although these findings underline the importance of working with the minimum flow velocities wherever possible, it is quickly shown that for a steel casting of height 1 m, giving a velocity of flow of $4.5\,\mathrm{m\,s^{-1}}$, for $\eta = 5.5 \times 10^{3}\,\mathrm{N\,s\,m^{-2}}$, and for a runner of 80 mm square, *Re* is over 100 000. Thus it seems that conditions for the deposition of solid materials such as sand and refractory particles in runners will not be easily met. Even so, every cast iron foundry worker knows that slag will accumulate on the tops of runners, where it is much to be preferred than in the casting. Separation in this case happens because of the great difference in density between the slag and the metal, and because of the large size of the slag droplets. Thus there are some conditions in which a slow runner speed is valuable to assist cleaning the metal.

If there is a choice, the runner should be moulded in the lower half of the mould (the drag). As emphasized previously, this will encourage the runner to fill completely prior to rising through the gates (moulded for preference in the cope) and into the mould cavity.

The basic plan of the filling design starts to become clear: the metal arrives in some chaos at the bottom of the sprue. Here, after this initial trauma, it is gathered together once again by the integrating action of a feature such as a filter to provide some delay and back-pressure, after which it is allowed to rise steadily against gravity, filling section after section of the running system, and finally arriving in the mould in good order at a speed below the critical velocity.

It should be noted that such a logical system and its consequential orderly fill is not to be taken for granted. For instance, a usual mistake is to mould the runner in the cope. This is mainly because the gates, which are in either the drag or the cope, will inevitably start to fill and allow metal into the mould cavity before the runner is full, as is clear from Figure 2.3a. The traditional running of cast iron in this way fails to achieve its potential in its intended separation of metal and slag. This is because the first metal and its load of slag enters the gates immediately, prior to the filling of the runner, and thus prior to the chance that the slag can be trapped against the upper surface of the runner. In short, the runner in the cope results in the violation of the fundamental 'no fall' criterion. The runner in the cope is not recommended for any type of casting—not even grey iron!

In gravity die castings the placing of the runner in the cope, and taking off gates on the die joint (Figure 2.3a), is especially bad. This is because the impermeable nature of the die prevents the escape of air and mould gases from the top of the runner. Thus the runner never properly fills. The entrapped gas floating on the surface of the metal will occasionally dislodge, as waves race backwards and forwards along the runner, and as the gases heat up and expand. Large bubbles will therefore continue to migrate through the gates from time to time throughout the pour, and possibly even afterwards. Because of their late arrival, it is likely that not only will bubble trails and splash problems occur, but also the advancing solidification front will trap whole bubbles.

This scenario is tempered if a die joint is provided along the top of the runner to allow the escape of air. Alternatively, a sand core sited above the runner can help to allow bubbles to diffuse away.

Even so, the complexity of behaviour of some filling system designs is illustrated by a runner in a gravity die, positioned in the cope, that acted to reduce the bubble damage in the casting (Figure 2.25). This result, apparently in complete contradiction to the behaviour described above, arose because of the exceptionally tall aspect ratio of the runner, which was shaped like a vertical slot. This shape retained bubbles high above the exits to the gates moulded below. In fact it seems that the reduction in bubbles into the casting by placing the gates low in this way only really resulted because of the extremely poor front end of the filling system. This was a bubble-producing design, so that almost any remedy had a chance to produce a better result.

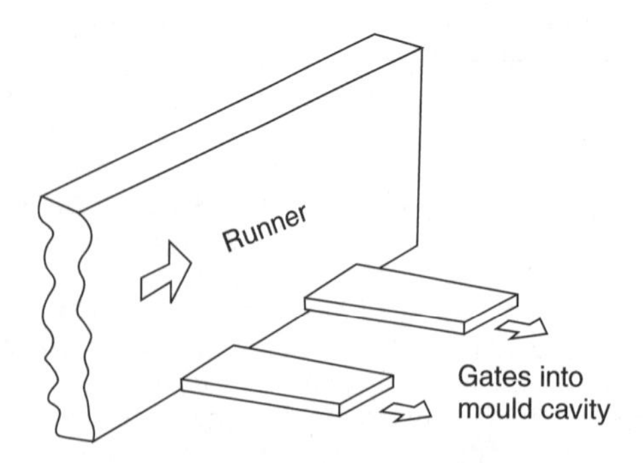

Figure 2.25 *Tall slot runner with bottom gates.*

However, there is a real benefit to be noted (running systems are perversely complicated) because the gates would prime slowly as a head of metal in the runner was built up, thus avoiding any early jetting through into the mould cavity. This is a benefit not to be underestimated, and highlights the problem of generalizing for complex geometries of castings and their filling systems that can sometimes contain not just liquid metal but sometimes emulsions of slag and/or air.

The tapered runner

It is salutary to consider the case where the runner has two or more gates, and where the stepping or tapering of the runner has been unfortunately overlooked. The situation is shown in Figure 2.26a. Clearly, the momentum of the flowing liquid causes the furthest gate, number 3, to be favoured. The rapid flow past the opening of gate 1 will create a reduced-pressure region in the adjacent gate at this point,

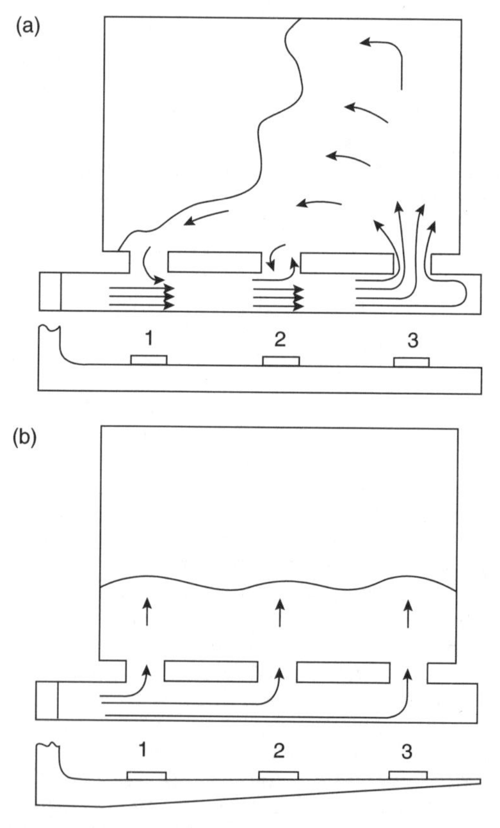

Figure 2.26 *(a) An unbalanced delivery of melt into the mould as a result of an incorrect runner design; (b) a tolerably balanced system.*

drawing liquid out of the casting! The flow may be either in or out of gate 2, but at such a reduced amount as to probably be negligible. In the case of a non-tapered runner it would have been best to have only gate 3.

Where more than one gate is attached to the runner, the runner needs to be reduced in cross-section as each gate is passed, as illustrated in Figure 2.26b. In the past such reductions have usually been carried out as a series of steps, producing the well-known stepped runner designs. For three ingates the runner would be reduced in section area by a step of one third the height of the runner as each gate was passed. However, real-time X-ray studies have noted how during the priming of such systems, because of the high velocity of the stream, the steps cause the flow to be deflected, leaping into the air, and ricocheting off the roof of the runner. Needless to say, the resulting flow was highly disturbed, and did not achieve its intended even distribution. It has been found that simply reducing the cross-section of the runner gradually, usually linearly, cures the deflection problem. A smooth, straight taper geometry does a reasonable job of distributing the flow evenly (Figure 2.26b).

Kotschi and Kleist (1979) allow a reduction in the runner area of just 10 per cent more than the area of the gate to give a slight pressurization bias to help to balance the filling of the gates. However, they used a highly turbulent non-pressurized system that will not have encouraged results of general applicability. In contrast, computer simulation of the narrow runners recommended in this work has shown that the last gate suffers some starvation as a result of the accumulation of friction along the length of the runner. Thus for slim systems the final gates require some additional area, not less. The author usually provides for this in an *ad hoc* way by simply extending the runner past the final gate, and providing a linear taper to this more distant point (Figure 2.26b). The taper can, of course, be provided horizontally or vertically (an important freedom of choice often forgotten).

Finally, avoid tapering the runner to zero. The thinning section adds no advantage but to provide points on which people keep stabbing themselves in the foundry. It aids safety in the workplace to stop the taper at about 5 mm section thickness.

The expanding runner

In an effort to slow the metal in its early progress in the runner, a number of methods of expanding the area of the runner have been tried. The simple expansion of the runner at an arbitrary location along the runner is of no use at all (Figure 2.27a). The melt progresses without noticing the expansion. Even expanding the runner directly from the near side of the sprue (shown as having a square section for clarity) is not helpful (Figure 2.27b). However, expanding the runner from the far side of the sprue (Figure 2.27c) does seem to work considerably better. Even here, however, the front tends to progress in two main streams on either side of the central

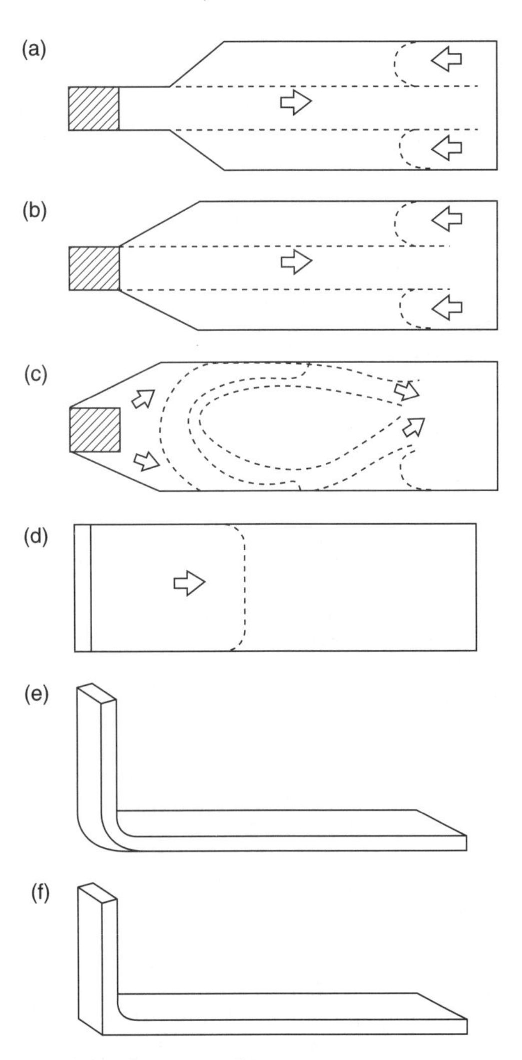

Figure 2.27 *Plan views of a square section sprue connected to a shallow rectangular runner showing attempts to expand the runner (a and b) that fail completely. Attempt (c) is better, but flow ricochets off the walls generates a central starved, low pressure region; (d) a slot sprue and slot runner produce a uniform flow distribution in the runner shown in (e) (recommended) and (f) (probably acceptable).*

axis of the runner, leaving the centre empty, or relatively empty, forming a low-pressure region some distance down-stream in the runner. This development of this double jet flow seems to be the result of the attempted radial expansion of the flow as it impacts on the runner, but finds itself constrained by and reflected from the walls of the runner. This situation for high-temperature liquids such as irons and steels leads to the downward collapse of the centre of the runner in sand moulds, since this becomes heated by radiation, and so expands, but is unsupported by the pressure of metal. The closing down of the runner in this way can be avoided by a central moulded support, effectively separating the runner into two separate, parallel runners. In practice I find that a slot runner about 100 mm wide for irons and steels is close to the maximum that can resist collapse.

A further pitfall for the unwary is the possible constricting effect that sometimes occurs as a result of attempting to connect a round or square section sprue on to a thin flat runner (Figure 2.28). For instance, if the runner were paper thin the constriction at the exit of the sprue would be nearly total; only a fraction of the flow would be able to squeeze into the narrow runner. To eliminate a constriction at this point the runner may need to be thickened, or, preferably, the fillet radius at the bend may require to be increased.

Even so, ultimately, it may come as some surprise to the reader to learn that the linking of a round or square section sprue to a slot runner, especially when attempting an expansion of the runner to reduce the velocity of the liquid, is not yet developed. To the author's knowledge, techniques for the satisfactory design of this junction do not yet exist. Some limited expansion in a horizontal plane might be achievable as indicated in Figure 2.28, but should probably be accompanied by at least a partial corresponding reduction in the vertical plane (not shown in Figure 2.28). The reduction in velocity would benefit from the friction provided by the extra surface area, but would probably not be successful to fill an expansion of a factor of 2. Thus the effect is of limited value. More research is required to evaluate what can be achieved by careful runner design.

What seems more certain, is that the distribution of flow would be simpler if a narrow slot sprue were simply to turn to link onto a horizontal slot type of runner (Figure 2.27d and e). The more uniform action of friction might assist better to achieve a modest expansion and corresponding speed reduction. This has yet to be tested. Even so, the use of slot sprues linked to slot runners promises to be a complete solution to the problem of the sprue/runner junction and deserves wider exploration.

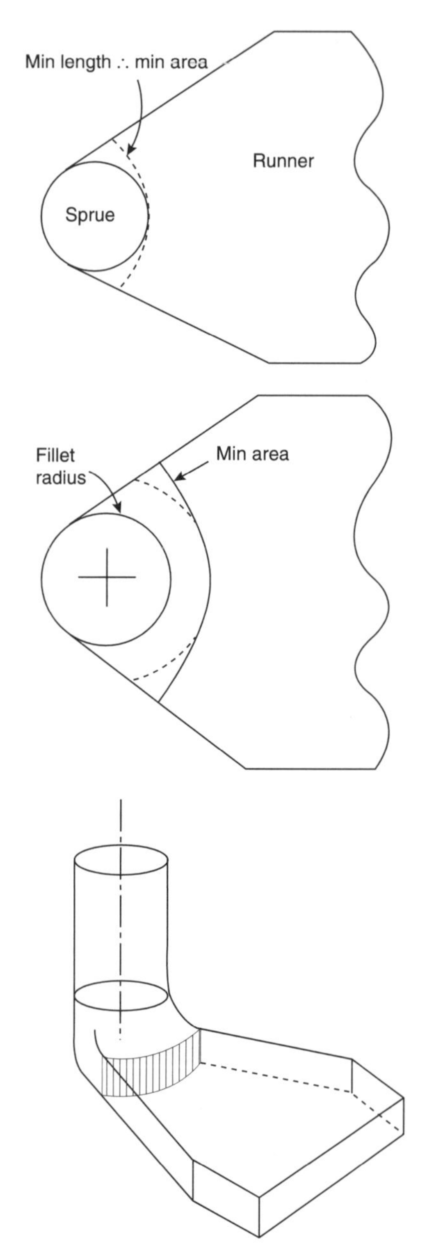

Figure 2.28 *Potential constriction to flow at the sprue to runner junction.*

2.3.2.6 Gates

Siting

When setting out the requirements for the site of a good gate, it is usual to start with the questions

'Where can we get the gate on?'

and

'Where can we get the gate off?'

Other practical considerations include

'Gate on a straight side if possible'

and

'Locate at the shortest flow distance to the key parts of the casting'.

This is a good start, but, of course, just the start. There are many other aspects to the design of a good gate.

Direct and indirect

In general, it is important that the liquid metal flows through the gates at a speed lower than the critical velocity so as to enter the mould cavity smoothly. If the rate of entry is too high, causing the metal to fountain or splash, then the battle for quality is probably lost. The turbulence inside the mould cavity is the most serious turbulence of all. Turbulence occurring early in the running system may or may not produce defects that find their way into the casting because many bifilms remain attached to the walls of the runners and many bubbles escape. However, any creation of defects in the mould cavity causes unavoidable damage to the casting.

One important rule therefore follows very simply:

Do not place the gate at the base of the down-runner so that the high velocity of the falling stream is redirected straight into the mould, as shown in Figures 2.3c, 2.5 and 2.7b. In effect, this *direct* gating is *too direct*. An improved, somewhat *indirect* system is shown in Figures 2.3b and 2.6, illustrating the provision of a separate runner and gate, and thus incorporating a number of right-angle changes of direction of the stream before it enters the mould. These provisions are all used to good effect in reorganizing the metal from a chaotic mix of liquid and gases into a coherent moving mass of liquid. Thus although we may not be reducing the entrainment of bifilms, we may at least be preventing bubble damage in the mould cavity.

As we have mentioned above, all of the oxides created in the early turbulence of the priming of the running system do not necessarily find their way into the mould cavity. Many appear to 'hang up' in the running system itself. This seems especially true when the oxide is strong as is known to be the case for Al alloys containing Be. In this case the film attached to the wall of the running system resists being torn away, so that such castings enjoy greater freedom from filling defects. The wisdom of lengthening the running system, increasing friction, especially by the use of right-angle bends, adds back-pressure for improved back-filling and reduces velocity. It also provides more surface to contain and hold the oxides generated during priming.

Total area of gate(s)

A second important rule concerns the sizing of the gates. They should be provided with sufficient area to reduce the velocity of the melt to below the critical velocity of about $0.5\,m\,s^{-1}$. The concept is illustrated in Figure 2.1. Occasionally, the author has permitted himself the risk of a velocity up to $1\,m\,s^{-1}$ and has usually achieved success. However, velocities above $1.2\,m\,s^{-1}$ for Al alloys always seem to give problems. Velocities of $2\,m\,s^{-1}$ in film-forming alloys, unless onto a core as explained below, would be expected to have consequences sufficiently serious that they could not be overlooked. With even higher velocities the problems simply increase.

Occasionally, there is a problem obtaining a sufficient size of gate to reduce the melt speed to safe levels before it enters the mould cavity. In such cases it is valuable if the gate opens at right angles onto a thin (thickness a few millimetres) wall. This is because the melt is now forced to spread sideways from the gate, and suffers no splashing problems because the section thickness of the casting is too small. As it spreads away from the gate it increases the area of the advancing front, thereby reducing its velocity. Thus by the time the melt arrives in a thicker section of the casting it is likely to be moving at a speed below critical. In a way, the technique uses the casting as an extension of the filling system.

This is a good reason for gating direct onto a core. This, once again, is contrary to conventional wisdom. In the past, gating onto a core was definitely bad because of the amount of air entrained in the flow. The air-assisted hammer action and oxidation of the binder thus led to sand erosion. With a good design of filling system, however, in which air is largely excluded, the action of the hot metal is safe. Little or no damage is done despite the high velocity of the stream, because the melt merely heats the core while exerting a steady pressure that holds the core material in place. Thus with a good filling system design, gating directly onto a core is recommended.

Returning to the usual gating problem whereby the gate opens into a large-section casting. If the area of the gate is too small then the metal will be accelerated through, jetting into the cavity as though from a hosepipe. Figure 2.1c shows the effect. In many castings the jet speed can be so high that the metal effectively blasts its way around the mould cavity. Historically, many castings have been gated in this way. At the present time most steels and grey cast iron appear to be cast with this technique. The approach has enjoyed tolerable success while greensand moulding has been employed, but it seems certain that better castings and lower scrap rates would have been achieved with less turbulent filling. In the case of cores and moulds made with resin binders that cause graphitic films on the liquid iron, the pressurized system is usually unacceptable. The same conclusion is true for ductile irons in all types of moulds.

We may define some useful quick rules for determining the total gate area that is needed. For an Al alloy cast at 1 $kg\,s^{-1}$ assuming a density of approximately 2500 $kg\,m^{-3}$ and assuming that we wish the metal to enter the gate at its critical speed of approximately 0.5 $m\,s^{-1}$ it means we need approximately 1000 mm^2 of gate area. The elegant way to describe this interesting ingate parameter is in the form of the units of area per mass per second; thus for instance '1000 $mm^2\,kg^{-1}\,s$'.

Clearly we may pro-rata this figure in different ways. If we wished to fill the casting at twice this rate (i.e. in half the time) we would require 2000 mm^2 and so on. It can also be seen that the area is quickly adjusted if it is decided that the metal can be allowed to enter at twice the speed, thus the 1 $kg\,s^{-1}$ would require only 500 mm^2, or if directly onto a core in a thin section casting, perhaps twice the rate once again, giving only 250 mm^2.

Allowing for the fact that denser alloys such as irons, steels and copper-based alloys, have a density approximately three times that of aluminium, but the critical velocity is slightly smaller at 0.4 $m\,s^{-1}$, the ingate parameter becomes, with sufficient precision, 500 $mm^2\,kg^{-1}\,s$.

The values of approximately 1000 $mm^2\,kg^{-1}\,s$ for light alloys and 500 $mm^2\,kg^{-1}\,s$ for dense alloys are useful parameters to commit to memory.

Gating ratio

In its progress through the running system the metal is at its highest velocity as it exits the sprue. If possible, we aim to reduce this in the runner, and further reduce as it is caused to expand once again into the gates. The aim is to reduce the velocity to below the entrainment threshold (the 0.4 or 0.5 $m\,s^{-1}$) at the point of entry into the mould cavity.

It is worth spending some time below describing an alternative method of defining running systems which is widely used, but erroneous. It is to be noted that it is not recommended!

It has been common to describe running systems in terms of ratios based on the area of the exit of the sprue. For instance, a widely used area ratio of the sprue/runner/gates has been 1:2:4. Note that in this abbreviated notation the ratios given for both the runner and the gates refer back to the sprue, so that for a sprue exit of 1, the runner area is 2 and the total area of the gates is 4. It is clear that such ratios cannot always be appropriate, and that the real parameter that requires control is the velocity of metal entering the mould. Thus on occasions this will result in ratios of 1:5:10 and other unexpected values. The design of running systems based on ratios is therefore a mistake.

Having said this, I do allow myself to use the ratio of the area of sprue exit to the (total) area of the gates. Thus if the sprue is 200 mm tall (measured of course from the top of the metal level in the pouring basin) the velocity at its base will be close to 2 $m\,s^{-1}$. Thus a gate of four times this area will be required to get to below 0.5 $m\,s^{-1}$. (Note therefore that the old 1:2:4 and 1:4:4 ratios can be seen to be applicable only up to 200 mm sprue height. Beyond this sprue height the ratios are insufficient to reduce the speed below 0.5 $m\,s^{-1}$.)

I am often asked what about the problem that occurs when the mould cross-sectional area reduces abruptly at some higher level in the mould cavity. The rate of rise of the metal will also therefore be increased suddenly, perhaps becoming temporarily too fast, causing jetting or fountaining as the flow squeezes through the constriction. Fortunately, and perhaps surprisingly, this is extremely rare in casting geometries. In forty years dealing with thousands of castings I have difficulty recalling whether this has ever happened. The most narrow area is usually the gate, so the casting engineer can devote attention to ensuring that the critical velocity is not exceeded at this critical location, and at the location just inside the mould because of the sideways spreading flow (see below). If the velocity in these two situations is satisfactory it usually follows that the velocity is satisfactory at all other levels in the casting.

Even in a rare situation where a narrowing of the mould is severe, it would still be surprising if the critical velocity were exceeded, because the velocity of filling is at its highest at the ingate, and usually decreases as the metal level rises, finally becoming zero when the net head is zero, as the metal reaches the top of the mould.

Once again, of course, counter-gravity filling wins outright. In principle, and usually with sufficient accuracy in practice, the velocities can be controlled at every level of filling.

Multiple gates

Premature filling problem via early gates Sutton (2002) applied Bernoulli's theorem to draw attention to the possibility that a melt travelling along a horizontal runner will partly enter vertical gates placed along the length of the runner, despite the fact that the runner may not yet have completely filled and pressurized (Figure 2.29). This arises as a result of the pressure gradient along the flow, and is proportional to the velocity of flow. In real casting conditions, the melt may rise sufficiently high in such gates that cavities attached to the gates might be partially filled with a slow dribble of upwelling metal prior to the filling of the runner, and therefore prior to the main flow up the vertical gates. These dribbles of metal in the cavity are poorly assimilated by the arrival of the main metal supply, and so usually constitute a lap defect resembling a misrun or part-filled casting.

This same effect would be expected to be even more noticeable in horizontal gates moulded in the cope, sited above a runner moulded in the drag (Figure 2.3b). The head pressure required to simply cross the parting line and start an unwanted early filling of part of the mould cavity would be relatively small, and easily exceeded.

Horizontal velocity in the mould When calculating the entry velocity of the metal through the gates, it is easy to overlook what happens to the melt once it starts to spread sideways into the mould cavity. The horizontal sideways velocity away from the gate can sometimes be high. In many castings where the ingate enters a vertical wall the transverse spreading speed inside the mould is higher than the speed through the gate, and causes a damaging splash as the liquid hits the far walls (Figure 2.30). We can make an estimate of this lateral velocity V_L in the following way.

The lateral travel of the melt will normally be at about the height h of a sessile drop. (In a thin wall the height of the flow might reach $2h$, reducing the problem considered below. We shall neglect this complication, and consider only the worst case.) We shall assume the section thickness t, for a symmetrical ingate, area A_i. The melt enters the ingate at the critical velocity V_C, and spreads in both directions away from the gate. Equating the volume flow rates through the gate and along the base of the casting gives

$$V_C \cdot A_i = 2\, V_L \cdot h \cdot t$$

If we limit the gate velocity and the transverse velocity to the same critical velocity V_C (for instance $0.5\,\mathrm{m\,s^{-1}}$) and adopt a gate thickness t the same as that of the casting wall, the relation simplifies to the fairly self-evident geometrical relation in terms of the length of the ingate L_i

$$L_i = 2h$$

The message from this simple formula is that if the length of the gate exceeds twice the height of the sessile drop, even if the gate velocity is below the critical velocity, the transverse velocity may still be too high, and surface turbulence will result from the impact of the transverse flow on the end walls of the mould cavity.

To be sure of meeting this condition, therefore, for aluminium alloys where $h = 13\,\mathrm{mm}$, gates must always be less than 26 mm wide (remembering that this applies only to gates that have the same thickness as the wall of the casting). For irons and steels gates should not exceed 16 mm wide. Since we often require areas

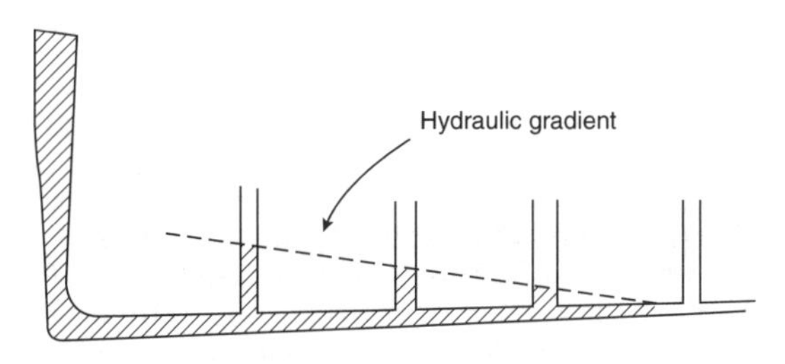

Figure 2.29 *The partial filling of vertical ingates along the length of a runner.*

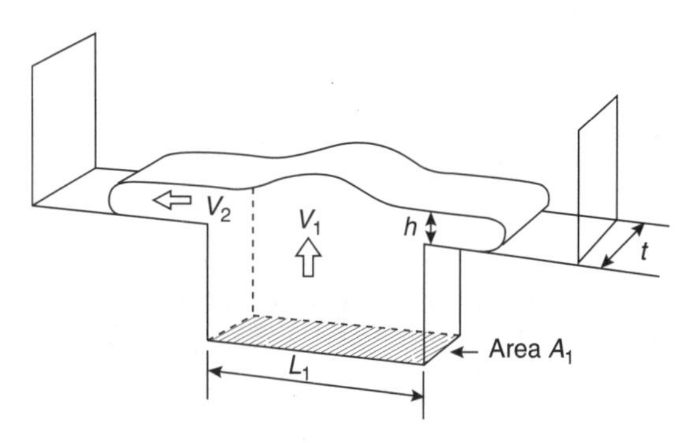

Figure 2.30 *Sideways flow inside mould cavity.*

considerably greater than can be provided by such a short gate, it follows that multiple gates are required to achieve the total ingate area to bring the transverse velocity below critical.

Clearly, it is a concern that in practice, gate lengths are often longer than these limits and may be causing quality problems from this unsuspected source.

For those conditions where more length (or area) is needed than the above formula will allow, the solution is the provision of more gates. Two equally spaced gates of half the length will halve the problem, and so on. In this way the individual gates lengths can be reduced, reducing the problem correspondingly. Our relation becomes simply for N ingates of total ingate length L_i

$$L_i/N = 2h$$

Or directly giving the number of ingates N that will be required

$$N = L_i/2h$$

These considerations based on velocity through the gate or in the casting take no account of other factors that may be important in some circumstances. For instance, the number of ingates might require to be increased to (i) distribute heat more evenly throughout the mould; (ii) avoid localized hot spots as a result of junction problems (see below); and (iii) provide liquid at all the lowest points in the mould cavity to avoid waterfall effects.

Junction effect When the gates are planted on the casting they create a junction. This self-evident statement requires explanation.

Some geometries of junction create the danger of a hot spot. The result is that a shrinkage defect forms in the pocket of liquid that remains trapped here at a late stage of freezing. Thus when the gate is cut off a shrinkage cavity is revealed underneath. This defect is widely seen in foundries. In fact, it is almost certainly the reason why most traditional moulders cut such narrow gates, causing the metal to jet into the mould cavity with consequent poor results to casting quality.

The magnitude of the problem depends strongly on what kind of junction is created. Figure 2.31 shows the different kinds of junctions. An in-line junction (c) is hardly more than an extension of the wall of the casting. Very little thermal problem is to be expected here. The T-junction (a) is the most serious problem. It is discussed below. The L-junction (b) is an intermediate case and is not further discussed.

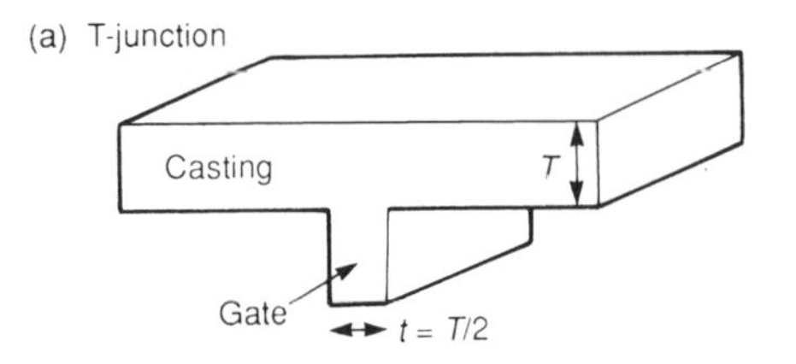

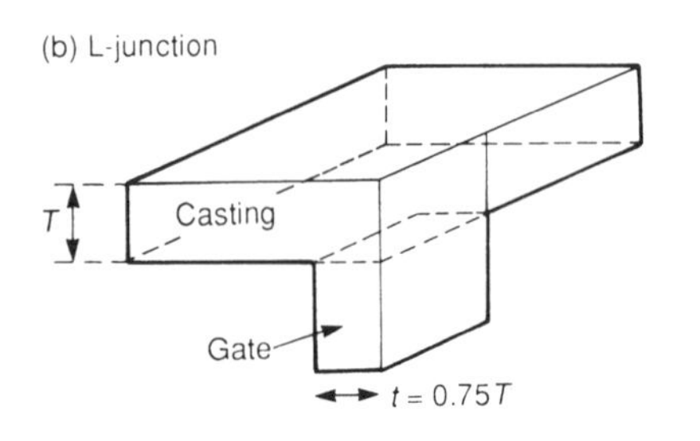

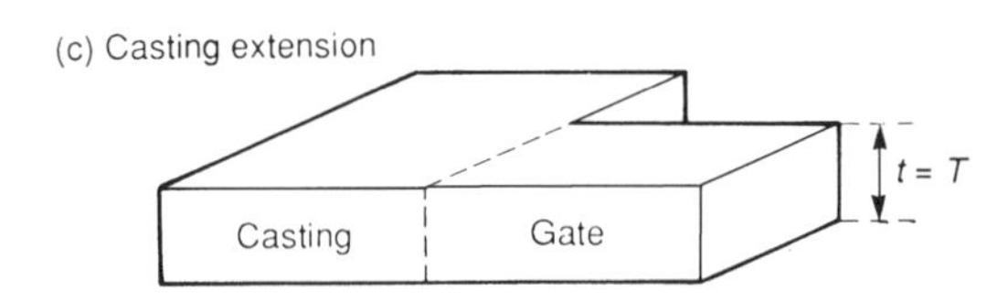

Figure 2.31 *Maximum allowable gate thickness to avoid a hot spot at the junction with the casting.*

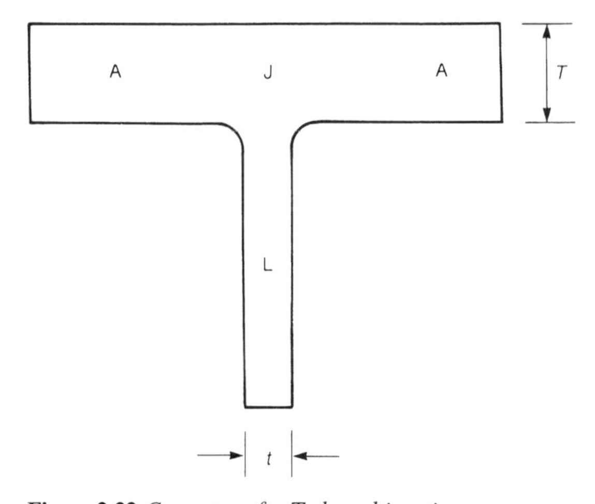

Figure 2.32 *Geometry of a T-shaped junction.*

The reader can make his or her own allowances assuming conditions intermediate between the zero (in-line junction) and T-junction cases.

To help to solve this problem it is instructive to examine the freezing patterns of T-sections. In the 1970s Kotschi and Loper carried out some admirable theoretical studies of T-junctions as shown in Figure 2.32 using only simple calculations based on modulus. These studies pointed the way for experimental work by Hodjat and Mobley in 1984 that broadly confirmed the predictions. The data are interpreted in Figure 2.33 simply as a set of straight lines of slopes 2/1, 1/1, and 1/2. (A study of the scatter in the data shows that the predictions are not infallibly correct in the transitional areas, so that some caution is required.)

Figure 2.34 presents a simplified summary of these findings. It is clear that a gate (the upright leg of the T) of 1 : 1 geometry, i.e. a section equal to the casting section (the horizontal arm of the T), has a hot spot in the junction, and so is undesirable. In fact, Figure 2.33 makes it clear that any medium-sized gate less than twice as thick, or more than half as thick, will give a troublesome hot spot. It is only when the gate is reduced to half or less of the casting thickness that the hot spot problem is removed. (Other lessons can be learned from the T-junction results: (1) an appendage of less than one half of the section thickness will act as a cooling fin, locally enhancing the rate of cooling in the manner of a metal chill in a sand mould; and (2) an appendage of section double that of the casting will freeze later without a hot-spot problem. This is the requirement for a feeder when planted on a plate-like casting, as will be discussed in Chapter 6.

In the case of gates forming T-junctions with the casting (Figure 2.34), the requirement to make the gate only half of the casting thickness ensures that under most circumstances no

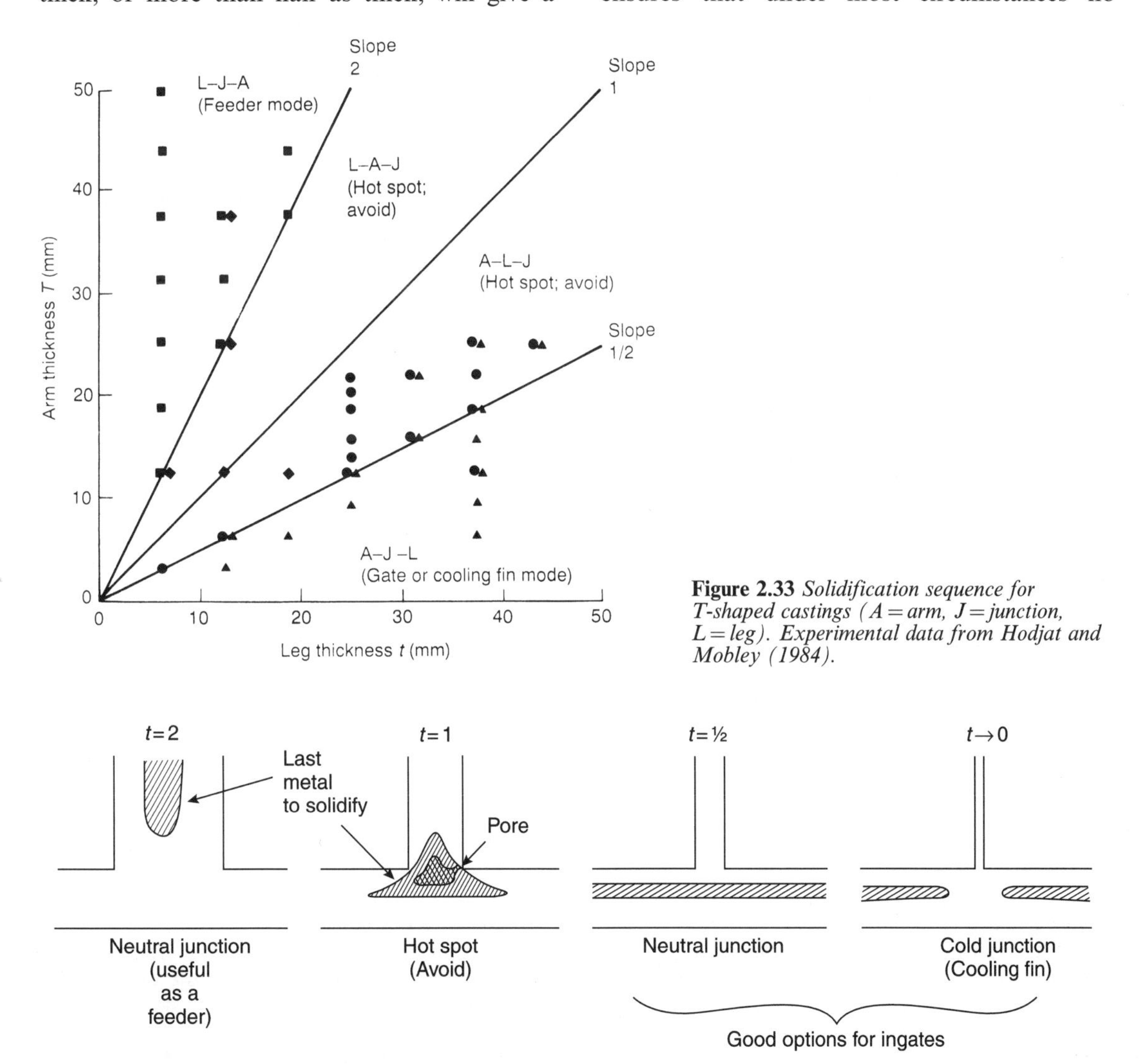

Figure 2.33 *Solidification sequence for T-shaped castings (A = arm, J = junction, L = leg). Experimental data from Hodjat and Mobley (1984).*

Figure 2.34 *Array of different T-junctions.*

localized shrinkage defect will occur, and almost no feeding of the casting will take place through the gate.

For slot gates much less than half the section of the casting that act as cooling fins the effect can be put to good use in setting up a favourable temperature gradient in the casting, encouraging solidification from the gate towards the top feeder. Such cooling fin gates have been used to good effect in the production of aluminium and copper-based alloys because of their high thermal conductivity (Wen *et al.* 1997). (The effect is much less useful in irons and steels.) Also, the common doubt that the cooling effect would be countered by the preheating because of the flow of metal into the casting is easily demonstrated to be, in most cases, a negligible problem. The preheating occurs only for a relatively short time compared to the time of freezing of the casting, and the thin gate itself has little thermal capacity. Thus following the completion of its role as a gate it quickly cools and converts to acting as a cooling fin. Where a gate cannot conveniently be made to act as a cooling fin, the author has planted a cooling fin on the sides of the gate. (This is simple if the slot gate is on a joint line.) By this means the gate is strongly cooled, and in turn, cools the local part of the casting.

The current junction rules have been stated only in terms of thickness of section. For gates and casting sections of more complex geometry it is more convenient to extend the rules, replacing section thickness by equivalent modulus. The more general rule that is inferred is 'the gate modulus should be half or less than the local casting modulus'.

It is worth drawing attention to the fact that not all gates form T-junctions with the casting. For instance, those that are effectively merely extensions of a casting wall may clearly be continued on at the full wall thickness without any hot-spot effect (Figure 2.31c).

Gates which form an L-junction with the wall of the casting are an intermediate case (Sciama 1974), where a gate thickness of 0.75 times the thickness of the wall is the maximum allowable before a hot spot is created at the junction (Figure 2.31b).

It is possible that these simple rules may be modified to some degree if much metal flows through the gate, locally preheating this region. In the absence of quantitative guidelines on this point it is wise to provide a number of gates, well distributed over the casting to reduce such local overheating of the mould. The Cosworth system devised by the author for the ingating of cylinder heads used ten ingates, one for every bolt boss. It contrasted with the two or three gates that had been used previously, and at least partly accounts for the immediate success of the gating design (although it did not help cut-off costs, of course!).

If the casting contains heavy sections that will require feeding then this feed metal will have to be provided from elsewhere. It is necessary to emphasize the separate roles of (i) the filling system and (ii) the feeding system. The two have quite different functions. In the author's experience attempts to feed the casting through the gate are to be welcomed if really possible; however, there are in practice many reasons why the two systems often work better when completely separate. They can then be separately optimized for their individual roles.

It is necessary to make mention of some approaches to gating that attempt to evaluate the action of running systems with gates that operate only partly filled (Davis 1977). The reader will confirm that such logic only applies if the gates empty downhill into the mould, like water spilling over a weir. This is a violation of one of our most important filling rules. Thus approaches designed for partially filled gates are not relevant to the technique recommended in this book. The placing of gates at the lowest point of the casting, and the runner below that, ensures that the runner fills completely, then the gates completely, and only then can the casting start to fill. The complete prior filling of the running system is essential; it avoids the carrying through of pockets of air as waves slop about in unfilled systems. Complete filling eliminates waves.

As in most foundrywork, curious prejudices creep into even the most logical approaches. In their otherwise praiseworthy attempt to formalize gating theory, Kotschi and Kleist (1979) omit to limit the thickness of their gates to reduce the junction hot spot, but curiously equalize the areas of the gates so as to equalize the flow into the casting. In practice, making the gates the same is rarely desirable because most castings are not uniform. For instance, a double flow rate might be required into part of the casting that is locally twice as heavy.

The design of gates may be summarized in these concluding paragraphs.

The requirement for gates to be limited to a maximum thickness naturally dictates that the gates may have to become elongated into a slot-type shape if the gate area is also required to be large. Limitations to the length of the slots to limit the lateral velocities in the mould may be required of course, dictating more than one gate as explained above. The limitation of lateral velocities in addition to ingate velocities is a vital feature.

The slot form of the gates is sometimes exasperating when designing the gating system because it frequently happens that there is not sufficient length of casting for the required length of slot! In such situations the casting engineer has to settle for the best compromise possible. In practice, the author has found that if the gate area is within a factor of 2 of the area required to give $0.5\,\mathrm{m\,s^{-1}}$, then an aluminium alloy casting is usually satisfactory. Any further deviation would be cause for concern. Grey irons and carbon steels are somewhat more tolerant of higher ingate velocities.

As a final part of this section on gating, it is worth examining some traditional gating designs.

The touch gate The touch gate, or kiss gate, is shown in Figure 2.35a. As its name suggests, it only just makes contact between the source of metal and the mould cavity. In fact there is no gate as such at all. The casting is simply placed so as to overlap the runner. The overlap is typically 0.8 mm for brass and bronze castings (Schmidt and Jacobson 1970; Ward and Jacobs 1962) although up to 1.2 mm is used. Over 2.5 mm overlap causes the castings to be difficult to break off, negating the most important advantage. The elimination of a gate in the case described by the authors was claimed to allow between 20 and 50 per cent more castings in a mould. Furthermore, the castings are simply broken off the runner, speeding production and avoiding cut-off costs and metal losses from sawing. The broken edge is so small that for most purposes dressing by grinding is not necessary; if anything, only shot blasting is required.

A further benefit of the touch gate is that a certain amount of feeding can be carried out through the gate. This happens because (1) the gate is preheated by the flow of metal through it, and (2) the gate is so close to the runner and casting that it effectively has no separate existence of its own; its modulus is not that of a tiny slot, but some average between that of the runner and that of the local part of the casting to which it connects. Investigations of touch gate geometry have overlooked this point, with confusing results. More work is needed to assess how much feeding can actually be carried out. The result is likely to be highly sensitive to alloy type so that any study would benefit from the inclusion of short and long freezing range alloys, and high and low conductivity metals.

Ward and Jacobs report a reduced incidence of mis-run castings when using touch gating. This observation is almost certainly the result of the beneficial effect of surface tension control in preventing the penetration of the gates before the runner is fully filled and at least partly pressurized. Only when the critical pressure to force the metal surface into a single curvature of 0.4 mm is reached (in the case of the 0.8 mm overlap) will the metal enter the mould cavity. This pressure corresponds to a head of 30 to 40 mm for copper-based alloys.

With such a thin gate, variations of only 0.1 mm in thickness have been found to change performance drastically. With the runner in one half of the mould and the casting in the other, this is clearly seen to be a problem from small variations in mismatch between the mould halves. The problem can be countered in practice by providing a small gate attached to the casting, i.e. in the same mould half as the casting cavity (Figure 2.35b), so that the gate geometry

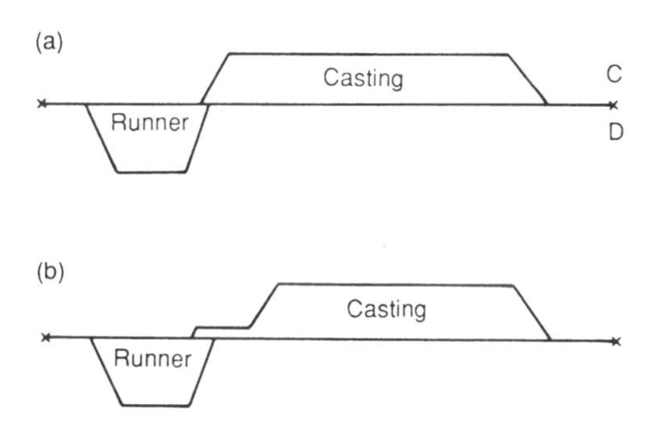

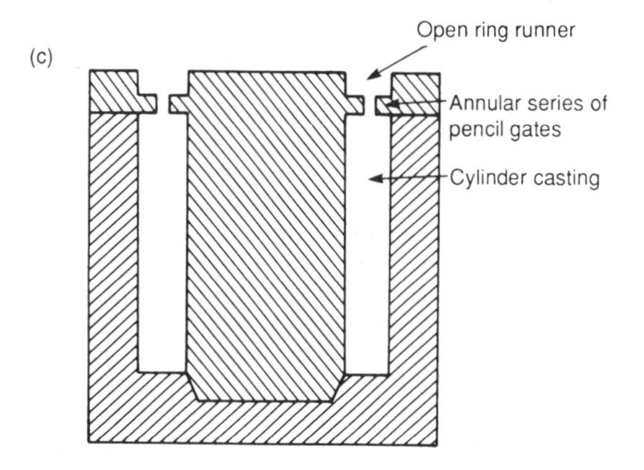

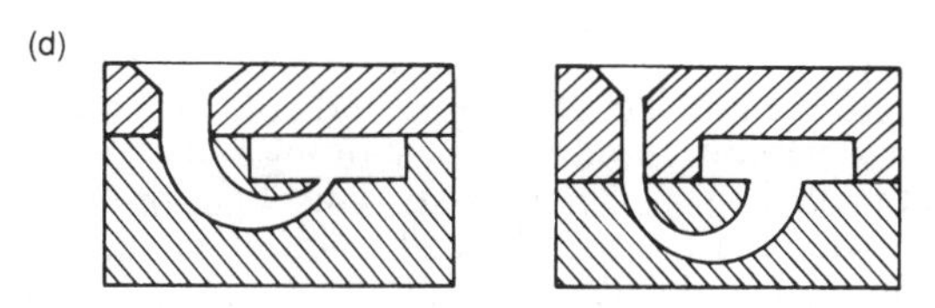

Figure 2.35 *(a) Touch gate, (b) knife gate, (c) pencil gate, (d) normal and reversed horn gates.*

is fixed regardless of mismatch. This is sometimes called a knife gate.

Although it is perhaps self-evident, touch and knife gates are not viable as knock-off gates on the modern designs of accurate, thin-walled, aluminium alloy castings. This is simply because the gate has a thickness similar to the casting, so that on trying to break it off, the casting itself bends! The breaking off technique works only for strong, chunky castings, or for relatively brittle alloys.

The system was said to be unsuitable for aluminium-bronze and manganese-bronze, both of which are strong film-forming alloys (Schmidt and Jacobson 1970), although this discouraging conclusion was probably the result of the runner being usually moulded in the cope and the castings in the drag and a consequence of their poor filling system, generating quantities of oxide films that would threaten to choke gates. The unfortunate fall into the mould cavity would further damage quality, as was confirmed by Ward and Jacobs (1962). They found that uphill filling of the mould was essential to providing a casting quality that would produce a perfect cosmetic polish.

The system has been studied for a number of aluminium alloys (Askeland and Holt 1975), although the poor gating and downhill filling used in this work appears to have clouded the results. Even so, the study implies that a better quality of filling system with runner in the drag and casting impressions in the cope could be important and rewarding.

The fundamental fear that the liquid may jet through the narrow gate may be unfounded. In fact, there may actually be no jetting problem at all. This appears to be a result of the high surface tension of liquid metals. Whereas water might be expected to jet through such a narrow constriction, liquid aluminium is effectively compressed when forced in to any section less than its natural sessile drop height of 12.5 mm. The action of a melt progressing through a thin gate, equipped with an even thinner section formed by a sharp notch was observed for aluminium alloys in the author's laboratory by Cunliffe (1994). The gate was 4 mm thick and the thickness under the various notches was only 1 to 2 mm. The progress of the melt along the section was observed via a glass window from above. The metal was seen to approach, cross the notch constriction, and continue on its way without hindrance, as though the notch constriction did not exist! This can only be explained if the melt immediately re-expands to fill the channel after passing the notch. It seems the liquid meniscus, acting like a compressed, doubled-over leaf spring, immediately expands back to fill the channel when the point of highest compression is passed.

If the surface turbulence through touch gates is tolerable, or minimal, then they deserve to be much more widely used. It would be so welcome to be able to end the drudgery of sawing castings off running systems, together with the noise and the waste. With good quality metal provided by a good front end to the filling system, and uphill filling of the mould cavity after the gate, it seems likely that this device could work well. It would probably not require much work to establish a proper design code for such a practice.

The pencil gate Many large rolls for a variety of industries are made from grey cast iron in greensand moulds. They often contain a massive proportion of grey iron chills around the roll barrel to develop the white iron wear surface of the roll. It is less common nowadays to cast rolls in *loam moulds* produced by strickling. (*Loam* is a sand mixture containing high percentages of clay and water, like a mud, which allow it to be formed by *sleeking* into place. It needs to be thoroughly dried prior to casting.) Steel rolls are similarly cast.

Where the roll is solid, it is often bottom-gated tangentially into its base. Where the roll or cylinder is hollow, it may be centrifugally cast, or it may be produced by a special kind of top gating technique using pencil gates.

Figure 2.35c represents a cross-section through a mould for a roll casting. Such a casting might weigh over 60 000 kg, and have dimensions up to 5 m diameter by 5 m face length, with a wall thickness 80 mm (Turner and Owen 1964). It is cast by pouring into an open circular runner, and the metal is metered into the mould by a series of pencil gates. The metal falls freely through the complete height of the mould cavity, gradually building up the casting. The metal–mould combination of grey iron in greensand is reasonably tolerant of surface turbulence. In addition, the heavy-section thickness gives a solidification time in excess of 30 minutes, allowing a useful time for the floating out and separation of much of the oxide entrained by splashing. The splashing is limited by the slimness of the falling streams from the narrow pencil gates.

The solidification geometry is akin to continuous casting. The slow, controlled build-up of the casting ensures that the temperature gradient is high, and thus favouring good feeding. The feeder head on top of the casting is therefore only minimal, since much of the casting will have solidified by the time the feeder is filled. This beneficial temperature gradient is encouraged by the use of pencil gates: the narrow falling streams have limited energy and so do not disturb the pool of liquid to any great depth (a single massive stream would be a disaster for this reason).

Top gating in this fashion using pencil gates is expected to be useful only for the particular conditions of: (i) grey iron; (ii) heavy sections; and (iii) greensand or inert moulds. It is not expected to be appropriate for any metal–mould combinations in which the metal is sensitive to the entrainment of oxide films, especially in thin sections where entrained material has limited opportunity to escape.

Even so, this top pouring, although occurring in the most favourable way possible as discussed above, still results in occasional surface defect in products that are required to be nearly defect-free. The use of bottom gating via an excellent filling system, entering the mould at a tangent to centrifuge defects away from the outer surface of the roll would be expected to yield a superior product. Even vertical-axis centrifugal casting would benefit from better filling design, applying the liquid metal to the rotating mould in a less turbulent fashion. No matter what the casting method, there is no substitute for a good filling system.

The horn gate The horn gate is a device used by a traditional greensand moulder to make a quick and easy connection from the sprue into the base of the mould cavity without the need to make and fit a core or provide an additional joint line (Figure 2.35d). The horn pattern could be withdrawn by carefully easing it out of the mould, following its curved shape. Although the ingenuity of the device can be admired, in practice it cannot be recommended. It breaks one of our fundamental rules for filling system design by allowing the metal to fall downhill. In addition, there are other problems. When used with its narrow end at the mould cavity it causes jetting of the metal into the mould. This effect has been photographed using an open-top mould, revealing liquid iron emerging from the exit of the gate, and executing a graceful arc through the air, before splashing into a messy, turbulent pool at the far side of the cavity (Subcommittee T535 1960). It has occasionally been used in reverse in an attempt to reduce this problem (Figure 2.35b). However the irregular filling of the first half of the gate by the metal running downhill in an uncontrolled fashion and slopping about in the valley of the gate is similarly unsatisfactory. Furthermore, the large end junction with the casting now poses the additional problem of a large hot spot that requires to be fed to avoid shrinkage porosity.

The horn gate might be tolerable for grey iron in greensand. Otherwise it is definitely to be avoided.

Vertical gate Sometimes it is convenient to place a vertical gate at the end of a runner. Whereas the slowing of the flow by expanding the channel was largely unsuccessful for the horizontal runner, an upward-oriented expanding fan-shaped gate can be extremely beneficial because of the aid of gravity. As always, the application is not completely straightforward. Figure 2.36a shows that if the fan gate is sited directly on top of a rectangular runner, the flow is constrained by the vertical sides of the runner, so that the liquid jets vertically, falling back to fill the fan gate from above. Figure 2.36b shows that if the expansion of the fan is started from the bottom of the runner, the flow expands nicely, filling the expanding volume and so reducing in speed before it enters the mould cavity. This result is valuable because it is one of the very few successful ways in which the speed of the metal entering the mould cavity can be reduced.

The work in the author's laboratory (Rezvani *et al.* 1999) illustrates that this form of gate produces castings of excellent reliability. Compared to conventional slot gates, the Weibull modulus of tensile test bars filled with the nicely diverging fan gate was raised nearly four times, indicating the production of castings of four times greater reliability.

Itamura and co-workers (2002) have shown by computer simulation that the limiting $0.5\,\mathrm{m\,s^{-1}}$ velocity is safe for simple vertical gates, but can be raised to $1.0\,\mathrm{m\,s^{-1}}$ if the gate is expanded as a fan. However, expansion does not continue to work at velocities of $2\,\mathrm{m\,s^{-1}}$ where the flow becomes a fountain. Similar results have been confirmed in the author's laboratory by Lai and Griffiths (2003) who used computer simulation to study the expansion of the vertical gate by the provision of a generous radius at the junction with the casting. All these desirable features involve additional cutting off and dressing costs of course.

Surge control systems The flowing of metal past the gates and into some kind of dump has been widely used to eliminate the first cold metal, diverting it away, together with any initial contamination by sand or oxide. When the dump is filled the gates can start to fill. If there is any raising of back-pressure as, for instance, the accumulation of friction along the length of the runner extension, particularly if

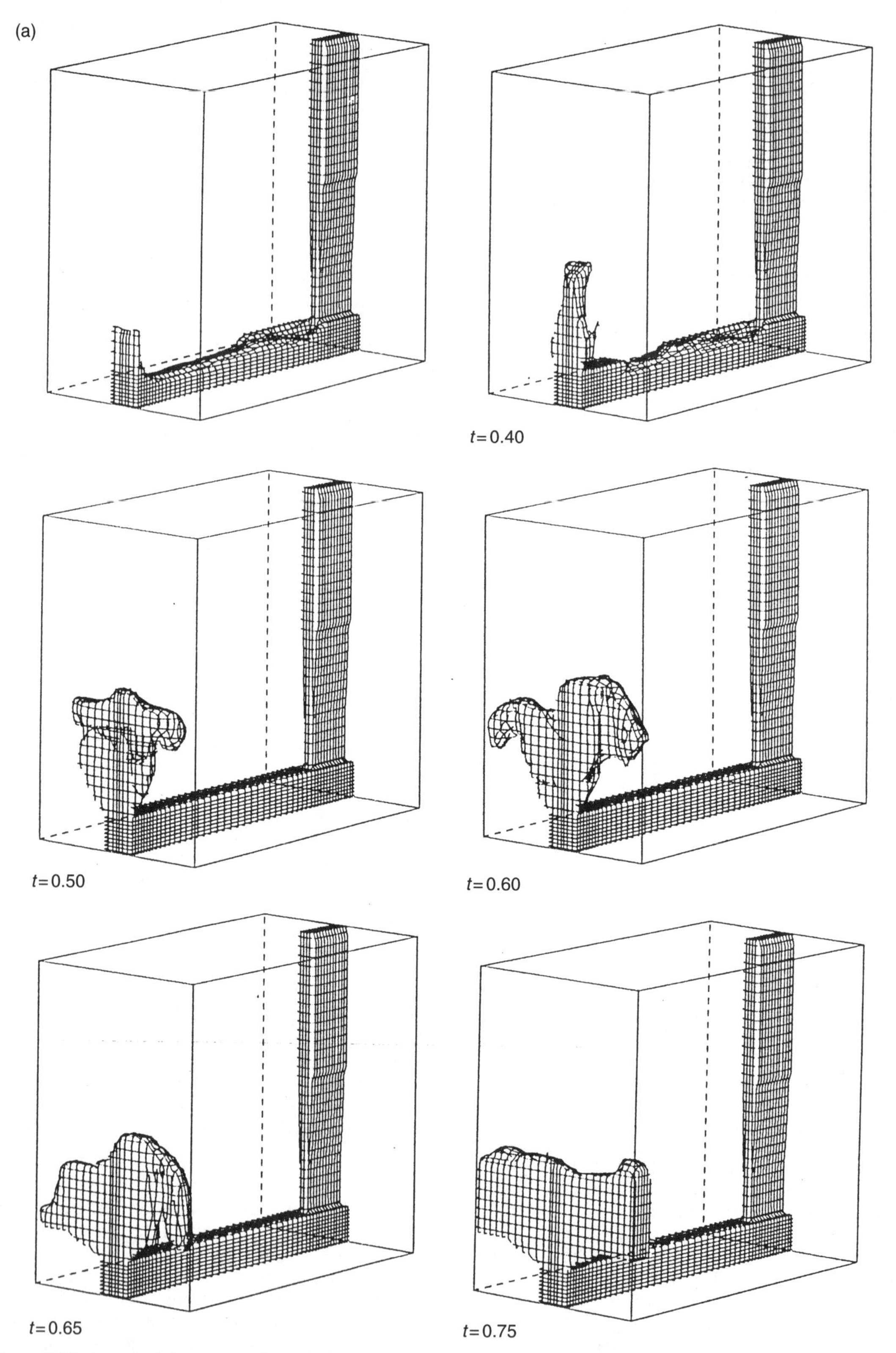

Figure 2.36 *A vertical fan gate at the end of a runner showing the difference in flow as a result of (a) top connection, and (b) bottom connection to the runner (courtesy X. Yang and Flow-3D).*

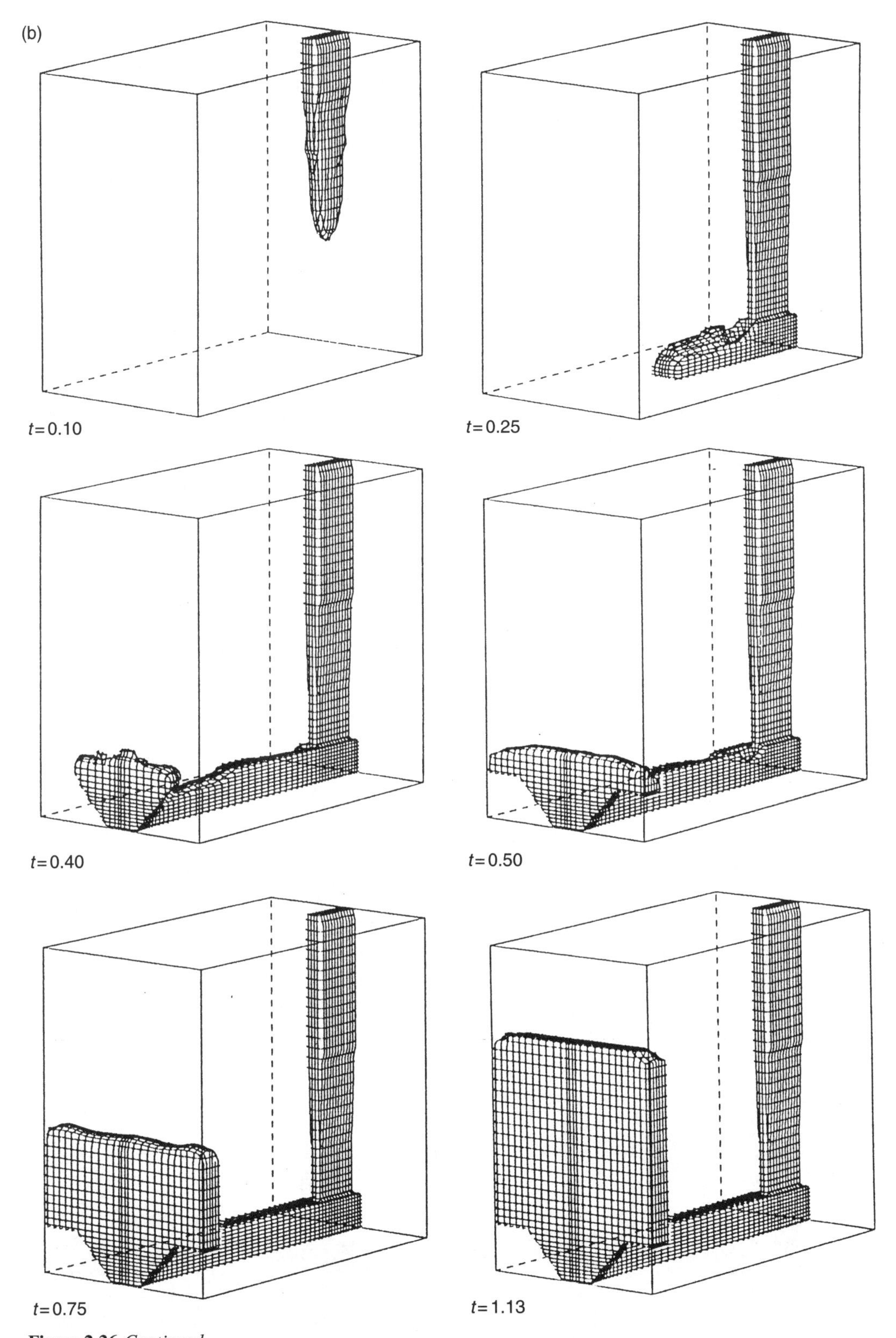

Figure 2.36 *Continued*

the runner is narrow, then the gates may start to fill earlier, before the dump is fully filled. This principle is nicely addressed by the use of the Bernoulli equation as used by Sutton (2002). The concept is developed further below.

The *by-pass* principle can be used to generate more important benefits. It can assist gaining control over the initial velocity of the melt through the gates. Usually, the first metal through the gate is a transient jet, the metal spurting through when the runner is suddenly filled. This is not a problem for castings of small height where the jet effect can be negligible, but becomes increasingly severe for those with a high head height. For a tall casting the velocity at the end of the runner is high so the momentum of the melt, shocked to an instant stop, causes the metal to explode through the far gate, and enter the cavity like a javelin. This damaging initial transient can occur despite the correct tapering of the runner, since the taper is designed to distribute the flow evenly into the mould only after the achievement of steady state conditions.

The problem can be greatly reduced by diverting the initial flow away from the casting. The provision of an additional gate at the end of the runner, beyond the casting, and not connected with the mould cavity, is a valuable technique for the reduction of the shock of the sudden filling of the runner and the impact of metal through the gates. The design of this *flow-off* device is capable of some sophistication, and promises to be a key ingredient, particularly for large, expensive one-off castings. This introduces the concept of surge control systems.

A gate that channels the initial metal into a dump *below* the level of the runner is probably the least valuable form of this technique (Figure 2.37a). The downward facing gate will continue to fill without generating significant back-pressure, the metal merely falling into the trap,

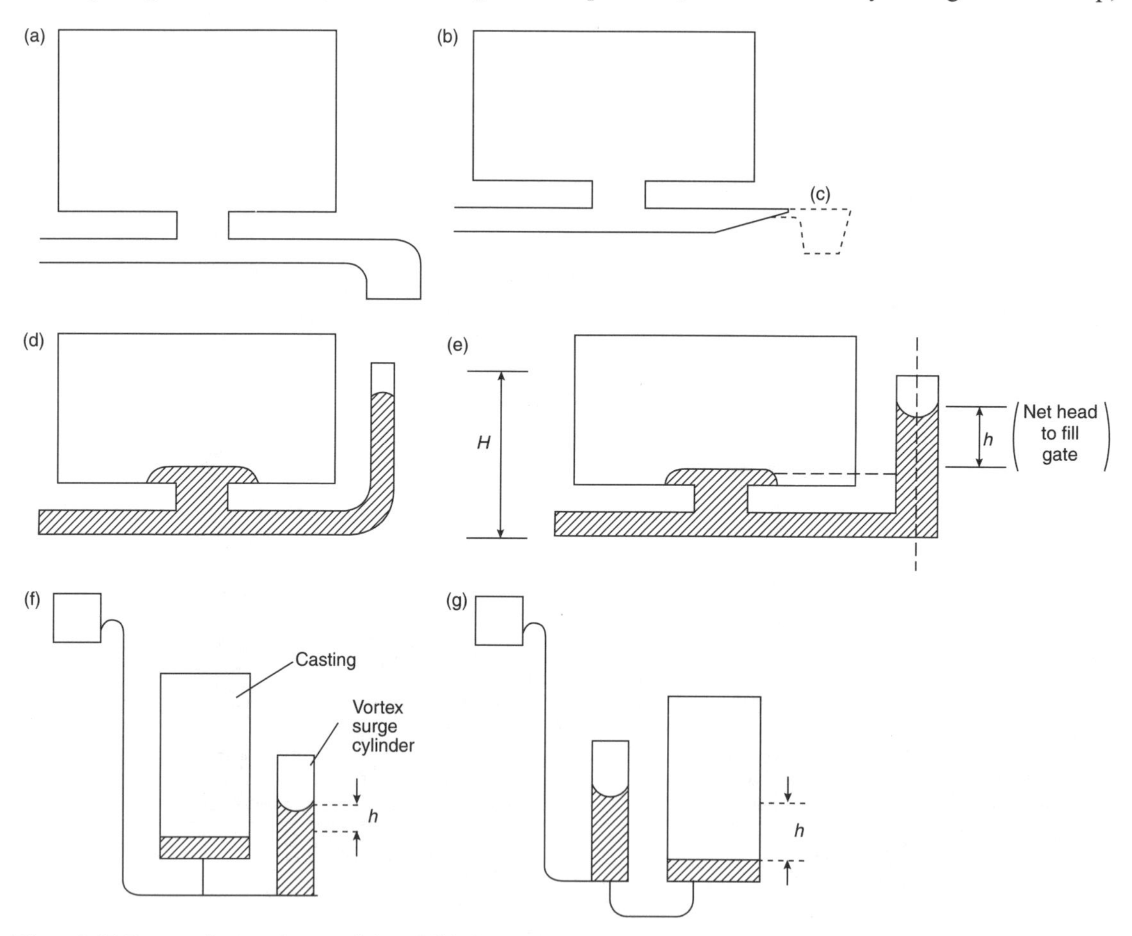

Figure 2.37 *By-pass designs showing (a) and (b) dross trap type (better than nothing, but not especially recommended); (c) non-return trap; (d) vertical runner extension for gravity deceleration; (e) and (f) surge control systems using a terminal vortex; (g) surge control system with in-line vortex with axial (central) outlet.*

until the instant the trap is completely filled. At that instant, the shock of filling is then likely to create, albeit at a short time later, the spurting action into the mould that it was designed to avoid. However, by this time the gates and perhaps even the mould are likely to contain some liquid, so there is a chance that any deleterious jetting action will to some extent be suppressed. This flow-off dump has the benefit of working as a classical dross trap, of course. A taper prior to the trap can prevent the back-wave reversing debris out of the trap, since there is only room for the inflow of metal (Figure 2.37c).

An improved form of the device is easily envisaged. A gate into a dump moulded *above* from the runner has a more positive action (Figure 2.37d). It provides a gradual reduction in flow rate along the runner because it generates a gradually increasing back-pressure as it fills, building up its head height. When placed at the end of the runner, the gate acts to reduce temporarily the speed into the (real) gates by providing additional gate area, and is valuable for reducing the unwanted final filling shock by some contribution to reducing speed. A simply upturned end to the runner as a runner extension (Figure 2.37b) will help in this way, but its limited area will mean that it generates its back-pressure too quickly, so any benefit of a slow increase in speed through the gates is limited. Additional volume of the dump is an advantage to delay the build-up of pressure to fill the gate.

The economically minded casting engineer might find that some castings could be made as 'free riders' in the mould at the end of such gates. The quality may not be high, especially because of the impregnation of the aggregate mould by the momentum of the metal. Even so, the part may be good enough for some purposes, and may help to boost earnings per mould.

A more sophisticated design incorporates all the desirable features of a fully developed surge control system. It consists of extending the runner into the base of an upright circular cylinder, entering tangentially (Figure 2.37e and f). The height and diameter of the cylinder are calculated to raise the back-pressure into the gates at a steady rate (avoiding the application of the full head from the filling system) for a sufficient time to ensure that the gates and the lower part of the casting are filled. When the cylinder (a kind of vortex dump) is completely filled only then does the full pressure of the filling system come into operation to accelerate the filling of the mould cavity. The final filling of the dump may still occur with a 'bang'—the water hammer effect—announced by the shock wave of the impact as it flashes back along the runner at the speed of sound. However, this final filling shock will be considerably reduced from that produced by the metal impacting the end of a simple closed runner.

Although the device actually controls the speed of metal through the ingate, it is not called a speed control since its role is over within the first few seconds of the pour. The name 'surge control' emphasizes its temporary nature.

An even more sophisticated variant that can be suggested is the incorporation of the surge control dump in line with the flow from the sprue (Figure 2.37g). The design of the dump as a vortex as before brings additional advantages: on arrival at the base of the sprue and turning into the runner at high speed, the speed creates a centrifugal action. This action is strongly organizing to the melt, retaining the integrity of the front rather than the chaotic splashing that would have occurred in an impact into a rectangular volume, for instance. The rotary action also centrifuges the entrained air, slag (and possibly some oxides) into the centre where they have opportunity to float if the cylinder is given sufficient height. The good quality melt is taken off from the centre of the base. The small fall down the exit of the surge cylinder is not especially harmful in this case because the rotational action assists the flow to progress with maximum friction down the walls of the exit channel. The system acts to take the first blast of high-velocity metal, gradually increasing the height in the surge cylinder. In this way a gradually increased head of metal is applied to the gates. Furthermore, of course, the metal reaching the gates should be free of air and other low density contaminants.

These surge control concepts promise to revolutionize the production of large steel castings, for which other good filling solutions are, in general, either not easy or not practical. The by-pass and surge control devices represent valuable additions to the techniques of controlling not only the initial surge through the gates, but if their action is extended, as seems possible, they can also make a valuable contribution to slowing velocities during the complete vulnerable early phase of filling.

The action of a by-pass to double as a classical dross trap is described further in Section 2.3.6.

2.3.2.7 Direct gating (from gates)

If the casting engineer has successfully designed the running system to provide bottom gating with minimal surface turbulence, then the casting will fill smoothly without the formation of film defects. However, the battle for a quality casting may not yet be won. Other defects can lie in wait for the unwary!

For the majority of castings the gate connects directly into the mould cavity. I call this simply 'direct gating'. In most cases it is allowable, or tolerable, but it sometimes causes other problems because of the effect it has on the solidification pattern of the casting.

Flow channel structure

Consider the direct-gated vertical plate shown in Figure 2.38a. Imagine this casting being filled slowly to reduce the potential for surface turbulence. If the filling rate was reduced to the point that the metal just reached the top of the mould by the time the metal had just cooled to its freezing point, then it might be expected that the top of the casting would be at its coldest, and freezing would then progress steadily down the plate, from the top to the gate. (At that time the gate would be assumed to be hot because of the preheating effect of the hot metal that would have passed through.) Nothing could be further from the truth.

In reality, the slow filling of the plate causes metal to flow sideways from the gate into the sides of the plate, cooling as it goes, and freezing near the walls. Layers of fresh hot metal would continue to arrive through the gate. The successive positions of the freezing front are shown in Figure 2.38. The final effect is a flow path kept open by the hot metal through a casting that by now has mainly solidified. Rabinovich (1969) describes these patterns of flow in thin vertical plates, calling them *jet streams. Flow channel* is suggested as a good name, if somewhat less dramatic. The final freezing of the flow channel is slow because of the preheated mould around the path, and so its structure is coarse and porous. The porosity will be encouraged by the enhanced gas precipitation under the conditions of slow cooling, and shrinkage may contribute if local feeding is poor because either the flow path is long or it happens to be distant from a source of feed metal.

Reducing the subsequent feeding problem in a flow channel by feeding down the channel from above, or by limited feeding uphill from below, is facilitated in thicker sections where the feeding distance is greater (see Chapter 6). Thus bottom gating into bosses can take advantage of the boss as a useful feed path (Figure 2.38b). However, this action increases the problems with slower cooling, leading to enhanced gas porosity and coarse structure.

The flow channel structure is a standard feature of castings that are filled slowly from their base. This serious limitation to structure control seems to have been largely overlooked.

Moreover, the defect is not easily recognizable. It can occasionally be seen as a region of coarse grain and fine porosity in radiographs of large plate-like parts of castings. The structure contrasts with the extensive areas of clear, defect-free regions of the plate on either side. It is possible that many so-called shrinkage problems (for which more or less fruitless attempts are made to provide a solution by extra feeders or other means) are actually residual flow channels that might be cured by changing ingate position or size, or raising fill rate. No research

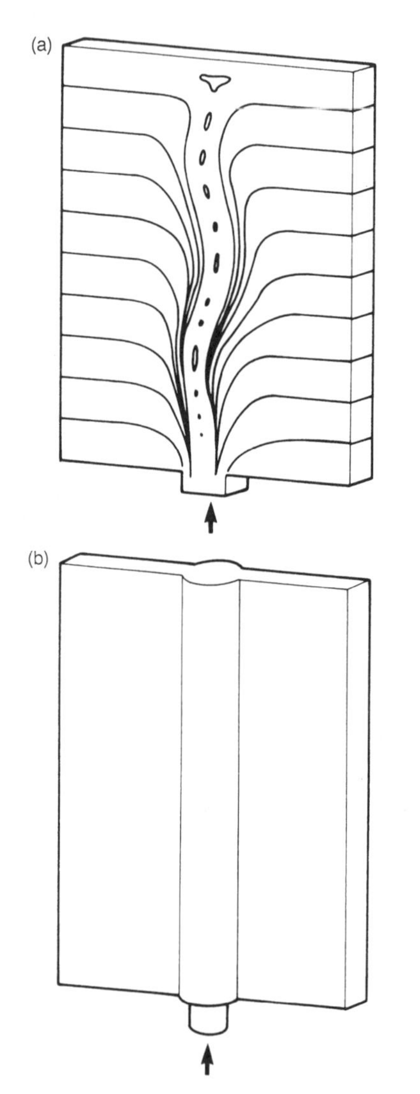

Figure 2.38 *(a) Direct bottom-gated vertical plate, and (b) the use of a boss to assist feeding after casting is filled.*

appears to have been carried out to guide us out of this difficulty.

Nevertheless, in general, the problem is reduced by filling faster (if that is possible without introducing other problems).

However, even fast filling does not cure the other major problem of bottom gating, which is the adverse temperature gradient, with the coldest metal being at the top and the hottest at the bottom of the casting. Where feeders are placed at the top of the casting this thermal regime is clearly unfavourable for effective feeding. In addition, particularly when solidification is slow, the problem of convection may become important. This serious problem is considered later under Rule 7.

2.3.2.8 Indirect gating (from gates)

There is an interesting gating system that solves the major features of the flow channel problem. The problem arises because the hot metal that is required to fill the casting is gated directly into the casting and has to travel through the casting to reach all parts.

The solution is *not* to gate into the casting: a main flow path is created *outside* the casting. It is called a *riser* or *up-runner*. Metal is therefore diverted initially away from the casting, through the riser, only entering subsequently by displacement sideways from the riser as fresh supplies of hot metal arrive. The fresh supplies flood up into the top of the riser, ensuring that the riser remains hot, and that the hottest metal is delivered to the top of the mould cavity. The system is illustrated in Figure 2.39. The system has the special property that the riser and slot gate combination acts not only to fill but also to feed. (The reader will notice that the use of the term 'riser' in this book is limited to this special form of feeder which also acts as an 'up-runner', in which the metal rises up the height of the casting. It is common in the USA to refer to conventional feeders placed on the tops of castings as risers. However, this terminology is avoided here; such reservoirs of metal are called feeders, not risers, following the simple logic of using a name that describes their action perfectly, and does not get confused with other bits of plumbing such as whistlers, up which metal also rises!)

The final parts of the casting to fill in Figure 2.39 will probably also require some feeding. This is easily achieved by planting a feeder on the top of the riser, as a kind of riser extension. This retains all the benefits of the system, since its metal is hot, and hotter metal below in the riser will convect into the feeder. The disadvantages of the riser and slot gate system are as follows:

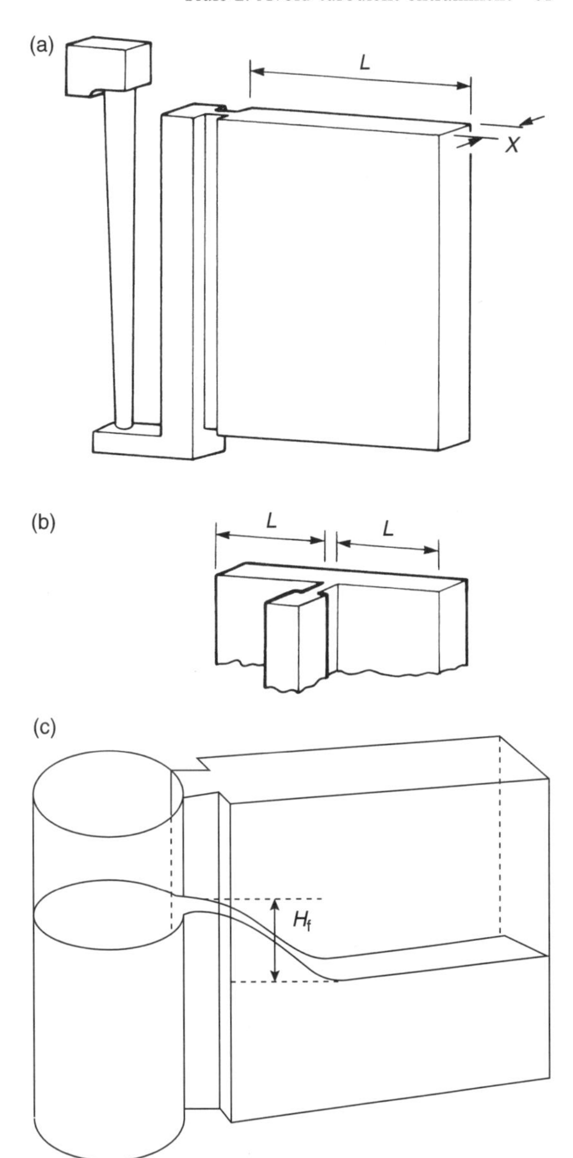

Figure 2.39 *Riser and slot gate to both gate and feed a vertical plate from (a) its side, or (b) its centre.*

1. The considerable cut-off and finishing problem, since the gate often has to be sited on an exterior surface of the casting, and so requires much subsequent dressing to achieve an acceptable cosmetic finish.
2. There appears to be no method of predicting the width and thickness of the gate at the present time. Further research is required here. In the normal gate where it is required to freeze before the casting section to avoid the hot-spot problem in the junction, the

thickness of the gate is held to half of the casting thickness or less. However, in this case the gate is really equivalent to a feeder neck, through which feed metal is required to flow until the casting has solidified. Whereas a thickness of double the casting thickness would then be predicted to avoid the junction hot spot under conditions of uniform starting temperature, the preheating effect of the gate due to the flow of metal through it might mean that a gate as narrow as half of the casting section may be good enough to continue feeding effectively. There are, unfortunately, no confirmatory data on this at the present time.

It is important to caution against the use of a gate which is too narrow for a completely different reason; if filling is reasonably fast then the resistance to flow provided by a narrow slot gate will cause the riser to fill up to a high level before much metal has had a chance to fill the mould cavity. The dynamics of filling and surface tension, compounded by the presence of a strong oxide film, will together conspire to retain the liquid in the riser for as long as possible. The metal will therefore spill through the slot into the casting from an elevated level (the height of fall H_f shown in Figure 2.39c). Again, our no-fall rule is broken. It is desirable, therefore, to fill slowly, and/or to have a gate sufficiently wide to present minimal resistance to metal flow. In this way the system can work properly, with the liquid metal in the riser and casting rising substantially together. Metal will then enter the mould gently.

It is also important for the gate not to be thinner than the casting when the casting wall thickness reduces to approximately 4 mm. A gate 2 mm thick would hold back the metal because of the effect of surface tension and the surface film allowing the metal head to build up in the riser. When the metal eventually breaks through, the liquid will emerge as a jet, and fall and splash into the mould cavity. For casting sections of 4 mm or less the gate should probably be at least as thick as the wall.

In general, it seems reasonable to assume that conditions should be arranged so that the fall distance h in Figure 2.39c should be less than the height of the sessile drop. The fall will then be relatively harmless.

For thinner-section castings (for instance, less than 2 mm thickness) made under normal filling pressure, the feeding of thin-section castings can probably be neglected (as will be discussed in Chapter 6). Thus any hot-spot problems can also be disregarded, with the result that the ingate can be equal to the casting thickness. Surface tension controls the entry through the gate and the further progress of the metal through the mould cavity, reducing the problems of surface turbulence. Fill speed can therefore be increased.

A further important point of detail in Figure 2.39 should be noted. The runner turns upward on entry to the riser, directing the flow upwards. A substantial upward step is required to ensure this upward direction to the flow. This is a similar feature to that shown in Figure 2.6 and contrasts with the poor system shown in Figure 2.5. If the provision of this step is neglected causing the base of the runner to be level with the gate and the base of the casting, metal rushing along the runner travels unchecked directly into the mould cavity. A flow path would then be set up so that the riser would receive no metal directly, only indirectly after it had circulated through the casting. The base of the casting would receive all the heated metal, and the riser would be cold. Such a flow regime clearly negates the reason for the provision of the system! Many such systems have failed through omitting this small but vital detail.

What rates are necessary to make the system work best? Again we find ourselves without firm data to give any guide. We can obtain some indication from the following considerations.

The first liquid metal to flow through the gate and along the base of the cavity travels as a stream. Being the first metal travelling over the cold surface of the mould, it is most at risk from freezing prematurely. Subsequent flow occurs over the top of this hot layer of metal, and therefore does not lose so much heat from its undersurface. Thus if we can ensure that conditions are right for the first metal to flow successfully, then all subsequent flow should be safe from early freezing. In the limiting condition where the tip of the first stream just solidifies on reaching the end of the plate, it will clearly have established the best possible temperature gradient for subsequent feeding by directional solidification back towards the riser. Subsequent layers overlying this initial metal will, of course, have slightly less beneficial temperature gradients, since they will have cooled less during their journey. Nevertheless, this will be the best that we can do with a simple filling method; further improvements will have to await the application of programmed filling by pumped systems.

Focusing our attention, therefore, on the first metal into the mould, it is clear that the problem is simply a fluidity phenomenon. We shall assume that the height of the stream corresponds to the height h of the meniscus which can be supported by surface tension (Figures 2.30).

If the distance to be run from the gate is L, and the solidification time of the metal is t_f in that section thickness x, then in the limiting condition where the metal just freezes at the limit of flow (thus generating the maximum temperature gradient for subsequent feeding):

$$t_f = L/V \quad \textbf{(2.1)}$$

where V is the velocity of flow (m s^{-1}) of the metal stream. The corresponding rate of flow **Q** (kg s^{-1}) for metal of density ρ (kg m^{-3}) is easily shown to be:

$$\begin{aligned} \mathbf{Q} &= Vhx\rho \\ &= Lhx\rho/t_f \end{aligned} \quad \textbf{(2.2)}$$

At constant filling rate the time t to fill a casting of height H is given by:

$$t = t_f(H/h)$$

A 10 mm thick bar (considering the first length of melt to travel along the base of the mould) in Al–7Si alloy would be expected to freeze completely in about 40 seconds giving a flow life for the solidifying alloy of perhaps 20 seconds. The meniscus height h is approximately 12.5 mm, and so for a casting $H = 100$ mm high the pouring time would be $8 \times 20 = 160$ seconds, or nearly 3 minutes. This is a surprisingly long time.

This conclusion is not likely to be particularly accurate, but does emphasize the important point that relatively thin cast sections do not necessarily require fast filling rates to avoid premature solidification. What is important is the steady, continuous advance of the meniscus. Naturally, however, it is important not to press this conclusion too far, and the above first-order approximation to the fill time probably represents a time that might be achievable in ideal circumstances: in fact, if the rate of filling is too slow, then the rate of advance of the liquid front will become unstable for other reasons:

1. Surface film problems may cause instability in the flow of some materials, as is explained in *Castings* (2003). Film-free systems will not suffer this problem, and vacuum casting may also assist.
2. Another instability that has been little researched is the flow of the metal in a pasty mode. Flow channels revealed in the radiograph in Figure 2.40 (Runyoro and Campbell 1990) show the curious behaviour in which channels take a line of least resistance through the casting, abandoning the riser. They adopt the form of magma vents in the earth's crust, and form volcano-like structures at the top surface. (Additionally in these radiographs a metal–mould reaction between A356 alloy and the furan resin binder has produced many minute bubbles that have floated to decorate the upper surfaces of flow channels, revealing the outline of the last regions to remain liquid.)

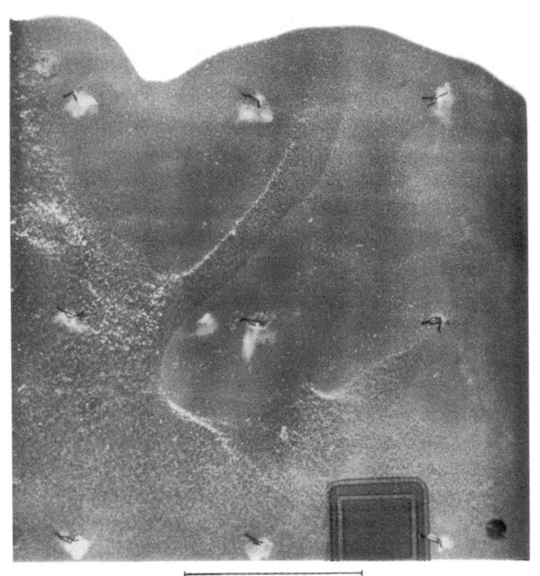

Figure 2.40 *Radiograph of an Al–7Si–0.4Mg alloy vertical plate filled via a riser and slot gate from the left-hand side, revealing unexpected filling behaviour when cast particularly cool. Remains of thermocouples can be seen (Runyoro 1992).*

2.3.2.9 Central versus external systems

Most castings have to be run via an external running system as shown in Figure 2.22. While this is satisfactory for the requirements of the running system, it is costly from the point of view of the space it occupies in the mould. This is especially noticeable in chemically bonded moulds, whose relatively high cost is, of course, directly related to their volume, and whose volume can be modified easily since the moulds are not contained in moulding boxes, i.e. they are boxless. Naturally, in this situation it would be far more desirable if the running system could somehow be incorporated inside the casting, so as to use no more sand than necessary. This ideal might be achieved in some castings by the use of direct gating in conjunction with a filter as discussed later (Section 2.3.6, Direct pour).

For castings that have an open base, however, such as open frames, cylinders or rings, an excellent compact and effective solution is possible. It is illustrated for the case of cylinder and ring castings in Figures 2.23 and 2.24. The runners radiate outwards from the sprue exit, and connect with vertical slot gates arranged as arcs around the base of the casting. Ruddle and Cibula (1957) describe a similar arrangement, but do not show how it can be moulded (with all due respect to our elder statesmen of the foundry world, their suggested arrangement looks unmouldable!), and omit the upward gates. The vertical gates are an important feature for success, introducing useful friction into the system, and making for easy cut-off.

Feeders can be sited on the top of the cylinder if required. Alternatively, if the casting is to be rolled through 180 degrees after pouring, the feeding of the casting can take the form of a ring feeder at the base (later to become the top, of course).

Experience with internal running has found it to be an effective and economical way to produce hollow shapes. It is also effective for the production of other common shapes such as gearboxes and clutch covers, where the sprue can be arranged to pass down through a rather small opening in one half of the casting and then be distributed via a spider of runners and gates on the open side.

However, it has been noted that aluminium alloy castings of 300 mm or more internal diameter exhibit a patternmaker's contraction considerably less than that which would have been expected for an external system. This seems almost certainly to be the result of the expansion of the internal core as a result of the extra heating from the internal running system. For a silica sand core this expansion can be between 1 and 1.5 per cent, effectively negating the patternmaker's shrinkage allowance, which is normally between 1 and 1.3 per cent.

2.3.2.10 Sequential filling

When there are multiple impressions on a horizontal pattern plate, it is usually unwise to attempt to fill all the cavities at the same time. (This is contrary to the situation with a vertically parted mould, in which many filling systems specifically target the filling of all the cavities at once to reduce pouring time. However, such vertically parted moulds have not been subjected to the same degree of study in terms of the defects probably introduced by this system. In the absence of data therefore, they are not described further here. We look forward to good data becoming available at some future date.)

The reasoning in the case of the horizontal mould is simple. The individual cavities are filling at a comparatively slow rate, and not necessarily in a smooth and progressive way. In fact, despite an otherwise good running system design, it is likely that filling will be severely irregular, with slopping and surging, because of the lack of constraint on the liquid, and because of the additional tendency for the flow to be unstable at low flow rates in film-forming alloys. The result will be the non-filling of a number of the impressions and doubtful quality of the others.

Loper (1981) provided a solution to this problem for multiple impressions on one plate as shown in Figure 2.41. He uses runner dams to retard the metal, allowing it time to build up a head of metal sufficient to fill the first set of impressions before overcoming the dam and proceeding to the next set of impressions, and so on.

The system has only been reported to have been used for grey iron castings in greensand moulds. It may give less satisfactory results for other metal–mould systems that are more susceptible to surface turbulence. However, the design of the overflow (the runner down the far side of each dam) could be designed as a miniature tapered down-runner to control the fall, and so reduce surface turbulence as far as possible. Probably, this has yet to be tested.

Another sequential-filling technique, 'horizontal' stack moulding, has also only so far been used with cast iron. This was invented in the 1970s by one of our great foundry characters from the UK, Fred Hoult, after his retirement at the age of 60. It is known in his honour as the 'H Process'. Figure 2.41 outlines his method. The progress of the metal across the top of those castings already filled keeps the feeders hot, and thus efficient. The length of the stack seems unlimited because the cold metal is repeatedly being taken from the front of the stream and diverted into castings. (The reader will note an interesting analogy with the up-runner and slot gate principle; one is horizontal and the other vertical, but both are designed to divert their metal into the mould progressively. The same effect is also used in the promotion of fluidity as described by Hiratsuka *et al.* 1966.) Stacks of 20 or more moulds can easily be poured at one time. Pouring is continued until all the metal is used up, only the last casting being scrapped because of the short pour, and the remaining unfilled moulds are usable as the first moulds in the next stack to be assembled.

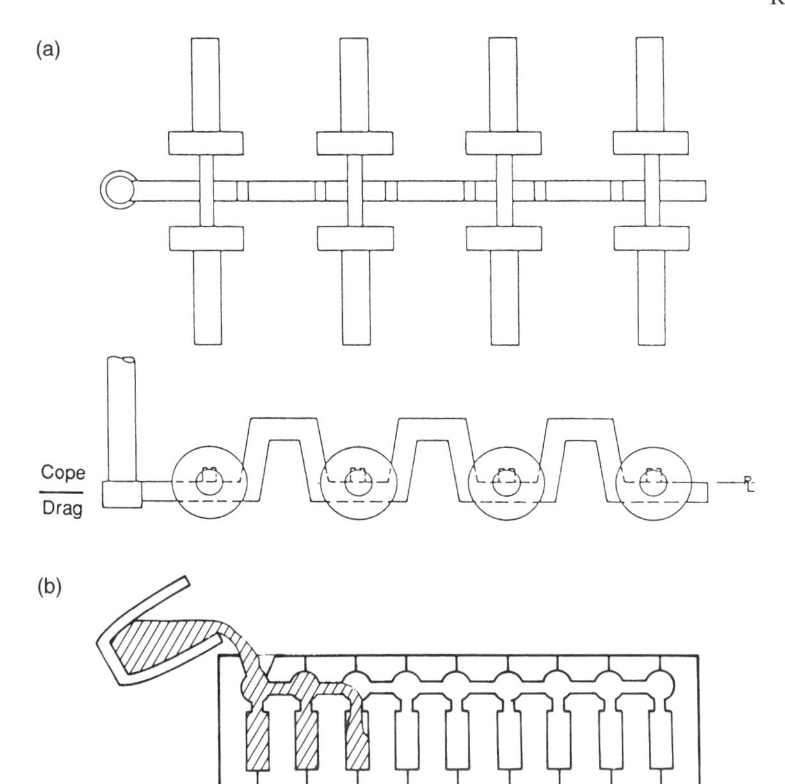

Figure 2.41 *(a) Sequential filling for a number of impressions on a pattern plate (after Loper 1981), and (b) sequential filling for horizontal stack moulded castings (H Process).*

The size of castings produced by the H Process is limited to parts weighing from a few grams to a few kilograms. Larger parts become unsuitable partly because of handling problems, since the moulds are usually stacked vertically during assembly, then clamped with long threaded steel rods, and finally lowered to the floor to make a horizontal line. Larger parts are also unsuitable because of the fundamental limitation imposed by the increase of defects as a necessary consequence of the increased distance of fall of the liquid metal inside the mould, and possibly greater opportunity to splash in thicker sections.

2.3.2.11 Two-stage filling

There have been a number of attempts over the years to introduce a two-stage filling process. The first stage consists of filling the sprue, after which a second stage of filling is started in which the runner and gates, etc. are allowed to fill.

The stopping of the filling process after the filling of the sprue brings the melt in the sprue to a stop, ensuring the exclusion of air. The melt is then allowed to start flowing once again. This second phase of filling has the full head H of metal in the sprue and pouring basin to drive it, but the column has to start to move from zero velocity. It reaches its 'equilibrium' velocity $(2gH)^{1/2}$ only after a period of acceleration. Thus the early phase of filling of the runner and gates starts from a zero rate, and has a gradually increasing velocity. The action is similar to our *'surge control'* techniques described earlier.

The benefits of the exclusion of air from the sprue, and the reduced velocity during the early part of stage 2, are benefits that have been recorded experimentally for semi-solid (actually partly solid) alloys. These materials are otherwise extremely difficult to cast without defects, almost certainly because their entrainment defects cannot float out but are trapped in suspension because of the high viscosity of the mixture.

Workers from Alcan (Cox *et al.* 1994) developed a system in which the advance of the melt was arrested at the base of the sprue by a layer of ceramic paper supported on a ceramic foam filter (Figure 2.42a). When the sprue was filled the paper was lifted from one corner by a rod, allowing the melt to flow through the filter and into the running system. These authors call their system 'interrupted pouring'. However, the

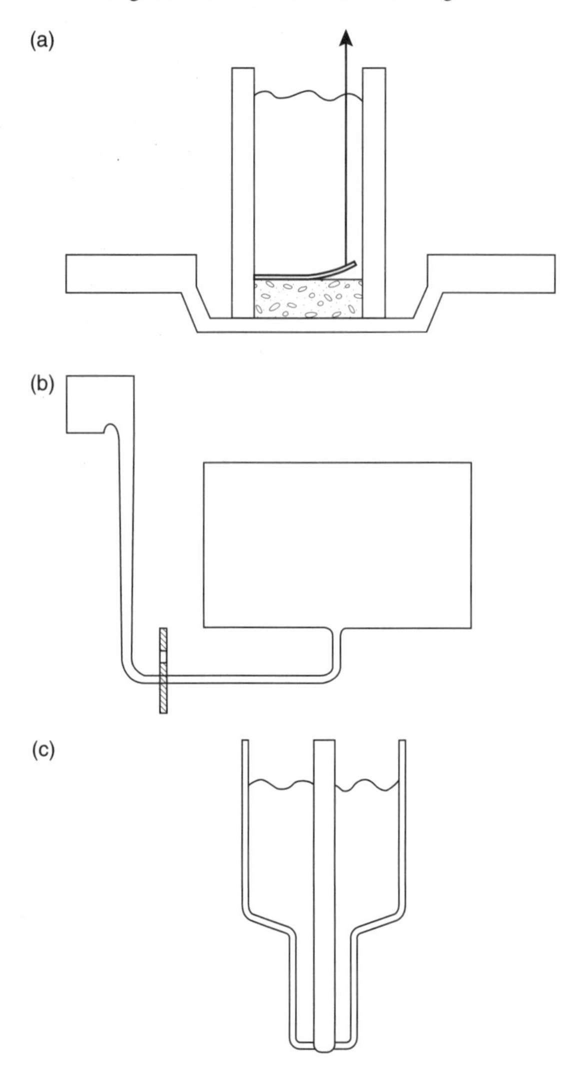

Figure 2.42 *Two-stage filling techniques: (a) paper seal on top of ceramic foam filter, lifted by wire; (b) steel slide gate at entrance to runner; (c) stopper in the base of a snorkel ladle.*

name 'two-stage pour' is recommended as being more positive, and less likely to be interpreted as a faulty pour as a result of an accident.

The two-stage pour was convincingly demonstrated as beneficial by Taghiabadi and colleagues (2003) for both partly solid and conventional aluminium casting alloys. These authors used Weibull statistics to confirm the reality of the benefits. They used a steel sheet to form a barrier, as a slide valve, in the runner (Figure 2.42b). After the filling of the sprue the sheet was withdrawn, allowing the mould to fill.

A second completely different incarnation of the two-stage filling concept is the *snorkel ladle*, sometimes known as the *eye-dropper ladle*. It is illustrated in Figure 2.42c. The device is used mainly in the aluminium casting industry, but would with benefit extend to other casting industries. Instead of transferring metal from a furnace via a ladle or spoon of some kind, and pouring into a pouring basin connected to a sprue, the snorkel dips into the melt, and can be filled uphill simply by dipping sufficiently deeply, or by a shallow dip and the melt sucked up by a reduced pressure applied in the body of the snorkel. The ladle is then transferred to the mould where it can deliver its contents into a conventional basin and sprue system, or, in the mode recommended here, lowered down through the mould to reach and engage with the runner. Only then is its stopper raised and the melt delivered to the start of the running system with minimal surface turbulence. The approach is capable of producing excellent products.

Two-stage filling in its various forms seems to offer real promise for many castings.

2.3.2.12 Vortex systems

The vortex has usually been regarded in foundries as a flow feature to be avoided at all costs. If the vortex truly swallowed air, and the air found its way into the casting, the vortex would certainly have to be avoided. However, in general, this seems to be not true.

The great value of the vortex is that it is a powerful organizer of the flow. Designers of water intakes for hydroelectric power stations are well aware of this benefit. Instead of the water being allowed to tumble haphazardly down the water intake from the reservoir, it is caused to spiral down the walls. At the base of the intake duct the loss of rotational energy allows the duct to back-fill to some extent. The central core of air terminates at the level surface of a comparatively tranquil pool, only gently circulating, near the base of the duct. (The spiralling central core of air does not extend indefinitely through the system.)

Several proposals to harness the benefits of vortices to running systems have originated in recent years from Birmingham, following the lead by Isawa (1994). They are potentially exciting departures from conventional approaches. Only initial results can be presented here. The systems merit much further investigation.

Vortex sprue

The benefits of the vortex for the action of a sprue were first explored by Campbell and

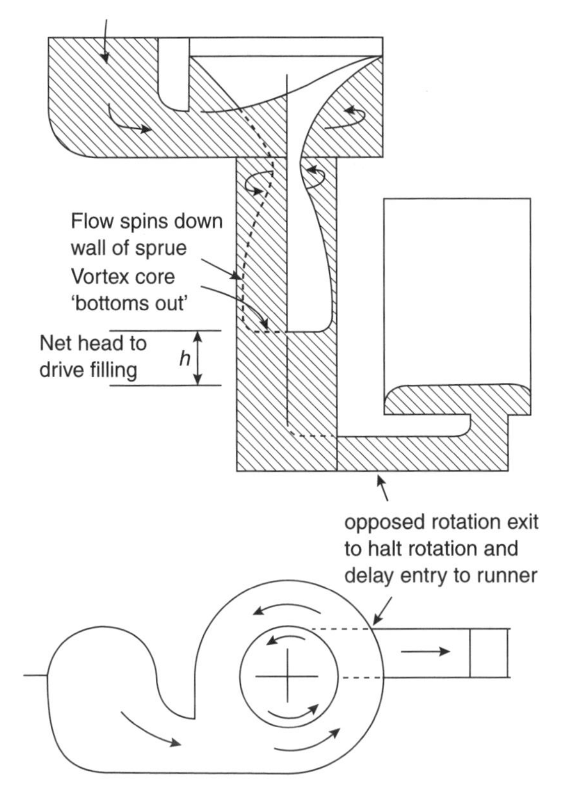

Figure 2.43 *Vortex sprue (after Isawa and Campbell 1994).*

Isawa (1994) as illustrated in Figure 2.43. An aluminium alloy was poured off-axis, being diverted tangentially into a circular pouring basin. The melt spun around the outside of the basin, gradually filling up and so progressing towards the central sprue entrance. As it progressed inwards, gradually reducing its radius, its rotation speeded up, conserving its angular momentum like the spinning ice skater who closes in outstretched arms. Finally, the melt reached the lip of the sprue, where, at maximum spinning speed, it started its downward fall.

The rationale behind this thinking is that the initial fall during the filling of the sprue is controlled by the friction of a spiral descent down the wall of the sprue. Once the sprue starts to fill, the core of the vortex terminates near the base of the sprue; it does not in fact funnel air into the mould cavity. The hydrostatic pressure generated by this system, driving the flow into the runner and gates, arises only from the small depth of liquid at the base of the vortex, so that the net pressure head h driving the filling of the mould is small. Because the level of the pool at the base of the vortex and the level of metal in the mould cavity both rise together, the net head to drive the filling of the mould remains remarkably constant during the entire mould filling process. Thus the filling of the mould is necessarily gentle at all times.

Despite some early success with this system, it seems that the technique is probably not suitable for sprues of height greater than perhaps 200 or 300 mm, because the benefits of the spiral flow are lost progressively with increasing fall distance. More research may be needed to confirm the benefits and limitations of this design. For instance, the early work has been conducted on parallel cylindrical sprues, since the taper has been thought to be not necessary as a result of the melt adopting its own 'taper' as it accelerates down the walls, becoming a thinner stream as it progresses. However, a taper may in any case be useful to favour the speeding up of the rate of spin, and so assist maintaining the spin despite losses from friction against the walls. Also of course, the provision of a taper will assist the sprue to fill faster, and increase yield. Much work remains to be done to define an optimum system.

Vortex well or gate

The provision of a cylindrical channel at the base of the sprue, entered tangentially by the melt, is a novel idea with considerable potential (Hsu *et al.* 2003). It gives a technique for dealing with the central issue of the high liquid velocity at the base of the sprue, and the problem of turning the right-angle corner and successfully filling the runner. What is even better, it promises to solve all of these problems without significant surface turbulence.

The vortex well can probably be oriented either horizontally or vertically as seen in Figure 2.44. The horizontal orientation may be useful for delivery to a single vertical gate. Alternatively, the vertical orientation is often convenient because the central outlet from the vortex can form the entrance to the runner, allowing the connection to many gates.

Notice that the device works exactly opposite to the supposed action of a spinner designed to centrifuge buoyant inclusions from a melt. In the vortex well the outlet to the rest of the filling system is the outlet that would normally be used to concentrate inclusions. Thus the device certainly does not operate to reduce the inclusion content. However, it should be highly effective in reducing the generation of inclusions by surface turbulence at the sprue base of poorly designed systems.

Once again, these are early days for this invention. Early trials on a steel casting of about 4 m height have suggested that the vortex is

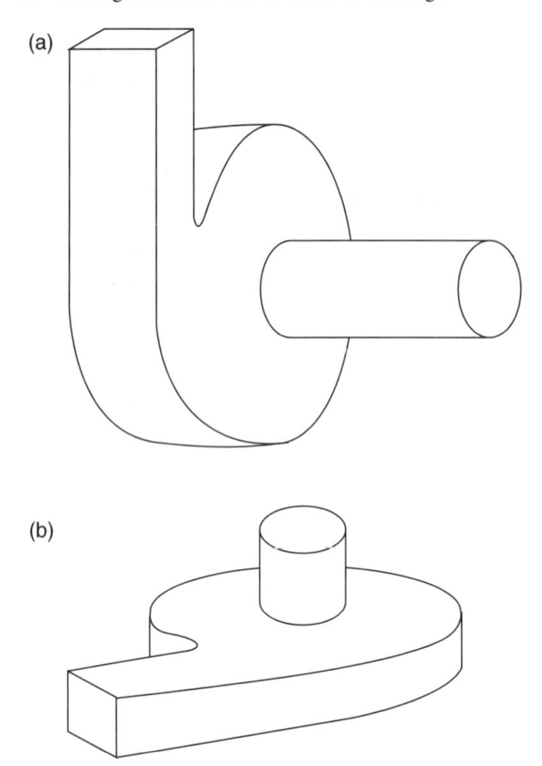

Figure 2.44 *(a) Vortex well (with horizontal axis) and (b) vortex gate (with vertical axis).*

extremely effective in absorbing the energy of the flow. In this respect its action resembles that of a ceramic foam filter. To enable the device to be used in routine casting production the energy absorbing behaviour would require to be quantified. There is no shortage of future tasks.

Vortex runner (the offset sprue)

The simple provision of an *offset sprue* causes the runner to fill tangentially, the melt spinning at high speed (Figures 2.19e and 2.45). The technique is especially suited to a vertically parted mould, where a rectangular cross-section sprue, moulded on only one side of the parting line, opens into a cylindrical runner moulded on the joint. The consequential highly organized filling of the runner is a definite improvement over many poor runner designs, as has been demonstrated by the Weibull statistics of strengths of castings produced by conventional in-line and offset (vortex) runners (Yang *et al.* 1998, 2000 and 2003). The technique produces convincingly more reliable castings than conventional in-line sprues and runner systems.

However, at this time it is not certain that the action of the vortex runner is better than other useful solutions such as the slot sprue/slot runner, or the vortex well, but it has the great benefit of simplicity. It promises to be valuable for vertically parted moulds; it deals effectively with the problem of high flow velocities in such moulds because of the great fall heights often encountered.

2.3.3 Horizontal transfer casting

The quest to avoid the gravity pouring of liquid metals has led to systems employing horizontal transfer and counter-gravity transfer. These solutions to avoid pouring are clearly seen to be key developments; both seem capable of giving competitive casting processes that offer products of unexcelled quality. The two major approaches to the first approach, horizontal transfer, are described below.

2.3.3.1 Level pour (side pour)

The 'level pour' technique was invented by Erik Laid (1978). At that time this clever technique delivered castings of unexcelled quality. It seems a pity that the process is not more generally used. This has partly occurred as the result of the process remaining commercially confidential for much of its history, so that relatively little has been published concerning the operational details that might assist a new user to achieve success. Also, the technique is limited to the type of castings, being applied easily only to plate, box or cylinder type castings where a long slot ingate can be provided up the complete height of the casting. In addition, of course, a fairly complex casting station is required.

The arrangement to achieve the so-called level filling of the mould is shown in Figure 2.46. An insulated pouring basin connects to a horizontal insulated trough that surrounds three of the four sides of the mould (a distribution system reminiscent of a Roman aqueduct). The melt enters the mould cavity via slot gates that extend vertically from the drag to the cope. Either side of each slot gate are guide plates that contain the melt between sliding seals as it flows out of the (stationary) trough and into the (descending) mould.

Casting starts with the mould sitting on the fully raised mould platform, so that the trough provides its first metal at the lowest level of the drag. The mould platform is then slowly lowered while pouring continues. The rate of withdrawal of the mould is such that the metal in the slot gate has time to solidify prior to its

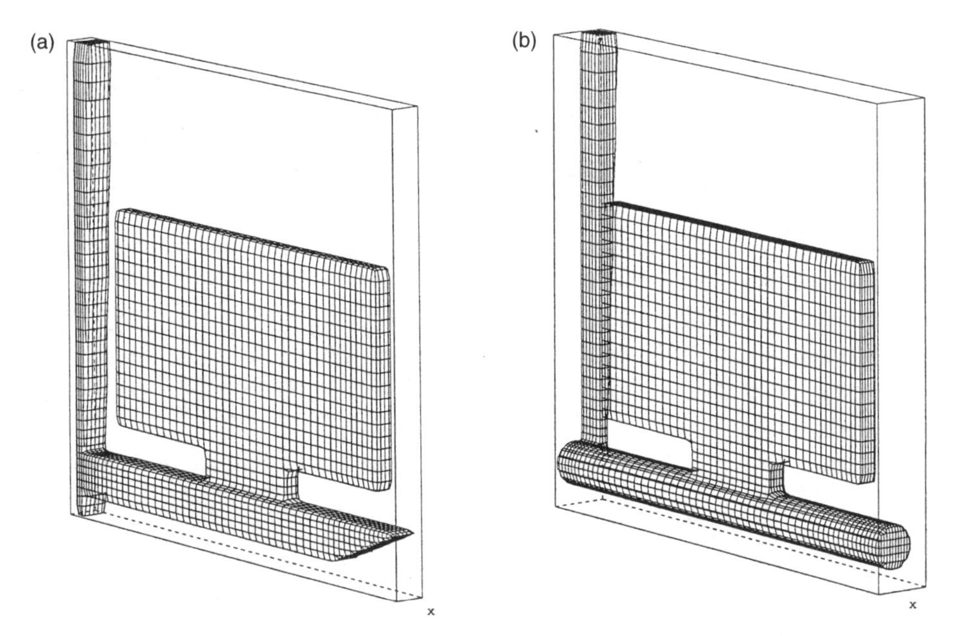

Figure 2.45 *(a) A conventional runner, and (b) a vortex runner, useful on a vertically parted mould (courtesy of X. Yang and Flow 3-D).*

appearance as it slides out below the level of the trough.

In one of the rare descriptions of the use of the process by Bossing (1982) the large area of melt contained in the pouring basin and distribution trough would materially help to smooth the rate of flow from the point of pour to delivery into the mould. Also in this description is an additional complicated distribution system inside the mould, in which tiers of runners are provided to minimize feeding distances and maximize temperature gradients. In general, such sophistication would not be expected to be necessary for most products.

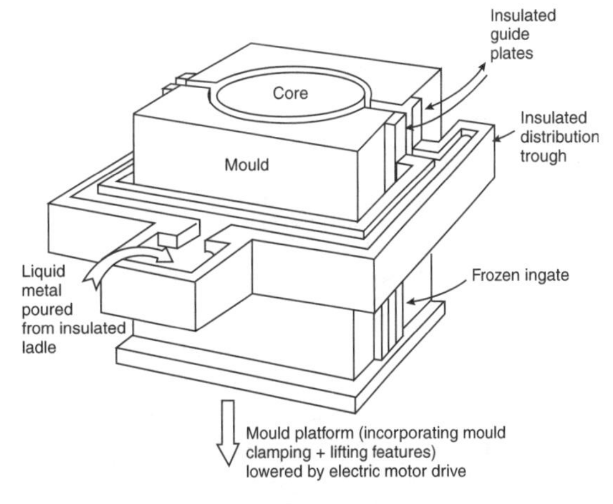

Figure 2.46 *Level pour technique.*

2.3.3.2 Controlled tilt casting

It seems that foundrymen have been fascinated by the intuition that tilt casting might be a solution to the obvious problems of gravity pouring. The result has been that the patent literature is littered with re-inventions of the process decade after decade.

Even so, the deceptive simplicity of the process conceals some fundamental pitfalls for the unwary. The piles of scrap seen from time to time in tilt-pour foundries are silent testimony to these hidden dangers. Generally, however, the dangers can be avoided, as will be discussed in this section.

Tilt casting is a process with the unique feature that, in principle, liquid metal can be transferred into a mould by simple mechanical means under the action of gravity, but without surface turbulence. It therefore has the potential to produce very high quality castings. This was understood by Durville in France in the 1800s and applied by him for the casting of aluminium-bronze in an effort to reduce surface defects in French coinage.

The various stages of liquid metal transfer in the Durville Process are schematically illustrated in Figure 2.47a. In the process as

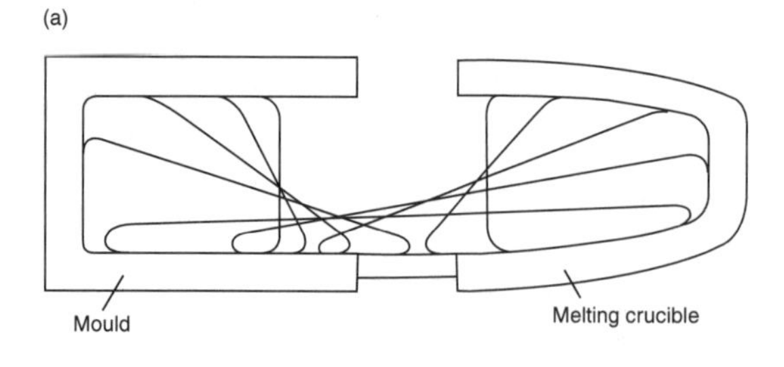

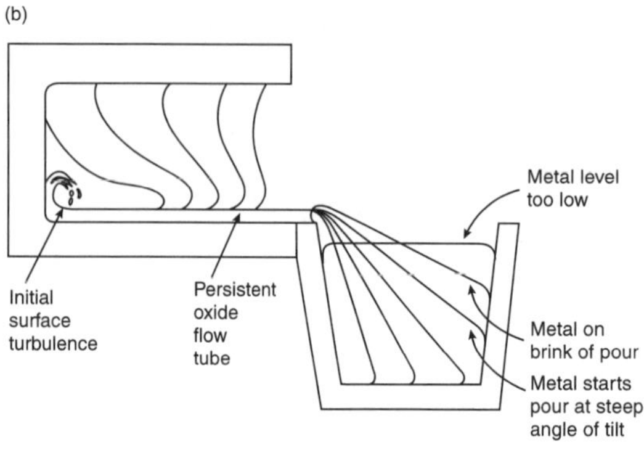

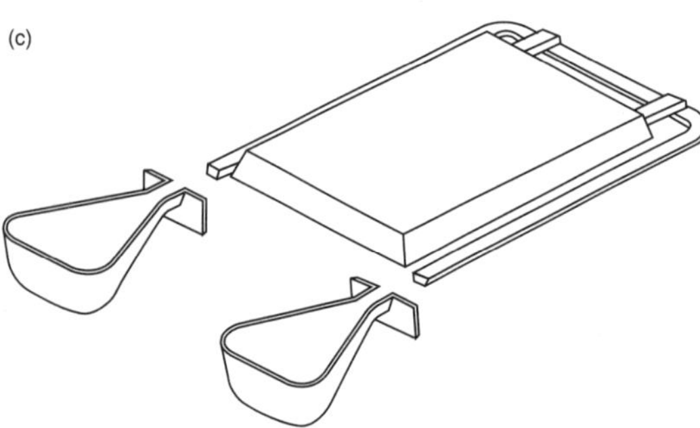

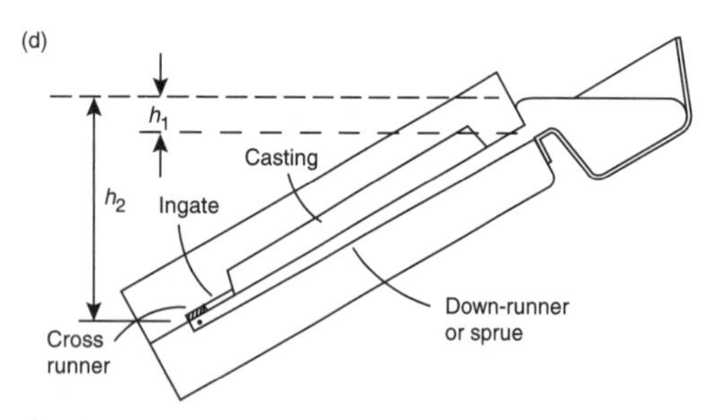

Figure 2.47 *Tilt casting process (a) Durville; (b) Semi-Durville; (c) twin-poured tilting die (adapted from Nyamekye et al. 1994); and (d) outline of tilt running system design at the critical moment that metal reaches the far end of the 'sprue'.*

originally conceived by Durville, the metal is melted in the same crucible as is used for the tilt machine. No pouring under gravity takes place at all. Also, since he was casting large, open-ended ingots for subsequent working, he was able to look into the crucible and into the mould, observing the transfer of the melt as the rotation of the mould progressed. In this way he could ensure that the rate of rotation was correct to avoid any disturbance of the surface of the liquid. During the whole process of the transfer, careful control ensured that the melt progressed by 'rolling' in its skin of oxide, like inside a rubber sack, avoiding any folding of its skin by disturbances such as waves. The most sensitive part of the transfer was at the tilt angle close to the horizontal. In this condition the melt front progresses by expanding its skin of oxide, while its top surface at all times remains horizontal and tranquil.

In the USA, Stahl (1961) popularized the concept of 'tilt pouring' for aluminium alloys into shaped permanent mould castings. The gating designs and the advantages of tilt pouring over gravity top pouring have been reviewed and summarized in several papers from this source (Stahl 1963, 1986, 1989).

A useful 'bottom-gated' tilt arrangement is shown in 2.47c, d. Here the sprue is in the drag, and the remainder of the running and gating system, and the mould cavity, is in the cope. Care needs to be taken with a tilt die to ensure that the remaining pockets of air in the die can vent freely to atmosphere. Also, the die side that retains the casting has to contain the ejectors if they are needed. The layout in Figure 2.47c illustrates a unique benefit enjoyed by tilt casting: a single operator can fill both pouring cups from a large ladle prior to starting the tilt. Static gravity casting would require two pourers to fill two pouring basins.

In an effort to understand the process in some depth, Nguyen and Carrig (1986) simulated tilt casting using a water model of liquid metal flow, and Kim and Hong (1995) carried out some of the first computer simulations of the tilt casting process. They found that a combination of gravity, centrifugal and Coriolis forces govern tilt-driven flow. However, for the slow rates of rotation such as are used in most tilt casting operations, centrifugal and Coriolis effects contribute less than 10 per cent of the effects due to gravitational forces, and could therefore normally be neglected. The angular velocity of the rotating mould also made some contribution to the linear velocity of the liquid front, but this again was usually negligible because the axis of rotation was often not far from the centre of the mould.

However, despite these studies, and despite its evident potential, the process has continued to be perfectly capable of producing copious volumes of scrap castings.

The first detailed study of tilt casting using the recently introduced concepts of critical velocity and surface turbulence was carried out in the author's laboratory by Mi (2002). In addition to the benefits of working within the new conceptual framework, he had available powerful experimental techniques. He used a computer controlled, programmable casting wheel onto which sand moulds could be fixed to produce castings in an Al–4.5%Cu alloy. The flow of the metal during the filling of the mould was recorded using video X-ray radiography, and the consequential reliability of the castings was checked by Weibull statistics.

Armed with these techniques, Mi found that at the slow rotation speeds used in his work the mechanical effect of surface tension and/or surface films on the liquid meniscus could not be neglected. For all starting conditions, the flow at low tilt speeds is significantly affected by surface tension (most probably aided by the effect of a strong oxide film). Thus below a speed of rotation of approximately 7 degrees per second the speed of the melt arriving at the end of the runner is held back. Gravity only takes control after tilting through a sufficiently large angle.

As with most casting processes, if carried out too slowly, premature freezing will lead to misrun castings. One interesting case was found in which the melt was transferred so slowly into the runner that frozen metal in the mouth of the runner acted as an obstructing ski jump to the remaining flow, significantly impairing the casting. At higher speeds, however, although ski jumps could be avoided, the considerable danger of surface turbulence increased.

The radiographic recordings revealed that the molten metal could exhibit tranquil or chaotic flow into the mould during tilt casting, depending on (i) the angle of tilt of the mould at the start of casting, and (ii) the tilting speed. The quality of the castings (assessed by the scatter in mechanical properties) could be linked directly to the quality of the flow into the mould.

We can follow the progress of the melt during the tilt casting process. Initially, the pouring basin at the mouth of the runner is filled. Only then is the tilting of the mould activated. Three starting positions were investigated:

(i) If the mould starts from some position in which it is already tilted downward, once the metal enters the sprue it is immediately unstable, and runs downhill. The melt accelerates under gravity, hitting the far end of the runner at a speed sufficient to cause splashing. The splash action entrains the melt surface. Castings of poor reliability are the result.

(ii) If the mould starts from a horizontal position, the metal in the basin is not usually filled to the brim, and therefore does not start to overflow the brim of the basin and enter the runner until a finite tilt angle has been reached. At this stage the vertical fall distance between the start and the far end of the runner is likely to be greater than the critical fall distance. Thus although slightly better castings can be made, the danger of poor reliability remains. This unsatisfactory mode of transfer typifies many tilt casting arrangements, particularly the so-called Semi-Durville type process shown in Figure 2.47b.

(iii) If, however, the mould is initially tilted slightly uphill during the filling of the basin, there is a chance that by the time the change of angle becomes sufficient to start the overflow of melt from the basin, the angle of the runner is still somewhat above the horizontal. The nature of the liquid metal transfer is now quite different. At the start of the filling of the runner the meniscus is effectively climbing a slight upward slope. Thus its progress is totally stable, its forward motion being controlled by additional tilt. If the mould is not tilted further the melt will not advance. By extremely careful control of the rate of tilt it is possible in principle to cause the melt to arrive at the base of the runner at zero velocity if required. (Such drastic reductions in speed would, of course, more than likely be counter-productive, involving too great a loss of heat, and are therefore not recommended.) Even at quite high tilting speeds of 30 degrees per second as used by Mi in his experimental mould, the velocity of the melt at the end of the runner did not exceed the critical value 0.5 m s^{-1}, and thus produced sound and repeatable castings.

The unique feature of the transfer when started above the horizontal in this way (mode iii above) is that the surface of the liquid metal is close to *horizontal* at all times during the transfer process. Thus in contrast to all other types of gravity pouring, this condition of tilt casting does not involve pouring (i.e. a free *vertical* fall) at all. It is a *horizontal* transfer process. It will be seen that in the critical region of tilt near to the horizontal, the nature of the transfer is the same as that employed originally by Durville.

Thus the optimum operational mode for tilt casting is the condition of horizontal transfer.

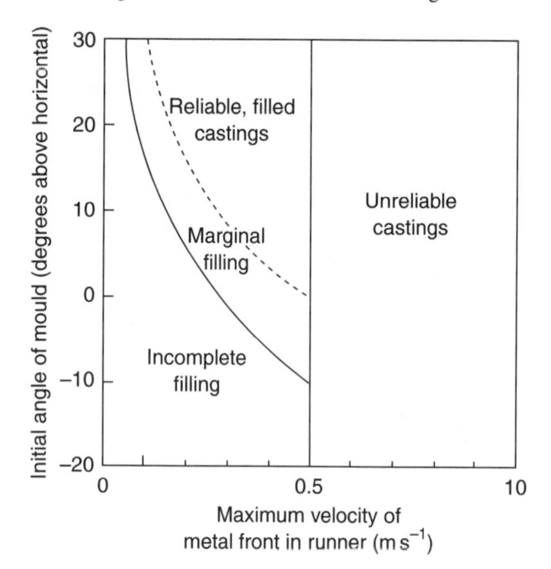

Figure 2.48 *Map of variables for tilt pouring, showing the operational window for good castings (Mi* et al. *2002).*

Horizontal transfer requires the correct choice of starting angle above the horizontal, and the correct tilting speed.

An operational map was constructed (Figure 2.48), revealing for the first time a window for the production of reliable castings. It is recognized that the conditions defined by the window are to some extent dependent on the geometry of the mould that is chosen. However, the mould in Mi's experiments was designed to be close to the size and shape of many industrial castings, particularly those for automotive applications. Thus although the numerical conclusions would require some adaptation for other geometries, the principles are of general significance and are clear: there are conditions, possibly narrowly restricted, but in which horizontal transfer of the melt is possible, and gives excellent castings.

The problem of horizontal transfer is that it is slow, sometimes resulting in the freezing of the 'ski jump' at the entrance to the runner, or even the non-filling of the mould. This can usually be solved by increasing the rate of tilt *after* the runner is primed. This is the reason for the extended threshold, denoted a marginal filling condition, on the left of the window shown on the process map (Figure 2.48). A constant tilt rate (as is common for most tilt machines at this time) cannot achieve this useful extension of the filling conditions to achieve good castings. Programmable tilt rates are required to achieve this solution.

A final danger should be mentioned. At certain critical rates of rise of the melt against an inclined surface, the development of the transverse travelling waves seems to occur to give lap problems on the cope surface of castings (illustrated later in Figure 2.60). In principle, such problems could be included as an additional threshold to be avoided on the operational window map (Figure 2.48). Fortunately, this does not seem to be a common fault. Thus in the meantime, the laps can probably be avoided by increasing the rate of tilt during this part of the filling of the mould. Once again, the benefits of a programmable tilt rate are clear.

In summary, the conclusions for tilt casting are:

1. If tilt casting is initiated from a tilt orientation at, or (even more especially) below the horizontal, during the priming of the runner the liquid metal runs downhill at a rate out of the control of the operator. The accelerating stream runs as a narrow jet, forming a persistent oxide flow tube. In addition, the velocity of the liquid at the far end of the runner is almost certain to exceed the critical condition for surface turbulence. Once the mould is initially inclined by more than 10 degrees below the horizontal at the initiation of flow, Mi found that it was no longer possible to produce reliable castings by the tilt casting process.
2. Tilt casting operations benefit from using a sufficiently positive starting angle that the melt advances into an upward sloping runner. In this way its advance is stable and controlled. This mode of filling is characterized by horizontal liquid metal transfer, promoting a mould filling condition free from surface turbulence.
3. Tilt filling is preferably slow at the early stages of filling to avoid the high velocities at the far end of the running system. However, after the running system is primed, speeding up the rate of rotation of the mould greatly helps to prevent any consequential non-filling of the castings.

2.3.4 Counter-gravity

There are some advantages to the use of gravity to action the filling of moulds. It is simple, low cost and completely reliable, since gravity has never been known to suffer a power failure. It is with regret, however, that the advantages finish here, and the disadvantages start. Furthermore, the disadvantages are serious.

Nearly all the problems of gravity pouring arise as a result of the velocity of the fall. After a trivial fall distance corresponding to the critical fall height, gravity has accelerated the melt to its critical velocity. Beyond this point there is the danger of entrainment defects. Because the critical fall distance is so small, being only the height of a sessile drop of the liquid, nearly all actual falls exceed this limit. In other words the energy content of the melt, when allowed to fall even only relatively small distances under gravity, is nearly always sufficiently high to lead to the break-up of the liquid surface. (It is of little comfort at this time to know that foundries on the moon would fare better.)

A second fundamental drawback of gravity filling is the fact that at the start of pouring, at the time the melt is first entering the ingates, the narrowest part of the mould cross-section where volume flow rate should be slowest, the speed of flow by gravity is highest. Conversely, at a late stage of filling, when the melt is at its coldest and approaching the top of the mould cavity, and the melt needs to be fastest, the speed of filling is slowest. Thus filling by gravity gives completely the wrong filling profile.

Thus to some extent, there are always problems to be expected with castings poured by gravity. The long section on filling system design in this book is all about reducing this damage as far as possible. It is a tribute to the dogged determination of the casting fraternity that gravity pouring, despite its severe shortcomings, has achieved the level of success that it currently enjoys.

Even so, over the last 100 years and more, the fundamental problems of gravity filling have prompted casting engineers to dream up and develop counter-gravity systems.

Numerous systems have arisen. The most common is low-pressure casting, in which air or an inert gas is used to pressurize an enclosed furnace, forcing the melt up a riser tube and into the casting (Figure 2.49). Other systems use a partial vacuum to draw up the metal. Yet others use various forms of pumps, including direct displacement by a piston, by gas pressure (pneumatic pumps), and by various types of electromagnetic action.

Clearly, with a good counter-gravity system, one can envisage the filling of the mould at velocities that never exceed the critical velocity, so that the air in the mould is pushed ahead of the metal, and no surface entrainment occurs. The filling can start gently through the ingates, speed up during the filling of the main part of the mould cavity, and finally slow down and stop as the mould is filled. The final deceleration is useful to avoid any final impact at the instant the mould is filled. If not controlled in this way, the transient pressure pulse resulting from the sudden loss of momentum of the melt can cause the liquid to penetrate any sand cores, open mould joints to produce flash, and generally impair surface finish.

When using a good counter-gravity system, good filling conditions are not difficult to achieve. In fact, in comparison with gravity pouring where it is sometimes difficult to achieve a good casting, counter-gravity is such a robust technique that it is often difficult to make a bad casting. This fundamental difference between gravity and counter-gravity filling is not widely

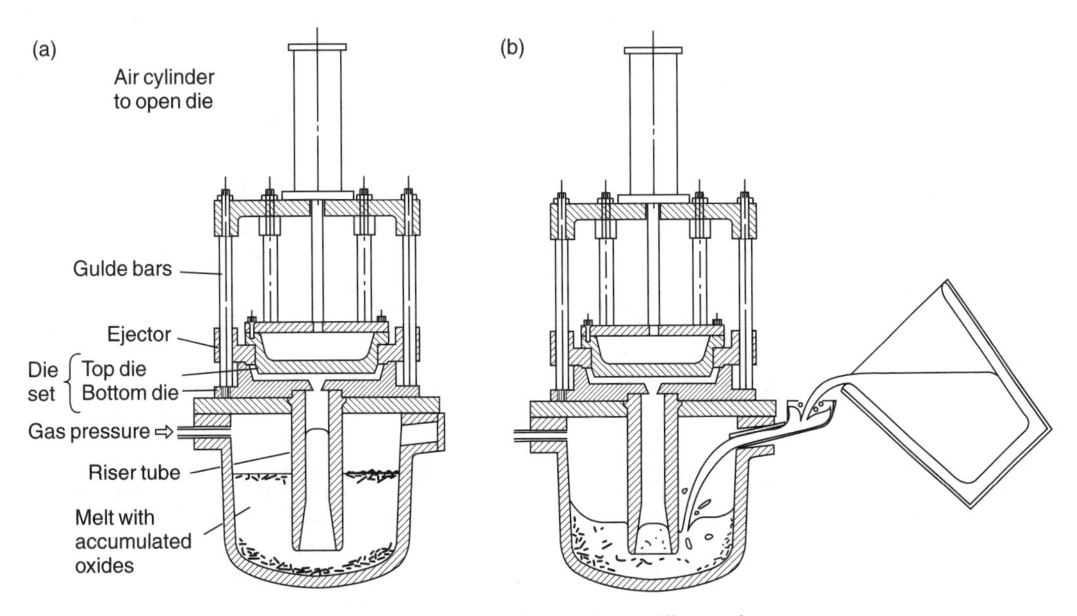

Figure 2.49 *(a) Low-pressure casting process, and (b) the usual poor filling technique.*

appreciated. In general, only those who have suffered gravity filling and finally accepted counter-gravity appreciate and are amazed by the powerful benefits.

That is not to say that the technique is not sometimes used badly. The usual failure is to keep the metal velocity under control. However, in principle it can be controlled, in contrast to gravity pouring where, in principle, control is often difficult or impossible.

A concern often expressed about counter-gravity is that the adoption of filling speeds below the critical speed of approximately $0.5\,m\,s^{-1}$ will slow the production rate. Such fears are groundless. For instance if the casting is 0.5 m tall (a tall casting) it can, in principle, be filled in one second. This would be a challenge!

In fact the unfounded fear of the use of low velocities of the melt leading to a sacrifice of production rate follows from the confusion of (i) flow velocity (usually measured in $m\,s^{-1}$) and (ii) melt volume flow rate (usually measured in $m^3\,s^{-1}$). For instance the filling time can be kept short by retaining a slow filling velocity but increasing the volume flow rate simply by increasing the areas of the flow channels. Worked examples to emphasize and clarify this point further will be given in Section 2.3.7 dealing with the calculation of the filling system.

2.3.4.1 Programmable control

The varying cross-sectional areas of the metal as it rises in the mould pose a problem if the fill rate through the bottom gate is fixed (as is approximately true for many counter-gravity filling systems that lack any sophistication of programmable control). Naturally, the melt may become too slow if the area of the mould increases greatly, leading to a danger of cold laps or oxide laps. Alternatively, if the local velocity is increased above the critical velocity through a narrow part of the mould, the metal may jet, causing entrainment defects.

Counter-gravity filling is unique in having the potential to address this difficulty. In principle, the melt can be speeded up or slowed down as required at each stage of filling. Even so, such programming of the fill rate is not easily achieved. In most moulds there is no way to determine where the melt level is at any time during filling. Thus if the pre-programmed filling sequence (called here the filling profile) gets out of step, its phases occurring either early or late, the filling can become worse than that offered by a constant rate system. The mis-timing problems can easily arise from splashes that happen to start timers early, or from blockage in the melt delivery system causing the time of arrival of the melt to be late.

2.3.4.2 Feedback control

The only sure way to avoid such difficulties is to provide feedback control. This involves a system to monitor the height of metal in the mould, and to feed this signal back to the delivery system, to force the system to adhere to a pre-programmed fill pattern.

One system for the monitoring of height is the sensing of the pressure of the melt in the melt delivery system. This has been attempted by the provision of a pocket of inert gas above the melt contained in the permanent plumbing of the liquid metal delivery system, and connected to a pressure transducer via a capillary.

A non-contact system used by the Cosworth sand casting process senses the change in capacitance between the melt and the mould clamp plate when the two are connected as a parallel plate condenser. Good feedback control solves many of the filling problems associated with casting production.

However, elsewhere, feedback control is little used at this time. The lack of proper control in counter-gravity leads to unsatisfactory modes of filling that explain some of the problems with the technique. The other problems relate to the remainder of the melting and melt handling systems in the foundry, that are often poor, involving multiple pouring operations from melt furnaces to ladles and then into the counter-gravity holding vessel. A widespread re-charging technique for a low-pressure casting unit is illustrated in Figure 2.49b; much of the entrainment damage suffered in such processes usually cannot be blamed on the counter-gravity system itself. The problem arises earlier because of inappropriate metal handling in the foundry before any casting takes place.

The lesson is that only limited success can be expected from a foundry that has added a counter-gravity system on to the end of a badly designed melting and melt-handling system. There is no substitute for an integrated approach to the whole production system. Some of the very few systems to achieve this so far have been the processes that the author has helped to develop; the Cosworth Process (see description later under Rule 7) and Alotech Processes. In these processes, when properly implemented, the liquid metal is never poured, never flows downhill, and is finally transferred uphill into the mould.

Finally, the concept of an integrated approach necessarily involves dealing with convection during the solidification of the casting. This serious problem is usually completely overlooked.

It has been the death of many otherwise good counter-gravity systems, but is specifically addressed in the Cosworth and Alotech systems. The problem is highlighted by the author as Rule 7.

The numerous forms of counter-gravity techniques will be discussed in detail in Volume 3 '*Casting Processes*'.

2.3.5 Surface tension controlled filling

This section starts with the interesting situation that the liquid may not be able to enter the mould at all. This is to be expected if the pressure is too low to force melt into a narrow section. It is an effect due to surface tension. If the liquid surface is forced to take up a sharp curvature to enter a non-wetted mould then it will be subject to a repulsive force that will resist the entry of the metal. Even if the metal enters, it will still be subject to the continuing resistance of surface tension, which will tend to reverse the flow of metal, causing it to empty out of the mould if there is any reduction in the filling pressure. These are important effects in narrow-section moulds (i.e. thin-section castings) and have to be taken into account.

We may usefully quantify our formulation of this problem with the well-known equation

$$Pi - Pe = \gamma(1/r_1 + 1/r_2) \quad \textbf{(2.3)}$$

where Pi, is the pressure inside the metal, and Pe the external pressure (i.e. referring to the local environment in the mould). The two radii r_1 and r_2 define the curvature of the meniscus in two planes at right angles. The equation applies to the condition when the pressure difference across the interface is exactly in balance with the effective pressure due to surface tension. To describe the situation for a circular-section tube of radius r (where both radii are now identical), the relation becomes:

$$Pi - Pe = 2\gamma/r \quad \textbf{(2.4)}$$

For the case of filling a narrow plate of thickness $2r$, one radius is, of course, r, but the radius at right angles becomes infinite, so the reciprocal of the infinite radius equates to zero (i.e. if there is no curvature there is no pressure difference). The relation then reduces to the effect of only the one component of the curvature, r:

$$Pi - Pe = \gamma/r \quad \textbf{(2.5)}$$

We have so far assumed that the liquid metal does not wet the mould, leading to the effect of capillary repulsion. If the mould is wetted then the curvature term γ/r becomes negative, so allowing surface tension to assist the metal to enter the mould. This is, of course, the familiar phenomenon of capillary attraction. The pores in blotting paper attract the ink into them; the capillary channels in the wick of a candle suck up the molten wax; and the water is drawn up the walls of a glass capillary. In general, however, the casting technologist attempts to avoid the wetting of the mould by the liquid metal. Despite all efforts to prevent it, wetting sometimes occurs, leading to the penetration of the melt into sand cores and moulds.

Continuing now in our assumption that the metal–mould combination is non-wetting, we shall estimate what head of metal will be necessary to force it into a mould to make a wall section of thickness $2r$ for a gravity casting made under normal atmospheric pressure. If the head of liquid is h, the hydrostatic pressure at this depth is ρgh, where ρ is the density of the liquid, and g the acceleration due to gravity. The total pressure inside the metal is therefore the sum of the head pressure and the atmospheric pressure, Pa. The external pressure is simply the pressure in the mould due to the atmosphere Pa plus the pressure contributed by mould gases Pm. The equation now is

$$(Pa + \rho gh) - (Pa + Pm) > \gamma/r \quad \textbf{(2.6)}$$

giving immediately

$$\rho gh - Pm > \gamma lr \quad \textbf{(2.7)}$$

The back-pressure due to outgassing in the mould lowers the effective head driving the filling of the mould. It is good practice, therefore, to vent narrow sections, reducing this resistance to practically zero if possible.

It is also clear from the above result that, provided the mould is permeable and/or well vented, atmospheric pressure plays no part in helping or resisting the filling of thin sections in air, since it acts equally on both sides of the liquid front, cancelling any effect. Interestingly, the same equation and reasoning applies to casting in vacuum, which, of course, can be regarded as casting under a reduced atmospheric pressure. Clearly, a vacuum casting is therefore not helpful in overcoming the resistance to filling provided by surface tension (although, to be fair, it may help by reducing Pm by outgassing the mould to some extent prior to casting, and it will help where the permeability of the mould is low, where residual gases may be compressed ahead of the advancing stream. Vacuum casting may also help to fill the mould by reducing—but not eliminating—the effect of the surface film of oxide or nitride).

The case of vacuum-assisted filling (not vacuum casting) is quite different, since the vacuum is not now applied to both the front and back of the liquid meniscus, thus cancelling any benefit as above, but applied only to the advancing front as illustrated in Figure 2.50. This application of a reduced pressure to one side of the meniscus creates a differential pressure that drives the flow. The differential pressure acts by atmospheric pressure continuing to apply to the liquid metal via the running system, but the atmospheric pressure in the mould is reduced by applying a (partial) vacuum in the mould cavity. This is achieved by drawing the air out either through the permeable mould, or through fine channels cut through to the section required to be filled (as is commonly applied to the trailing edge of an aerofoil blade section). In this way Pm is guaranteed to be zero or negligible, and Pa remains a powerful pressure to assist in overcoming surface tension as the equation indicates:

$$Pa + \rho gh > \gamma / r \quad \textbf{(2.8)}$$

It is useful to evaluate the terms of this equation to gain a feel for the size of the effects involved. Taking, roughly, g as $10\,\mathrm{m\,s^{-2}}$, and the liquid aluminium density ρ as $2500\,\mathrm{kg\,m^{-3}}$ and γ as $1.0\,\mathrm{N\,m^{-1}}$ (for steels and high-temperature alloys the corresponding values are approximately $7000\,\mathrm{kg\,m^{-3}}$ and $2.0\,\mathrm{N\,m^{-1}}$), the resistance term γ/r works out to be 2 kPa for a 1 mm section (0.5 mm radius) and 10 kPa for a 0.1 mm radius trailing edge on a turbine blade.

For a head of metal $h = 100$ mm the head pressure ρgh is 2.5 kPa, showing that the 1 mm section might just fill. However, the 0.1 mm trailing edge has no chance; the head pressure being insufficient to overcome the repulsion of surface tension. However, if vacuum assistance were applied (NB not vacuum casting, remember) then the additional 100 kPa of atmospheric pressure normally ensures filling. In practice it should be noted that the full value of atmospheric pressure is not easily obtained in vacuum-assisted casting; in most cases a value nearer half an atmosphere is more usual. Even so, the effect is still important: one atmosphere pressure corresponds to 4 m head of liquid aluminium, and approximately 1.5 m head of denser metals such as irons, steels and high-temperature alloys. In modest-sized castings of overall height around 100 mm or so, these valuable pressures to assist filling are not easily obtainable by other means. The pressure delivered by a feeder placed on top of the casting may only apply the additional head corresponding to its height of perhaps 0.1 to 0.4 m; only one tenth of the pressures that can be applied by the atmosphere.

For those castings that have sections of only 1 or 2 mm or less, the surface tension wields strong control over the tight radius of the front. Filling is only possible by the operation of additional pressure, such as that provided by the jeweller's centrifuge, or the application of vacuum assistance. Filling can occur upwards or downwards without problems, being always

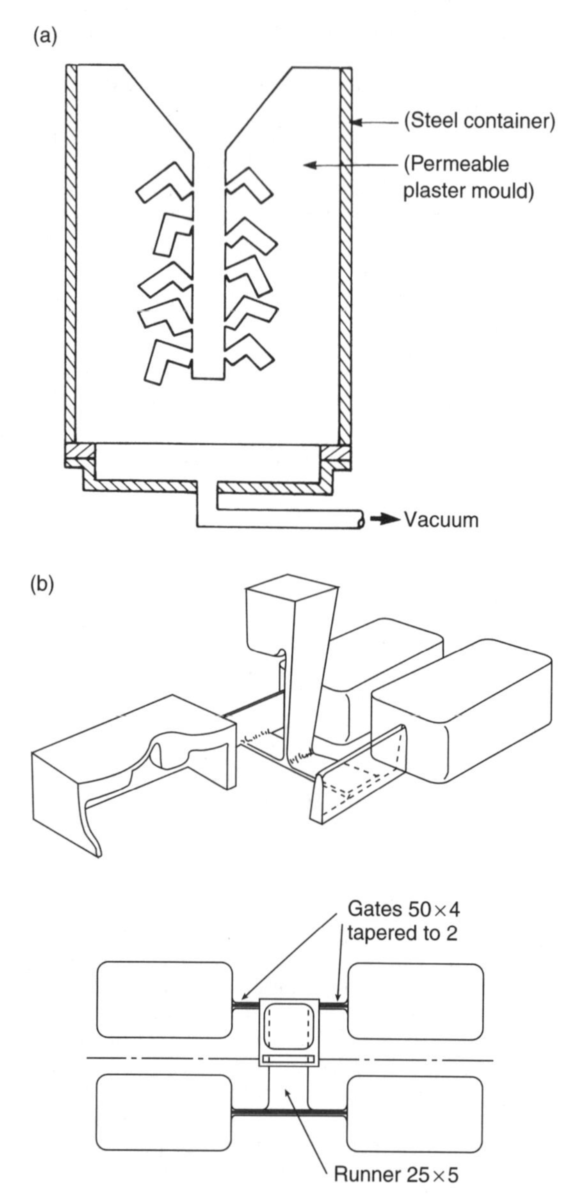

Figure 2.50 *(a) A plaster mould encased in a steel box using vacuum-assisted filling through the base of the mould. No formal running system is required for such small thin-walled castings. (b) Sand mould to make four cover castings, using narrow slot filling system to maximize benefits from surface tension and wall friction.*

under the control of the surface tension, which is effectively so strong in such thin sections that it keeps the surface intact. Surface turbulence is thereby suppressed. The liquid has insufficient room to break up into drops, or to jet or splash. The integrity of the front is under the control of surface tension at all times. This special feature of the filling of very thin-walled castings means that they do not require formal running systems. In fact, such thin-walled investment castings are made successfully by simply attaching wax patterns in any orientation directly to a sprue (Figure 2.50). The metal flows similarly with either gravity or counter to gravity, and no 'runner' or 'gate' is necessary.

To gain an idea of the head of metal required to force the liquid metal into small sections, from Equation 2.8 we have:

$$\rho g h = \gamma / r$$

$$h = \gamma / r \rho g \qquad \textbf{(2.8a)}$$

Using the values for aluminium and steel given above, we can now quickly show that to penetrate a 1 mm section we require heads of at least 80 and 60 mm respectively for these two liquids.

If the section is halved, the required head for penetration is, of course, doubled. Similarly, if the mould shape is not a flat section that imposes only one curvature on the meniscus, but is a circular hole of diameter 1 mm, the surface then has an additional curvature at right angles to the first curvature. Equation 2.4 shows the head is doubled again.

In general, because of the difficulty of predicting the shape of the liquid surface in complex and delicate castings, the author has found that a safety factor of 2 is not excessive when calculating the head height required to fill thin sections. This safety factor is quickly used up when allowances for errors in the wall thickness, and the likely presence of surface films is taken into account.

The resistance to flow provided by surface tension can be put to good effect in the use of slot-shaped filling systems. In this case the slots are required to be a maximum thickness of only 1 or 2 (perhaps 3 at the most) mm for engineering castings (although, clearly, jewellery and other widget type products might require even thinner filling systems). Figure 2.50b shows a good example of such a system. A similar filling system for a test casting designed by the author, but using a conical basin (not part of the author's original design!), was found to perform tolerably well, filling without the creation of significant defects (Groteke 2002). It is quite evident, however, when filling is complete such narrow filling channels offer no possibility of significant feeding. This is an important issue that should not be forgotten. In fact, in these trials, this casting never received the proper attention to feeding, and as a consequence suffered surface sinks and internal microporosity (the liquid alloy was clearly full of bifilms that were subsequently opened by the action of solidification shrinkage).

Finally, however, in some circumstances there may be fundamental limitations to the integrity of the liquid front in very thin sections.

(i) There is a little-researched effect that the author has termed microjetting (*Castings 2003*). This phenomenon has been observed during the filling of liquid Al–7Si–0.4Mg alloy into plaster moulds of sections between 1 and 3 mm thickness (Evans *et al.* 1997). It seems that the oxide on such small liquid areas temporarily restrains the flow, but repeatedly splits open, allowing jets of liquid to be propelled ahead of the front. The result resembles advancing spaghetti. The mechanical properties are impaired by the oxide films around the jets that become entrained in the maelstrom of progress of the front. Whether this unwelcome effect is common in thin-walled castings is unknown, and the conditions for its formation and control are also unknown. Very thin walled castings remain to be researched.

(ii) In pressure die-castings a high velocity v of the metal through the gate is necessary to fill the mould before too much heat is lost to the die. Speeds of between 25 and 50 $m\,s^{-1}$ are common, greatly exceeding the critical velocity of approximately 0.5 $m\,s^{-1}$ that represents the watershed between surface tension control and inertial control of the liquid surface. The result is that entrainment of the surface necessarily occurs on a huge scale. The character of the flow is now dictated by inertial pressure, proportional to v^2, that vastly exceeds the restraining influences of gravity or surface tension. This behaviour is the underlying reason for the use of PQ^2 diagrams as an attempt to understand the filling of pressure die castings. In this approach a diagram is constructed with vertical axis denoting pressure P, and horizontal axis denoting flow rate Q. The parabolic curves are linearized by squaring the scale of the Q values on the horizontal axis. The approach is described in detail in much of the pressure die-casting literature (see, for instance, Wall and Cocks 1980). In practice, it is not certain how valuable this technique is, now that computer simulation is beginning to be accepted as an accurate tool for the understanding of the process.

2.3.6 Inclusion control: filters and traps

The term 'inclusion' is a shorthand generally used for 'non-metallic inclusion'. However, it is to be noted that such defects as tungsten droplets from a poor welding technique can appear in some recycled metals; these, of course, constitute 'metallic inclusions'. Furthermore, one of the most common defects in many castings is the bubble, entrained during pouring. This constitutes an 'air inclusion' or 'gas inclusion.'

The fact that bubbles are trapped in the casting from the filling stage is remarkable in itself. Why did the bubble not simply rise to the surface, burst and disappear? This is a simple but important question. In most cases the bubble will not have been retained by the growth of solid, because solid will, in general, not have time to form. The answer in practically all cases is that oxide films will also be present. In fact the bubbles themselves are simply sections of the oxide films that have not perfectly folded back together. The bubbles decorate the double films, as inflated islands in the folds. Thus many bubbles, entangled in a jumble of films, never succeed to reach the surface to escape. Even those that are sufficiently buoyant to power their way through the tangle may still not burst at the surface because of the layers of oxide that bar its final escape.

This close association of bubbles and films (since they are both formed by the same turbulent entrainment process; they are both entrainment defects) is called by me bubble damage. We need to keep in mind that the bubble is the visible part of the total defect. The surrounding region of bifilms to which it is connected act as cracks, and can be much more extensive and often invisible. However, the presence of such films is the reason that cracks will often appear to start from porosity, despite the porosity having a nicely rounded shape that would not in itself appear to be a significant stress raiser.

Whereas inclusions are generally assumed to be particles having a compact shape, it is essential to keep in mind that the most damaging inclusions are the films (actually always double, unbonded films, remember, so that they act as cracks), and are common in many of our common casting alloys. Curiously, the majority of workers in this field have largely overlooked this simple fact. It is clear that techniques to remove particles will often not be effective for films, and vice versa. The various methods to clean metals prior to casting have been reviewed in Chapter 1 as a fulfilment of Rule 1. The various methods to clean metals travelling through the filling systems of castings will be reviewed here.

2.3.6.1 Dross trap (or slag trap)

The dross trap is used in light alloy and copper-based alloy casting. In ferrous castings it is called a slag trap. For our purposes we shall consider the devices as being one and the same.

It is good sense to include a dross trap in the running system. In principle, a trap sited at the end of the runner will take the first metal through the runner and keep it away from the gates. This first metal is both cold, having given up much of its heat to the running system en route, and will have suffered damage by oxide or other films during those first moments before the sprue is properly filled.

In the past, designs have been along the lines of Figure 2.37a. This type of trap was sized with a view to accommodating the total volume of metal through the system until the down-runner and horizontal runner were substantially filled. This was a praiseworthy aim. In practice, however, it was a regular joke among foundrymen that the best quality metal was concentrated in the dross trap and all the dross was in the casting! What had happened to lend more than an element of truth to this regrettable piece of folk-law?

It seems that this rather chunky form of trap sets up a circulating eddy during filling. Dross arriving in the trap is therefore efficiently floated out again, only to be swept through the gates and into the casting a few moments later! Ashton and Buhr (1974) have carried out work to show that runner extensions act poorly as traps for dirt. They observed that when the first metal reached the end of the runner extension it rose, and created a reflected wave which then travelled back along the top surface of the metal, carrying the slag or dirt back towards the ingates. Such observations have been repeated on iron and steel casting by Davis and Magny (1977) and on many different alloys in the author's laboratory using real-time radiography of moulds during casting. The effect has also been simulated in computer models. It seems, therefore, to be real and universal in castings of all types. We have to conclude that this design of dross trap cannot be recommended!

Figure 2.26b shows a simple wedge trap. It was thought that metal flowing into the narrowing section was trapped, with no rebound wave from the end wall, and no circulating eddy can form. However, video radiographic studies have shown that such traps can reflect a backward wave if the runner is sufficiently deep.

Also, of course, the volume of melt that they can retain is very limited.

A useful design of dross trap appears to be a volume at the end of the runner that is provided with a narrow entrance (the extension shown in broken outline in Figure 2.37b) to suppress any outflow. It is a kind of wedge trap fitted with a more capacious end. In the case of persistent dross and slag problems, the trap can be extended, running around corners and into spare nooks and crannies of the mould. If the entrance section is less than the height of a sessile drop, it will be filled by the entering liquid, thus being too narrow to allow a reflected wave to exit. It should therefore retain whatever material enters. In addition, depending on the narrowness of the tapered wedge entrance, to some extent the device should be capable of filling and pressurizing the runner in a progressive manner akin to the action of a gate. This is a useful technique to reduce the initial transient momentum problems that cause gates to fill too quickly during the first few seconds. This potentially useful benefit has yet to be researched more thoroughly so as to provide useful guidelines for mould design.

The device can be envisaged to be useful in combination with other forms of by-pass designs such as those shown in 2.37d and e.

Slag pockets

For iron and steel castings the term 'slag pocket' is widely used for a raised portion of the runner that is intended to collect slag. The large size of slag particles and their large density difference with the melt encourages such separation. However, such techniques are not the panacea that the casting engineer might wish for.

For instance the use of traps of wedge-shaped design, Figure 2.51, is expected to be almost completely ineffective because the circulation pattern of flow would take out any material that happened to enter. On the other hand, a rectangular cavity has a secondary flow into which buoyant material can transfer if it has sufficient time, and so remain trapped in the upper circulating eddy. This consideration again emphasizes the need for relatively slow flow for its effectiveness. Also, of course, none of these traps can become effective until the runner and the trap become filled with metal. Thus many filling systems will have passed much if not all such unwanted material before the separation mechanisms have a chance to come into operation. A further consideration that causes the author to hesitate to recommend such traps is that they locally remove the constraint on the flow of the metal, allowing surface turbulence. Thus the traps might cause more problems than they solve.

Davis and Magny (1977) observed the filling of iron and steel castings by video radiography. They confirmed that most slag retention devices either do not work at all, or work with only partial effectiveness. These authors made castings with different amounts of slag, and tested the ability of slag pockets sited above runners to retain the slag. They found that rectangular pockets were tolerably effective only if the velocity of flow through the runner was below $0.4\,\mathrm{m\,s^{-1}}$ (interesting that this is precisely the critical speed at which surface turbulence will occur, and so cause surface phases to be turbulently stirred back into the bulk liquid). For a casting only 0.1 m high the metal is already travelling four times too fast. For such reasons the experience with slag pockets has been somewhat mixed in practice.

In defence of the historical use of such traps it must be borne in mind that they were traditionally used with pressurized filling systems, heavily choked at the gate, so that the runner was encouraged to fill as quickly as possible, making the trap effective at an early stage of filling. Also, if the choking action was sufficiently severe the speed of flow in the runner may be sufficiently slow to ensure slag entrapment. However, this text does not recommend pressurized filling systems mainly because of the problems that follow from the necessarily high ingate velocities.

Perhaps, therefore, the slag trap has come to the end of its useful life.

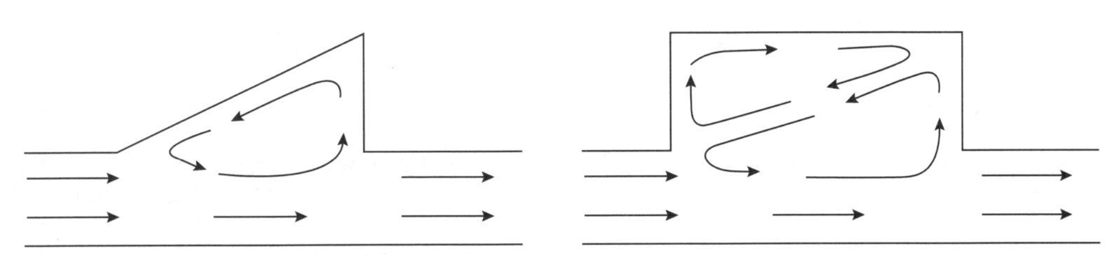

Figure 2.51 *Various designs of slag pockets: (a) relatively ineffective self-emptying wedge; (b) rectangular trap stores buoyant phases in upper circulating flow.*

2.3.6.2 Swirl traps

The centripetal trap is an accurate name for this device, but rather a mouthful. It is also known as a whirl gate, or swirl gate, which is shorter, but inaccurate since the device is not really a gate at all. Choosing to combine the best of both names, we can call it a swirl trap. This is conveniently short, and accurately indicates its main purpose for trapping rubbish.

The idea behind the device is the use of the difference in density between the melt and the various unwanted materials which it may carry, either floating on its surface or in suspension in its interior. The spinning of the liquid creates a centrifugal action, throwing the heavy melt towards the outside where it escapes through the exit, to continue its journey into the casting. Conversely, the lighter materials are thrown towards the centre, where they coagulate and float. The centripetal acceleration a_c is given by:

$$a_c = V^2/r \qquad \textbf{(2.9)}$$

where V is the local velocity of the melt, and r is the radius at that point. For a swirl trap of 50 mm radius and sprue heights of 0.1 m and 1 m, corresponding to velocities of 1.5 and 4.5 m s^{-1} respectively we find that accelerations of 40 and 400 m s^{-2} respectively are experienced by the melt. Given that the gravitational acceleration g is 9.81 m s^{-2}, which we shall approximate to 10 m s^{-2}, these values illustrate that the separating forces within a swirl trap can be between 4 and 40 times that due to gravity. These are, of course, the so-called 'g' forces experienced in centrifuges.

So much for the theory. What about the reality?

Foundrymen have used swirl traps extensively. This popularity is not easy to understand because, unfortunately, it cannot be the result of their effectiveness. In fact the traps have worked so badly that Ruddle (1956) has recommended not to use them on the grounds that their poor performance does not justify the additional complexity. One has to conclude that their extraordinarily wide use is a reflection of the fascination we all have with whirlpools, and an unshakeable belief, despite all evidence to the contrary, that the device should work.

Regrettably, the swirl trap is expected to be completely useless for film-forming alloys where inclusions in the form of films will be too sluggish to separate. Since some of the worst inclusions are films, the swirl trap is usually worse than useless, creating more films than it can remove. Worse still, in the case of alloys of aluminium and magnesium, their oxides are denser than the metal, and so will be centrifuged outwards, into the casting! Swirl traps are therefore of no use at all for light alloys (however, notice that the vortex sprue base, although not specifically designed to control inclusions, might have some residual useful effect for light alloys since the outlet is central). Finally, swirl traps seem to be difficult to design to ensure effective action. In the experience of the author, most do much more harm than good.

The inevitable conclusion is that swirl traps should be avoided.

The remainder of this section on swirl traps is for those who refuse to give up, or refuse to believe. It also serves as a mini-illustration of the real complexity of apparently simple foundry solutions. Such illustrations serve to keep us humble.

It is worthwhile to examine why the traditional swirl trap performs so disappointingly. On examination of the literature, the textbooks, and designs in actual use in foundries, three main faults stand out immediately:

1. The inlet and exit ducts from the swirl traps are almost always opposed, as shown in Figure 2.52a. The rotation of the metal as a result of the tangential entry has, of course, to be brought to a stop and reversed in direction to make its exit from the trap. The disorganized flow never develops its intended rotation and cannot help to separate inclusions with any effectiveness.

   Where the inlet and exit ducts are arranged in the correct tangential sense, then Trojan *et al.* (1966) have found that efficiency is improved in their model results using wood chips in water. Even so, efficiencies in trapping the chips varied between the wide limits of 50 and 100 per cent.
2. The inlet is nearly always arranged to be higher than the exit. This elementary fault gives two problems. First, any floating slag or dross on the first metal to arrive is immediately carried out of the trap before the trap is properly filled (Figure 2.52c). Second, as was realized many years previously (Johnson and Baker 1948), the premature escape of metal hampers the setting up of a properly developed spinning action. Thus the trap is slow to develop its effect, perhaps never achieving its full speed in the short time available during the pour. This unsatisfactory situation is also seen in the work of Jirsa (1982), who describes a swirl trap for steel casting made from preformed refractory sections. In this design the exit was again lower than the entrance, and filling of the trap was encouraged merely by making the exit smaller than the entrance.

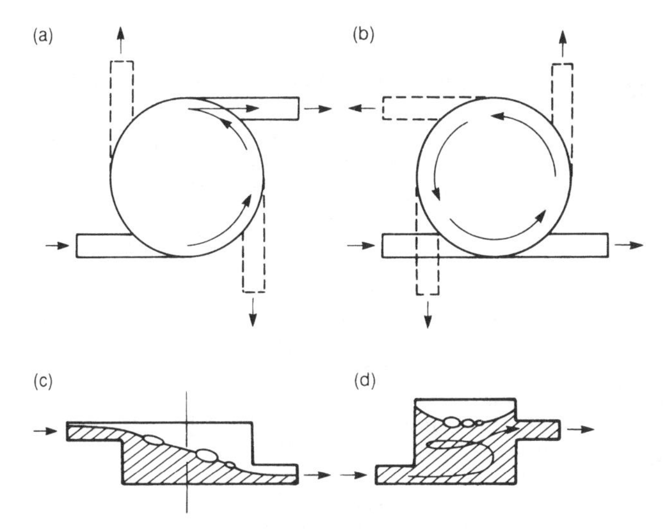

Figure 2.52 *Swirl traps showing (a) incorrect opposed inlet and exit ducts; (b) correct tangential arrangements; (c) incorrect low exit; (d) correct high exit.*

Much metal and slag almost certainly escaped before the trap could be filled and become fully operational. Since 90 per cent efficiency was claimed for this design it seems probable that all of the remaining 10 per cent which evaded the trap did so before the trap was filled (in other words, the trap was working at zero efficiency during this early stage). Jeancolas *et al.* (1971) report an 80 per cent efficiency for their downhill swirl trap for bronze and steel casting, but admit that the trap does not work at all when only partly full.

3. In many designs of swirl trap there is insufficient attention paid to providing accommodation for the trapped material. For instance, where the swirl trap has a closed top the separated material will collect against the centre of the ceiling of the trap. However, work with transparent models illustrates clearly how perturbations to the flow cause the inclusions, especially if small, to ebb and flow out of these areas back into the main flow into the casting (Jeancolas *et al.* 1971; Trojan *et al.* 1966). Also, of course, traps of such limited volume are in danger of becoming completely overwhelmed, becoming so full of slag or dross that the flow into the casting becomes necessarily contaminated.

Where the trap has an open top the parabolic form of the liquid surface assists the concentration of the floating material in the central 'well' as shown in Figure 2.52d. The extra height for the separated materials to rise into is useful to keep the unwanted material well away from the exit, despite variations from time to time in flow rate. Some workers have opened out the top of the trap, extending it to the top of the cope, level with the pouring bush. This certainly provides ample opportunity for slag to float well clear, with no danger of the trap becoming overloaded with slag. However, the author does not recommend an open system of this kind, because of the instability which open-channel systems sometimes exhibit, causing surging and slopping between the various components comprising the 'U'-tube effect between the sprue, swirl trap and mould cavity.

It is clear that the optimum design for the swirl trap must include the features:

(i) the entrance at the base of the trap;
(ii) the exit to be sited at a substantially higher level;
(iii) both entrance and exit to have similar tangential direction, and
(iv) an adequate height above the central axis to provide for the accumulation of separated debris.

In most situations the inlet will be moulded in the drag, and the exit in the cope, which is the most marginal difference in level between the two. At the high speeds at which the metal can be expected to enter the trap the metal will surge over this small ledge with ease, taking inclusions directly into the casting, particularly if the inlet and outlet are in line as shown in Figure 2.52b. This simplest form of cope/drag parting line swirl trap cannot be expected to work.

The trap may be expected to work somewhat more effectively as the angle of the outlet progresses from 90–180 degrees. (The 270 degree option would be more effective still, except that some reflection will show it to be unmouldable on this single joint line; the exit will overlay the entrance ports! Clearly, for the 270 degree option to be possible, the entrance and exits have to be moulded at different levels, necessitating a second joint line provided by a core or additional mould part.)

When using preformed refractory sections, or pre-formed baked sand cores, as is common for larger steel castings, the exit can with advantage be placed considerably higher than the entrance (Figure 2.52d).

These simple rules are designed to assist the trap to spin the metal up to full speed before the exit is reached, and before any floating or emulsified less-dense material has had a chance to escape.

For the separation of particulate slag inclusions from some irons and steels, *Castings 1991* showed that a trap 100 mm diameter in the running system of moulds 0.1 to 1 m high would be expected to eliminate inclusions of 0.2 to 0.1 mm respectively. The conclusion was that, when correctly designed, the swirl trap could be a useful device to divert unwanted buoyant particles away from ferrous castings.

We have to remember, however, that it is not expected to work for film type inclusions. Compared to particles, films would be expected to take between 10 and 100 times longer to separate under an equivalent field force. Thus most of the important inclusions in a large number of casting alloys will not be effectively trapped. Thus the alloys that need the technique most are least helped.

This damning conclusion applies to other field forces such as electromagnetic techniques that have recently been claimed to remove inclusions from melts. It is true that forces can be applied to non-conducting particles suspended in the liquid. However, whereas compact particles move relatively quickly, and can be separated in the short time available while the melt travels through the field. Films experience the same force, but move too slowly because of their high drag, and so are not removed.

In summary, we can conclude that apart from certain designs of by-pass trap, other varieties of traps are not recommended. In general they almost certainly create more inclusions than they remove.

2.3.6.3 Filters

Filters take many forms: as simple strainers, woven ceramic cloths, and ceramic blocks of various types. Naturally, their effectiveness varies from application to application, as is discussed here.

Strainers

A sand or ceramic core may be moulded to provide a coarse array of holes, of a size and distribution resembling a domestic colander. A typical strainer core might be a cylinder 30 to 50 mm diameter, 10 to 20 mm long, containing 10 or more holes, of diameter approximately 3–5 mm (Figure 2.53a). These devices are mainly used to prevent slag entering iron castings. The domestic colander is usually used to strain aggregates such as peas. These represent solid spherical particles of the order of 5 mm diameter. Thus, when applied to most metal castings, the rather open design of strainers means that they can hardly be expected to perform any significant role as filters.

In fact, Webster (1967) has concluded that the strainer works by reducing the rate of flow of metal, assisting the upstream parts of the filling system to prime, and thereby allowing the slag to float. It can be held against the top surface of the runner, or in special reservoirs placed above the strainers to collect the retained slag. Webster goes on to conclude that if the strainer only acts to reduce the rate of flow, then this can be carried out more simply and cheaply by the proper design of the running system.

This may not be the whole story. The strainer may be additionally useful to laminize the flow (i.e. cause the flow to become more streamlined).

However, whatever benefits the strainer may have, its action to create jets downstream of the strainer is definitely not helpful. The placing of a strainer in a geometry that will quickly fill the region at the back of the strainer would be a

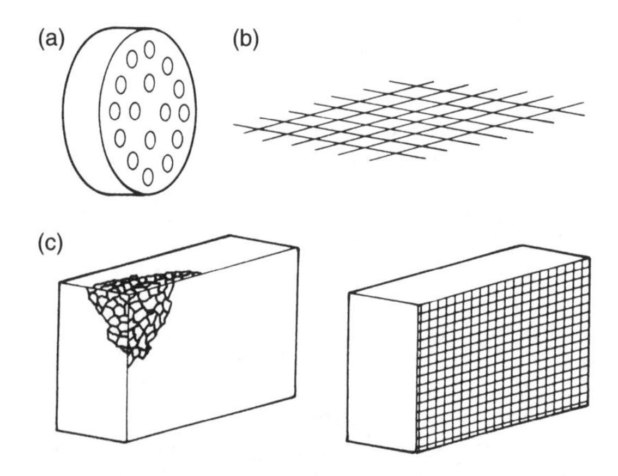

Figure 2.53 *Various filters showing (a) a strainer core (hardly a filter at all); (b) woven cloth or mesh, forming a two-dimensional filter; (c) ceramic foam and extruded blocks, constituting three-dimensional filters.*

great advantage. A geometry to suppress jetting is provided by the tangential filter print to be discussed later (Figure 2.56). The extruded or pressed ceramic filters with their arrays of parallel pores are, of course, equivalent to strainers with a finer pore size. They also benefit from the tangential placement to the oncoming flow as will be described.

Over the years there has been much work carried out to quantify the benefits of the use of filters. Nearly all of these have shown measurable, and sometimes important, gains in freedom from defects and improvements in mechanical properties. These studies are too numerous to list here, but include metals of all types, including Al alloys, irons and steels. The relatively few negative results can be traced to the use of unfavourable siting or geometry of the filter print. For positive and reliable results, these aspects of the use of filters cannot be overlooked. Special attention is devoted to them in what follows.

Woven cloth or mesh

For light alloys, steel wire mesh or glass cloth (Figure 2.53b) is used to prevent the oxides from entering the casting. Cloth filter material has the great advantage of low cost.

The surprising effectiveness of these rather open meshes is the result of the most important inclusions being in the form of films, which appear to be intercepted by and wrap around the strands of the mesh. Openings in the mesh or weave are typically 1–2 mm; this gives good results, being highly effective in retaining films down to this size range. Significantly, it is also a confirmation of the large size of the majority of films that cause problems in castings, particularly in light alloys.

The use of steel wire mesh is also useful to retain films. The steel does not have time to go into solution during the filling of aluminium alloy castings, so that the material of the casting is in no danger of contamination. However, of course, the steel presents a problem of iron contamination during the recycling of the running system. Even the glass cloth can sometimes cause problems during the break-up of the mould, when fragments of glass fibre can be freed to find their way into the atmosphere of the foundry, and cause breathing problems for operators. Both materials therefore need care in use.

Some glass cloth filters are partially rigidized with a ceramic binder, and some by impregnation with phenolic resin. (The outgassing of the resin can cause the evolution of large bubbles when contacted by the liquid metal. Provided the bubbles do not find their way into the casting the overall effect of the filters is definitely beneficial in aluminum alloys.) Both types soften at high temperature, permitting the cloth to stretch and deform.

A woven cloth based on a high silica fibre has been developed to avoid softening at these temperatures, and might therefore be very suitable for use with light alloys. In fact at the present time its high-temperature performance usually confines its use to copper-based alloys and cast irons. There are few data to report on the use of this material. However, it is expected that its use will be similar to that of the other meshes, so that the principles discussed here should still apply.

Despite the attraction of low cost, it has to be admitted that, in general, the glass cloth filters are not easy to use successfully.

For instance, as the cloth softens and stretches there is a strong possibility that the cloth will allow the metal to by-pass the filter. It is essential to take this problem into account when deciding on the printing of the filter. Clearly, it is best if it can be firmly trapped on all of its edges. If it can be held on only three of its four edges the vulnerability of the unsupported edge needs careful consideration. For instance, even though a cross-joint filter may be properly held on the edges that are available, the filter is sometimes defeated by the leading edge of the cloth bending out of straight, bowing like a sail, and thus allowing the liquid to jet past. All the filters shown in Figure 2.54 show this problem. It may be better to abandon cloth filtration if there is any danger of the melt jetting around a collapsing filter.

When sited at the point where the flow crosses a joint as in Figure 2.54 a greensand mould will probably hold the cloth successfully, the sand impressing itself into the weave, provided sufficient area of the cloth is trapped in the mould joint. In the case of a hard sand mould or metal die, the cloth requires a shallow print which must be deep enough to allow it room if the joint is not to be held apart. Also, of course, the print must not be too deep, otherwise the cloth will not be held tight, and may be pulled out of position by the force of the liquid metal. Some slight crushing of a hard sand mould is desirable to hold the cloth as firmly as possible.

A rigidized cloth filter can be inserted across the flow by simply fitting it into the pattern in a pre-moulded slot across the runner and so moulding it integral with the mould (Figure 2.54c). However, this is only successful for relatively small castings. Where the runner area becomes large and the time and temperature become too

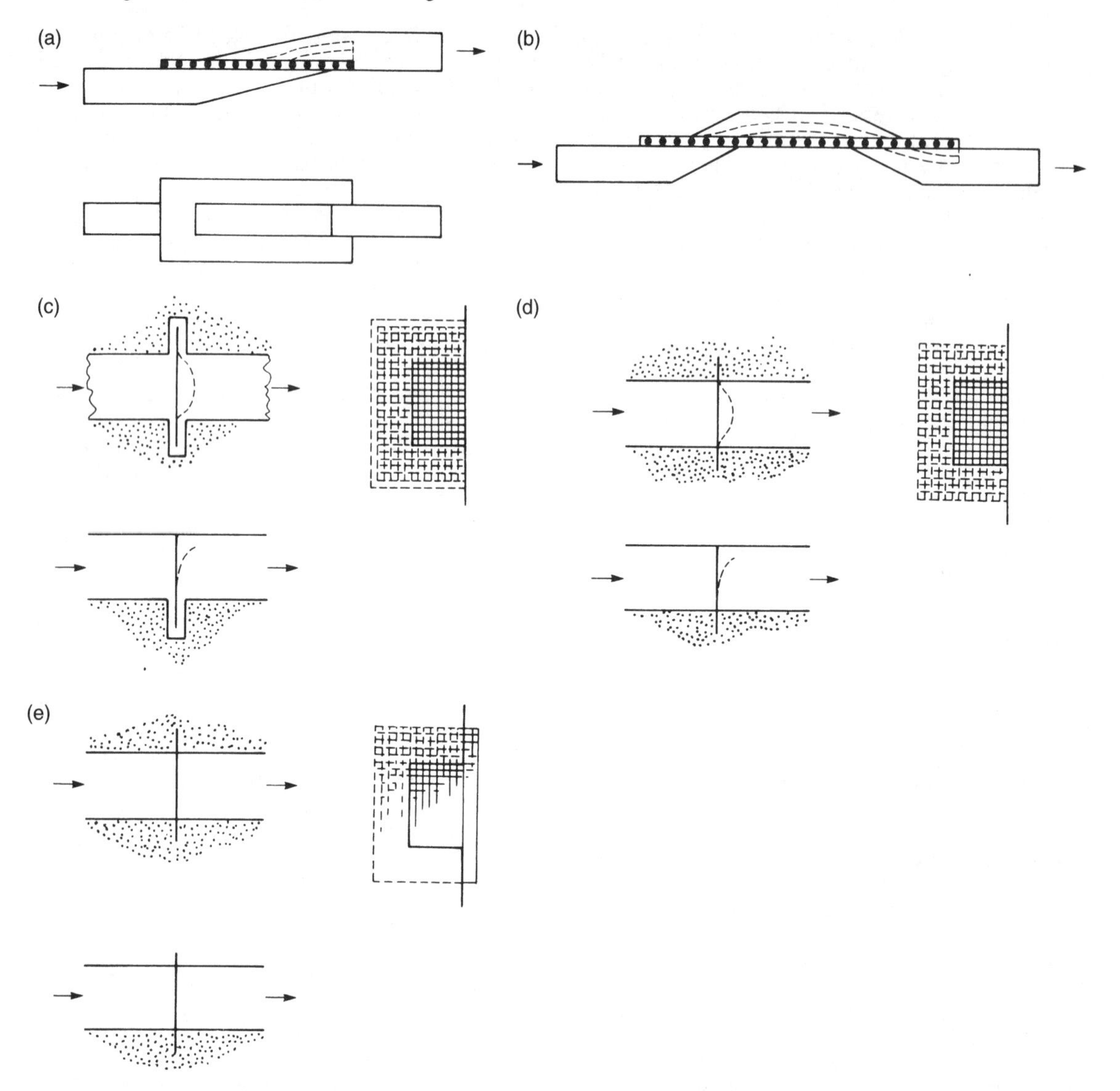

Figure 2.54 *Siting of cloth filters (a) in the mould joint; (b) in a double crossing of the joint; (c) in a slot moulded across the runner; (d) in a slot cut in the runner pattern; (e) with an additional upstand across the joint plane to assist sealing.*

high, the filter softens and bows in the force of the flow. Even if it is not entirely pulled out of position it may be deformed to sag like a fence in a gale, so that metal is able to flow over the top. This is the reason for the design shown in Figure 2.54e. The edge of the filter crosses the joint line, either to sit in a recess accurately provided on the other half of the mould, or if the upstand is limited to a millimetre or so, to be simply crushed against the other mould half. (The creation of some loose grains of sand is of little consequence in the running system, as has been shown by Davis and Magny (1977); loose material in the runner is never picked up by the metal and carried into the mould. The author can confirm this observation as particularly true for systems that are not too turbulent. The laminizing action of the filter itself is probably additionally helpful.)

If the filter is introduced at an earlier stage of manufacture during the production of a sand mould, it can be placed in position in a slot cut in the runner pattern. When the sand is introduced the filter is automatically bonded into the mould (Figure 2.54d). Again, an upstand above the level of the joint may be useful

(Figure 2.54e). In any case, when using filters across runners, it also helps to arrange for the selvage (the reinforced edge of the material) of the cloth to be uppermost to give the unsupported edge most strength; the ragged cut edge has little strength, letting the cloth bend easily, and allowing some, or perhaps all, of the flow to avoid the filter. All the cloth filters used as shown in Figure 2.54 are defeatable, since they are held only on three sides. The fourth side is the point of weakness. Failure of the filter by the liquid overshooting this unsupported edge can result in the creation of more oxide dross than the filter was intended to prevent! Increasing the trapped area of filter in the mould joint can significantly reduce the problem.

Geometries that combine bubble traps (or slag or any other low density phase) are shown in Figure 2.55 for in-line arrangements, and for those common occasions when the runner is required to be divided to go in opposite directions. With shallow runners of depth of a few millimetres there is little practical difference in whether the metal goes up or down through the filter. Thus several permutations of these geometries can be envisaged. Much depends on the links to the gates, and how the gates are to be placed on the casting. In general, however, I usually aim to have the runner exit from the filter below the joint.

Cloth filters are entirely satisfactory where they can be held around all four sides. This is the case at the point where gates are taken vertically upwards from the top of the runner. This is a relatively unusual situation, where, instead of a two-parted mould, a third mould part forms a base to the mould and allows the runner system to be located under the casting. Alternatively, a special core can be used to create an extra joint beneath the general level of the casting.

Another technique for holding the filter on all sides is the use of a 'window frame' of strong paper or cardboard that is bonded or stapled to the cloth. The frame is quickly dropped into its slot print in the mould, and gives a low cost rigid surround that survives sufficiently long to be effective.

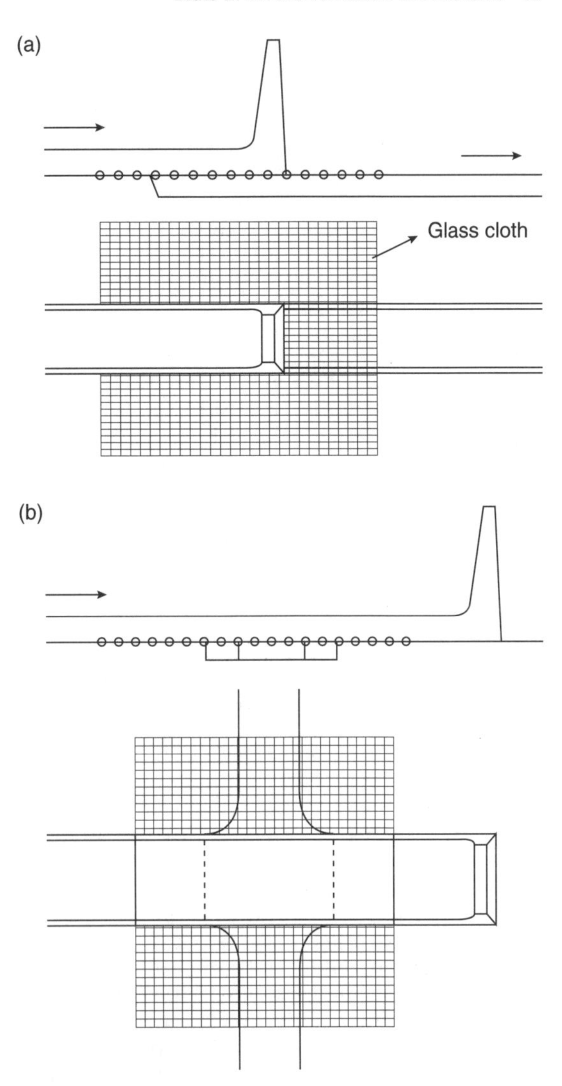

Figure 2.55 *Uses of glass cloth filtration (a) for an in line runner; (b) for transverse runners.*

Ceramic block filters

Ceramic block filters of various types introduced in about 1980 have become popular, and have demonstrated impressive effectiveness in many applications in running systems.

Unfortunately, much that has been written about the mechanisms by which they clean up the cast material appears to be irrelevant. This is because most speculation about the filtration mechanisms has considered only particulate inclusions. As has become quite clear over recent years, the most important and widespread inclusions are actually films. Thus the filtration mechanism at work is clearly quite different, and, in fact, easily understood.

In aluminium alloys the action of a ceramic foam filter to stop films has, in general, not been recognized. This is probably because the films are so thin (a new film may be only 20 nm thick, making the doubled-over entrained bifilm still only about 40 nm) that they cannot be detected when wrapped around sections of the ceramic filter. This explains part of the curious experience of finding that a filter has cleaned up a casting, but on sectioning the filter to examine it under the microscope, not a single inclusion can be found.

The contributing effect, of course, is that the filter acts to improve the filling behaviour of the casting, so reducing the number of inclusions that are created in the mould during the filling process. This behaviour was confirmed by Din *et al.* (2003) who found only about 10 per cent of the action of the filter was the result of filtration, but 90 per cent was the result of improved flow.

A further widespread foundry experience is worth a comment. On occasions the quantity of inclusions has been so great that the filter has become blocked and the mould has not filled completely. Such experiences have caused some users to avoid the use of filters. However, in the experience of the author, such unfortunate events have resulted from the use of poor front ends of filling systems (poor basin and sprue designs) that create huge quantities of oxide films in the pouring process. The filter has therefore been overloaded, leading either to its apparently impressive performance, or its failure by blockage. The general advice given to users by filter manufacturers that filters will only pass limited quantities of metal is seen to be influenced by similar experience. The author has not found any limit to the volume of metal that can be put through a filter without danger of blockage, provided the metal is sufficiently clean and the front end of the filling system is designed to perform well.

Thus the secret of producing good castings using a filter is to team a good front end of the filling system together with the filter. (If the remainder of the filling system design is good, this will, of course, help additionally.) Little oxide is then entrained, so that the filter appears to do little filtration. However, it is then fully enabled to serve a valuable role as a flow rate control device. The beneficial action of the filter in this case is probably the result of several factors:

(i) The reduction in velocity of the flow (provided an appropriately sized cross-section area channel is provided downstream of course). This is probably the single most important action of the filter. However, there are other important actions listed below.
(ii) Reduces the time for the back-filling of the sprue, thereby reducing entrainment defects from this source.
(iii) Smooths fluctuations in flow.
(iv) Laminizes flow, and thus aids fluidity a little.
(v) The freezing of part of the melt inside the filter by the chilling action of the filter (as predicted for Al alloys in ceramic foam filters by computer simulation and found by experiment by Gebelin and Jolly 2003) may be an advantage, because this may act to restrict flow, and so to reduce delivery from the filter in its early moments. The subsequent re-melting of the metal as more hot metal continues to pass through the filter will allow the flow to speed up to its full rate later during filling. (Interestingly, this advantage did not apply to preheated ceramic moulds where the preheat was sufficient to prevent any freezing in the filter).

There are different types of ceramic block filter.

(i) Foam filters made by impregnation of open-cell plastic foams with a ceramic slurry, squeezing out the excess slurry, and firing to burn out the plastic and develop strength in the ceramic. The foam structure consists of a skeleton of ceramic filaments and struts defining a network of interconnecting passageways.
(ii) Extruded forms that have long, straight, parallel holes. They are sometimes referred to as cellular filters.
(iii) Pressed forms, again with long, straight but slightly tapered holes. The filters are made individually from a blank of mouldable clay by a simple pressing operation in a two-part steel die.
(iv) Sintered forms, in which crushed and graded ceramic particles are mixed with a ceramic binder and fired.

In all types the average pore size can be controlled in the range 2–0.5 mm approximately, although the sintered variety can achieve at least 2–0.05 mm. Insufficient research (other than that funded by the filter manufacturers!) has been carried out so far to be sure whether there are any significant differences in the performance between them. An early result of Khan *et al.* (1987) found that the fatigue strength of ductile iron was improved by extruded cellular filters, but that the foam filters were unpredictable, with results varying from the best to the worst. Their mode of use of the filters was less than optimum, being blasted by metal in the entrance to the runner, and with no back protection for the melt. (We shall deal with these aspects below.) The result underlines the probable unrealized potential of both types, and reminds us that both would almost certainly benefit from the use of recent developments. In general, we have to conclude that the published comparisons made so far are, unfortunately, often not reliable.

For aluminium alloys the results are less controversial, because the filters are highly effective in removing films which have, of course, a powerful effect on mechanical properties. Mollard and Davidson (1978) are typical in their findings that the strength of Al–7Si–Mg alloy is improved by 50 per cent, and elongation to failure is doubled. This kind of result is now common experience in the industry.

For some irons and steels, where a high proportion of the inclusions will be liquid, most filter materials are expected to be wetted by the inclusion so that collection efficiency will be high for those inclusions. Ali *et al.* (1985) found that for alumina inclusions in steel traversing an alumina filter, once an inclusion made contact with the filter it became an integral part of the filter. It effectively sintered into place; despite the fact that both inclusion and filter are solid at the temperature of operation, they behave as though they are 'sticky'. This behaviour is likely to characterize many types of inclusion at the temperature of liquid steel.

In contrast with this, Wieser and Dutta (1986) find that whereas alumina inclusions in steel are retained by an alumina filter, even up to the point at which it will clog, deoxidation of steel with Mn and Si produces silica-containing products that are not retained by an extruded zirconia spinel filter. These authors also tested various locations of the filter, discovering that placing it in the pouring basin was of no use, because it was attacked by the slag and dissolved!

Although these results might have been influenced by the rapid flow rates that appear to have been used in this work, it is a warning that filtration efficiency is likely to be strongly dependent on inclusion and filter types. Ali *et al.* (1985) confirms this strong effect of velocity, finding only at very low velocities measured in mm s^{-1} was a high level (96 per cent) of filtration achieved in steel melts.

Block filters are more expensive than cloth filters. However, they are easier to use and more reliable. They retain sufficient rigidity to minimize any danger of distortion that might result in the by-passing of the filter. It is, however, important to secure a supply of filters that are manufactured within a close size tolerance, so that they will fit immediately into a print in the sand mould or into a location in a die, with minimal danger of leakage around the sides of the filter. Although all filter types have improved in this respect over recent years, the foam filter seems most difficult to control, the extruded is intermediate, whereas the pressed filter exhibits good accuracy and reproducibility as a result of it being made in a steel die; residual variation seems to be result of poor control of shrinkage on firing.

Leakage control

It is essential to control the leakage past the filter. There are various techniques.

(i) A seating of a compressible gasket of ceramic paper. This approach is useful when introducing filters into metal dies, where the filter is held by the closing of the two halves of the die. The variations in size of the filters, and the variability of the size and fit of the die parts with time and temperature, which would otherwise cause occasional cracking or crushing of the rather brittle filter, are accommodated safely by the gasket.
(ii) Moulding the filter directly into an aggregate (sand) mould. This is achieved simply by placing the filter on the pattern, and filling the mould box or core box with aggregate in the normal way. The filter is then perfectly held. In greensand systems or chemically bonded sands the mould material seems not to penetrate a ceramic foam filter more than the first pore depth. This is a smaller loss than would be suffered when using a normal geometrical print. However, the technique often requires other measures such as the moulding of the filter into a separate core, or the provision of a loose piece in the pattern to form the channel on the underside of the filter.

There are other aspects of the siting of filters in running systems that are worth underlining.

(i) Siting a filter so that some metal can flow by (into a slag trap for instance) prior to priming the filter is suggested to have the additional benefit that the preheat of the filter and the metal reduces the priming problem associated with the chilling of the metal by the filter (Wieser and Dutta 1986). An example is seen in Figure 2.56d.
(ii) The area of the filter needs to be adequate. There is much evidence to support the fact that the larger the area (thereby giving a lower velocity of flow through the filter) the better the effectiveness of the filter. For instance, if the filter area is too small in relation to the velocity of flow then the filter will be unable to retain foreign matter: the force of the flow will strip away retained films like sheets from the washing line in a hurricane; particles and droplets will follow a similar fate.

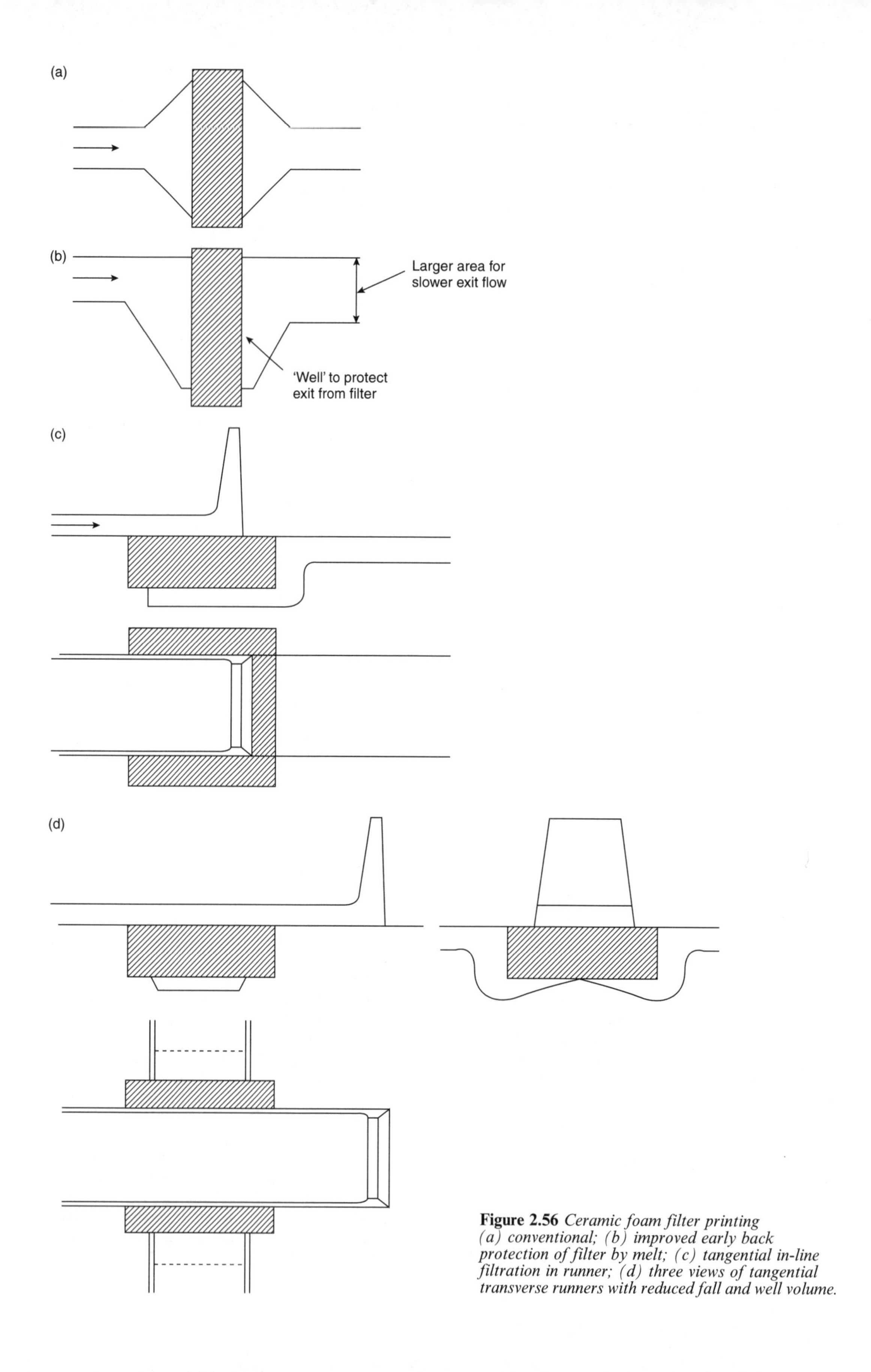

Figure 2.56 *Ceramic foam filter printing (a) conventional; (b) improved early back protection of filter by melt; (c) tangential in-line filtration in runner; (d) three views of tangential transverse runners with reduced fall and well volume.*

(iii) Many filter placements do not distribute the flow evenly over the whole of the filter surface. Thus a concentrated jet is unhelpful, being equivalent to reducing the active area of the filter. The tangential placement of a filter can also be poor in this respect, since the flow naturally concentrates through the farthest portion of the filter. This is countered by tapering the tangential entrance and exit flow channels as illustrated in Figure 2.57b. The provision of a bubble trap reduces the effectiveness of the taper, but the presence of the trap is probably worth this sacrifice. (If the trap is not provided, bubbles arriving from entrainment in the basin or sprue gather on the top surface of the filter. When they have accumulated to occupy almost the whole of the area of the filter the single large bubble is then forced through the filter, and travels on to create severe problems in the mould cavity. The trap is expected to be similarly useful for the diversion of slag from the filter face during the pouring of irons and steels.)

Mutharasan *et al.* (1981) find that the efficiency of removal of TiB_2 inclusions from liquid aluminium increased as the velocity through the filter fell from about $10\,mm\,s^{-1}$ to $1\,mm\,s^{-1}$. Later, the same authors found identical behaviour for the removal of up to 99 per cent of alumina inclusions from liquid steel (Mutharasan *et al.* 1985). However, it is to be noted that these are extremely low velocities, lower than would be found in most casting systems. In the work by Wieser and Dutta (1986) on the filtration of alumina from liquid steel, somewhat higher velocities, in the range $30–120\,mm\,s^{-1}$, are implied despite the use of filter areas up to ten times the runner area in an attempt to obtain sufficient slowing of the rate of flow. Even these

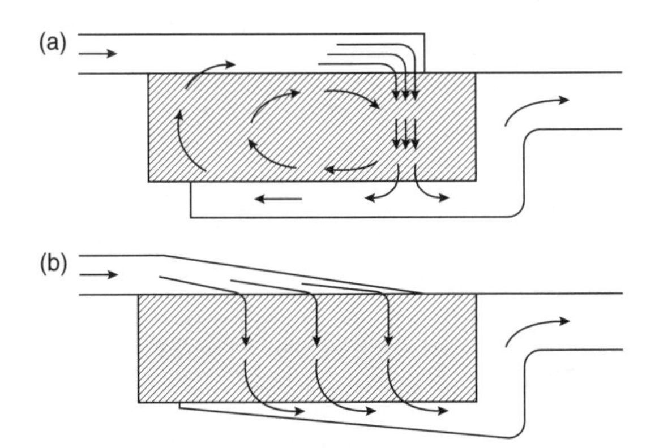

Figure 2.57 *(a) Concentration and reverse flow in a foam filter; (b) tapered inlet and exit ducting to spread flow.*

flow velocities will not match most running systems. These facts underline the poverty of the data that currently exists in the understanding of the action of filters.

Wieser and Dutta go on to make the interesting point that working on the basis of providing a filter of sufficient size to deal with the initial high velocity in a bottom-gated casting, the subsequent fall in velocity as the casting fills and the effective head is reduced implies that the filter is oversize during the rest of the pour. However, this effect may be useful in countering the gradual blockage of the filter in steel containing a moderate amount of inclusions.

Use of filters in running systems

In general the correct location for the filter is near the entrance to the runner, immediately following the sprue. The resistance to penetration of the pores of the filter by the action of surface tension is an additional benefit, delaying the entry into the filter until the sprue has at least partially filled. The frictional resistance to flow through the filter once it is operational provides a further contribution to the reduction in speed of the flow. This frictional resistance has been measured by Devaux (1987). He finds the head loss to be large for filters of area only one or two times the area of the runner. He concludes that whereas a filter area of twice the runner area is the minimum size that is acceptable for a thick-section casting, the filter area has to be increased to four times the runner area for thin-section castings. The pressure drop through filters is a key parameter that is not known with the accuracy that would be useful. Midea (2001) has attempted to quantify this resistance to flow but used only low flow velocities useful for only small castings. A slight improvement is available with Lo and Campbell (2000) who study flow up to $2.5\,m\,s^{-1}$. Even so, at this time the author regrets that it remains unclear how these measurements can be used in a design of a running system. A clear worked example would be useful for us all.

The filter positioned at the entrance to the runner also serves to arrest the initial splash of the first metal to arrive at the base of the sprue. At the beginning of the runner the filter is ideally positioned to take out the films created before and during the pour. The clean liquid can be maintained relatively free from further contamination so long as no surface turbulence occurs from this point onwards. This condition can be fulfilled if:

(i) The melt proceeds at a sufficiently low velocity and/or is sufficiently constrained by geometry to prevent entrainment (this particularly includes the provision to eliminate jetting from the rear face of the filter). A low velocity will be achieved if the cross-sectional area of the runner downstream of the filter is increased in proportion to the reduction in speed provided by the filter.

(ii) Every part of the subsequent journey for the liquid is either horizontal or uphill. The corollary of this condition is that the base of the sprue and the filter should always be at the lowest point of the running system and the casting. This excellent general rule is a key requirement.

Tangential placement

Filters have been seen to be open to criticism because of their action in splitting up the flow, thereby, it was thought, probably introducing additional oxide into the melt. There is some truth in this concern. A preliminary exploration of this problem was carried out by the author (Din and Campbell 1994). Liquid Al alloy was recorded on optical video flowing through a ceramic foam filter in an open runner. The filter did appear to split the flow into separate jets; a tube of oxide forming around each jet. However, close observation indicated that the jets recombined about 10 mm downstream from the filter, so that air was excluded from the stream from that point onwards. The oxide tubes around the jets appeared to wave about in the eddies of the flow, remaining attached to the filter, like weed attached to a grill across a flowing stream. The study was repeated and the observations confirmed by X-ray video radiography. The work was carried out at modest flow velocities in the region of 0.5 to 2.0 $m s^{-1}$. It is not certain, however, whether the oxides would continue to remain attached if speeds were much higher, or if the flow were to suffer major disruption from, for instance, the passage of bubbles through the filter.

What is certain is the damage that is done to the stream after the filter if the melt issuing from the filter is allowed to jet into the air. Loper and co-workers (1996) call this period during which this occurs *the spraying time*. This is so serious a problem that it is considered in some detail below.

Unfortunately, most filters are placed transverse to the flow, simply straight across a runner (Figure 2.56a) and in locations where the pressure of the liquid is high (i.e. at the base of the sprue or entrance into the runner). In these circumstances, the melt shoots through a straight-through-hole type filter almost as though the filter was not present, indicating the such filters are not particularly effective when used in this way. When a foam filter suffers a similar direct impingement, penetration occurs by the melt seeking out the easiest flow paths through the various sizes of interconnected channels, and therefore emerges from the back of the filter at various random points. Jets of liquid project from these exit points, and can be seen in video radiography. The jets impinge on the floor of the runner, and on the shallow melt pool as it gradually builds up, causing severe local surface turbulence and so creating dross. If the runner behind the filter is long or has a large volume, the jetting behaviour can continue until the runner is full, creating volumes of seriously damaged metal.

Conversely, if the volume of the filter exit channel is kept small, the volume of damaged melt that can be formed is now reduced correspondingly. Although this factor has been little researched, it is certain to be important in the design of a good placement for the filter. Loper *et al.* (1996) realized this problem, describing the limited volume at the back of the filter as a hydraulic lock, the word lock being used in a similar sense to a lock on an inland waterway canal.

Figure 2.56b shows an improved geometry that enables the back of the filter to be covered with melt quickly. Figure 2.56c shows an improved technique, placing the block filter tangentially to the direction of flow. The tangential mode has the advantage of the limitation of the exit volume from the filter, and providing a geometrical form resembling a sump, or lowest point, so that the exit volume fills quickly. In this way the opportunity for the melt to jet freely into air is greatly reduced so that the remainder of the flow is protected. A further advantage of this geometry is the ability to site a bubble trap over the filter, providing a method whereby the flow of metal and the flow of air bubbles can be divided into separate streams. The air bubbles in the trap are found to diffuse away gradually into sand moulds. For dies, the traps may need to be larger.

An additional benefit is that the straight-through-hole extruded or pressed filters seem to be effective when used tangentially in this way. A study of the effectiveness of tangential placement in the author's laboratory (Prodham *et al.* 1999) has shown that a straight-through-pore filter could achieve comparable reliability of mechanical properties as could be achieved by a relatively well-placed ceramic foam filter (Sirrell and Campbell 1997).

Adams (2001) draws attention to the importance of the flow directed downwards through the filter. In this way buoyant debris such as dross or slag can float clear. In contrast, with upward flow through the filter the buoyant debris collects on the intake face of the filter and progressively blocks the filter.

The tapering of both the tangential approach and the off-take from the filters further reduces the volume of melt, and distributes the flow through the filter more evenly. In the absence of these wedge-like features, only the far side of the filter carries the main flow, whereas the side nearest the upstream end is redundant, experiencing a circulating flow in the reverse direction (Figure 2.57).

Direct pour

Sandford (1988) showed that a variety of top pouring could be used in which a ceramic foam filter was used in conjunction with a ceramic fibre sleeve. The sleeve/filter combination was designed to be sited directly on the top of a mould to act as a pouring basin, eliminating any need for a conventional filling system. In addition, after filling, the system continued to work as a feeder. This simple and attractive system has much appeal.

Although at first sight the technique seems to violate the condition for protection of the melt against jetting from the underside of the filter, jetting does not seem to be a problem in this case. Jetting is avoided almost certainly because the head pressure experienced by the filter is so low, and contrasts with the usual situation where the filter experiences the full blast of flow emerging from the base of the sprue.

Sandford's work illustrated that without the filter in place, direct pour of an aluminium alloy resulted in severe entrainment of oxides in the surface of a cast plate. The oxides were eliminated if a filter was interposed, and the fall after the filter was less than 50 mm. Even after a fall of 75 mm after the filter relatively few oxides were entrained in the surface of the casting. The technique was further investigated in some detail (Din *et al.* 2003) with fascinating results illustrated in Figure 2.58. It seems that under conditions used by the authors in which the melt emerging from the filter fell into a runner bar and series of test bars, some surface turbulence was suffered, and was assessed by measuring the scatter of tensile test results. The effect of the filter acting purely to filter the melt was seen to be present, but slight. The castings were found to be repeatable (although not necessarily free from defects) for fall distances after the filter of up to about 100 mm, in agreement with Sandford. Above a fall of 200 mm reproducibility was lost (Figure 2.58a, b).

This interesting result explains the mix of success and failure experienced with the direct pour system. For modest fall heights of 100 mm or so, the filter acts to smooth the perturbations to flow, and so confers reproducibility on the casting. However, this may mean 100 per cent good or 100 per cent bad. The difference was seen by video radiography to be merely the chance flow of the metal, and the consequential chance location of defects.

The conclusion to this work was a surprise. It seems that direct pour should not necessarily be expected to work first time. If the technique were found to make a good casting it should be used, since the likelihood would be that all castings would then be good. However, if the first casting was bad, the site of the filter and sleeve should simply be changed to seek a different pattern of filling. This could mean a site only a few centimetres away from the original site. The procedure could be repeated until a site was found that yielded a good casting. The likelihood is that all castings would subsequently be good.

However, the technique will clearly not be applicable to all casting types. For instance, it is difficult to see how the approach could reliably produce extensive relatively thin-walled products in film-forming alloys where surface tension is not quite in control of the spread of metal in the cavity. For such products the advance of the liquid front is required to be steady, reproducible and controlled. Bottom gating in such a case is the obvious solution. Also, the technique works less well in thicker section castings where the melt is less constrained after its fall from the underside of the filter. Figure 2.58c illustrates the fall in reliability of products as the diameter of the test bars increases above 20 mm. Equivalent results would be expected for the increase of plate sections above about 10 mm.

Even though the use and development of the direct pour technique will have to proceed with care, it is already achieving an important place in casting production. A successful application to a permanent moulded cylinder head casting is described by Datta and Sandford (1995). Success here appears to be the result of the limited, and therefore relatively safe fall distance.

Flow rate data through the filter/sleeve combination is necessary to predict the pour times of castings. Such data has been measured by Bird (1989). His results presented here (Figure 2.58d) have been rationalized to apply to 50 mm diameter ceramic foam filters, and relate to Al–Si alloys cast at 720 °C. Clearly, filters of different sizes will pass correspondingly more or less melt per second proportional to their areas, assuming their thickness and pore sizes are sufficiently close. The pores' diameters of approximately 1 and 2 mm in Figure 2.58d refer to the 'pore per inch' categories 20 ppi and 10 ppi respectively.

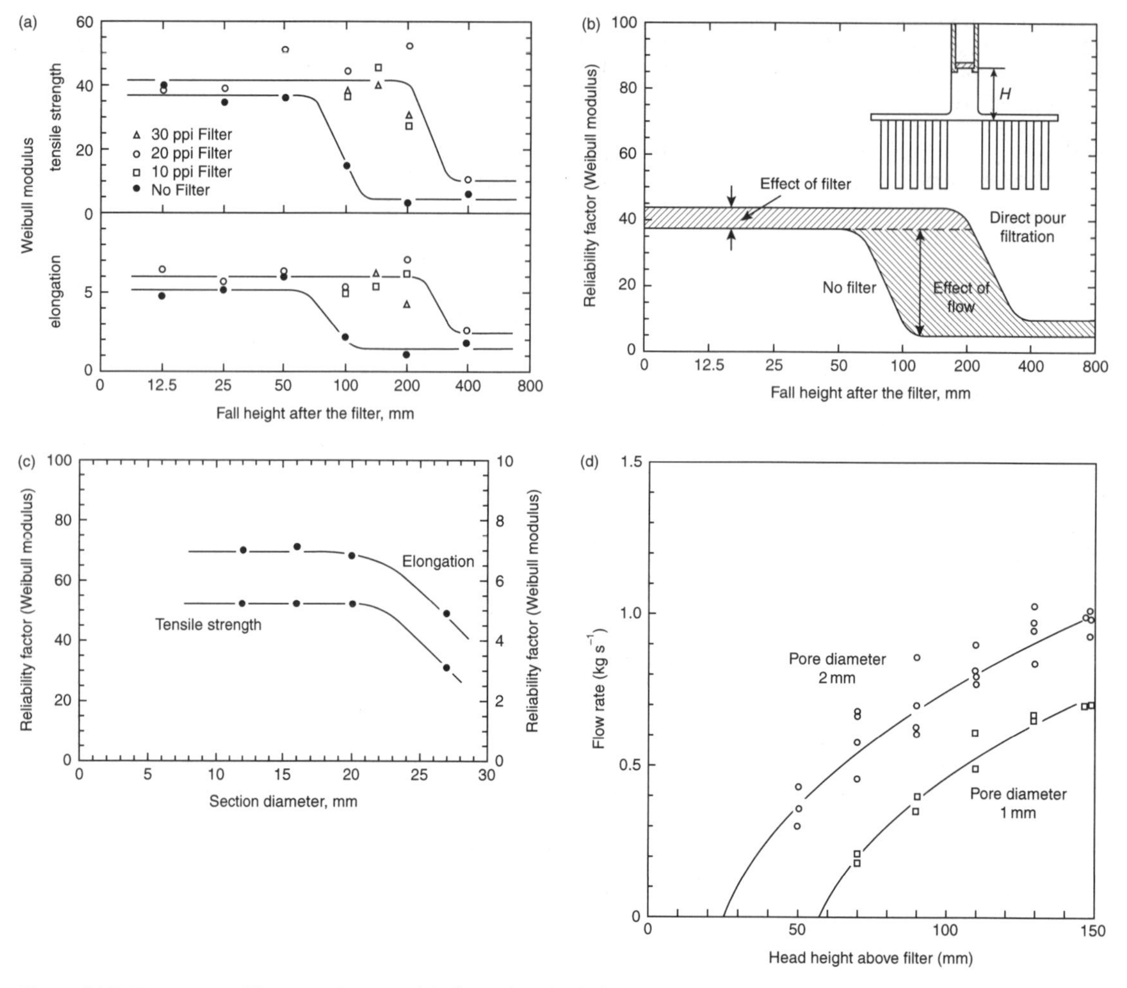

Figure 2.58 *Direct pour filtration showing (a) the reduced reliability as the fall increases with and without a filter in place; (b) the interpretation of 'a'; (c) the reduced reliability as diameter of test bars is increased; (d) the rate of flow of Al–Si alloys at 720°C by direct pour through a 50 mm diameter filter (Bird 1989, courtesy of Foseco).*

A recent development of the direct pour technique is described by Lerner and Aubrey (2000). For the direct pouring of ductile iron they use a filter that is a loose fit in the ceramic sleeve. It is held in place by the force of impingement of the melt. When pouring is complete the filter then floats to the top of the sleeve and can be lifted off and discarded, avoiding contamination of remelted material.

Sundry aspects

1. The dangers of using ceramic block filters in the direct impingement mode is illustrated by the work of Taylor and Baier (2003). They found that a ceramic foam filter placed transversely in the down-sprue worked better at the top of the sprue rather than at its base. This conclusion appears to be the result of the melt impact velocity on the filter causing jetting out of the back of the filter. Thus the high placement was favoured for the reasons outlined in the section above on the direct pour technique. This result is unfortunate, because if the filter exit volume had been limited to a few millilitres (a depth beneath the filter limited to 2 or 3 mm) the lower siting of the filter would probably have performed in the best way.
2. It is essential for the filter to avoid the contamination of the melt or the melting equipment. Thus for many years there appeared to be a problem with Al–Si alloys that appeared to suffer from Ca contamination

from an early formulation of the filter ceramic. This problem now seems to be resolved by modification of the chemistry of the filter material. In addition, modern filters for Al alloys are now designed to float, so that they can be skimmed from the top of the melting furnace when the running systems are recycled. This avoids the costly cutting out and separation of spent filters from recycled rigging to avoid them collecting in a mass at the bottom of the melting furnace.

Interestingly, the steel gauzes used for Al alloys do not contaminate the alloy entering the casting. This is almost certainly the result of the alloy wrapping a protective alumina film over the wires of the mesh as the meniscus passes through. However, the steel will dissolve later if recycled via a melting furnace of course.

The recent introduction of carbon-based filters for steel does add a little carbon to the steel, but this seems negligible for most grades. Whether the use of such filters will be suitable for ultra-low carbon steels is being decided as I write.

Xu and Mampaey (1997) report the additional benefit of a ceramic foam filter in an impressive 12-fold increase in the fluidity of grey iron poured at about 1400 °C in sand moulds. They attribute this unlooked-for bonus to the effect of the filter in (i) laminizing the flow and so reducing the apparent viscosity due to turbulence, and (ii) reducing the content of inclusions. One would imagine that films would be particularly important.

Summary

So far as can be judged at this time, among the many requirements to achieve a clean casting, the key practical recommendations for the casting engineer can be summarized as:

(i) Do not allow slag and dross to enter the filling system. This task is best solved by eliminating the conical pouring basin and substituting an offset stepped basin.
(ii) Use a good early part of the filling system to avoid the creation of additional slag or dross that may block the filter.
(iii) Use filters together with a buoyant phase trap. The bubble trap described earlier should also work as a slag trap. The presence of the filter significantly aids the separation of the two fluids. Where particularly dirty metals are in use, the trap will, of course, require the provision of sufficient volume and height on its upstream side to accommodate retained material, allowing slag and dross to float clear, and leaving the filter area to continue working without blockage.
(iv) Avoid the great danger of by-passing the filter by poor printing. Mould-in the filter if possible.
(v) Provide protection of the melt at the exit side of the filter, by rapid fill of this volume with liquid metal. A useful geometry to achieve this is the tangential placement of the filter, followed by a shallow well that can be quickly filled.

2.3.7 Practical calculation of the filling system

In view of all the information listed under Rule 2 in the previous part of this chapter, this section attempts to gather this together, to see how we might achieve a complete, practical solution to a filling system design. The ability to design a quantitative solution, yielding precise dimensions of the filling channels at all points, is a key responsibility, perhaps the *key* responsibility, of the casting engineer.

Naturally, computers are beginning to have some capability of optimizing the design of filling systems (McDavid and Dantzig 1989; Jolly *et al.* 2000). Even so, until the time that the computer is fully proficient, it will be necessary for the casting engineer to undertake this duty. The complication of the procedure is not to be underestimated (if it were easy the procedure would have been developed years ago). Many factors need to be taken into account. This short outline cannot cover all eventualities, but will present a systematic approach that will be generally applicable.

2.3.7.1 Background to the methoding approach

If a computer package is available to simulate the solidification of the casting, it is best to carry this out first. Most software packages are sufficiently accurate when confined to the simulation of solidification (it is the filling simulation, and other sophisticated simulations such as that of stress, strain and distortion that are more difficult, and the results often less accurate). A solidification simulation with the addition of no filling or feeding system will illustrate whether there are special problems with the casting. Figure 2.59 illustrates the formal logic of this approach. It lays out a powerful methodology that is strongly recommended.

If in fact there are no special problems it is good news. Otherwise, if problems do appear for a long-running part, and if they can be eliminated by discussion with the designer of the

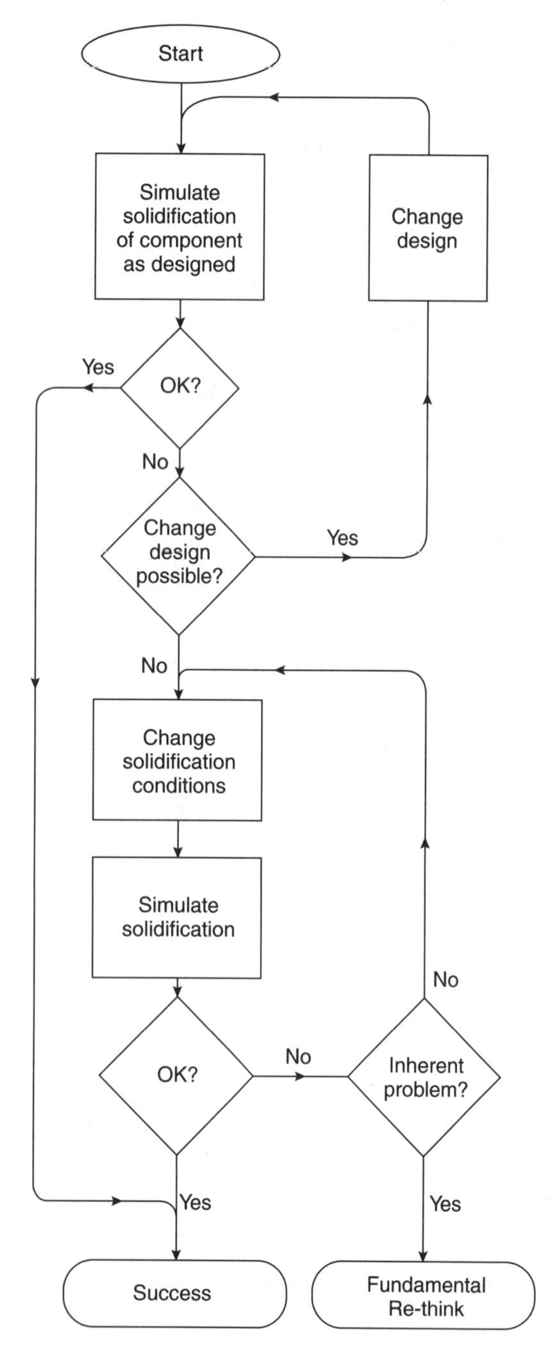

Figure 2.59 *Methoding procedure for computer simulation.*

component at this initial stage, this is usually the most valuable strategy. Such actions often include the shift of a parting line, or the coring-out of a heavy section or boss. The purpose of a modest one-time design change is to avoid, so far as possible, the ongoing expense for the life of the product of special actions such as the provision of chills or feeders, or an extra core, etc.

If, despite these efforts, a problem remains, the various options including additional chills, feeders or cores will require detailed study to limit, so far as possible, the cost penalties. The following section provides the background for the next steps of the procedure.

2.3.7.2 Selection of a layout

First, it will be necessary to decide which way up the part is to be cast (this may be changed later in the light of many considerations, including problems of core assembly, desirable filling patterns, subsequent handling and de-gating issues, etc.). If a two-part mould is to be used, the form of the casting should preferably be mainly in the cope, allowing gating at the lowest parts of the casting. This may prove so difficult that a third box part may be selected through which the running system could be sited under the casting. If some solution to the challenge of lowest point gating cannot easily be found, the risk of filling the casting at some slightly higher point may need to be assessed. Some filling damage might have to be accepted for some castings. Even so, it is unwelcome to have to make such decisions because the extent of any such damage is difficult to predict.

A heavy section of the casting needs special attention. This may most easily be achieved by orienting this part of the casting at the top and planting a feeder here. Alternatively, other considerations may dictate that the casting cannot be oriented this way, so arrangements may have to be made to provide chills and/or fins to this section if it has to be located in the drag.

When a general scheme is decided, including the approximate siting of gates and runners, the provision of feeders, if any, and the location of the sprue, a start can be made on the quantification of the system.

2.3.7.3 Weight and volume estimate

The weight of the casting will be known, or can be estimated. This is added to an estimate of the weight of the rigging (the filling and feeding system) to give an estimate of the total poured weight. Dividing this by the density of the liquid metal will give the total poured volume. Unfortunately, of course, the weight of the rigging is clearly not accurately known at this early stage because it has not yet been designed. However, an approximate estimate is nearly always good enough. Although a revised value can always be used in a subsequent iteration of the rigging design calculations to obtain an accurate value for the weight of the rigging, after some experience an additional iteration will be found to be hardly ever necessary.

2.3.7.4 Selection of a pouring time

The selection of a pouring time is always an interesting moment in the design of a filling system for a new casting.

A common concern is how can production rates be maintained high if metal velocities in the filling system need to be kept below the critical 0.5 m s^{-1}? Fortunately, this is not usually a problem because the time to fill a casting is dependent on the rate of mould filling measured as a volume per second, and can be fixed at a high level. At the same time the velocity through the ingate can be independently lowered simply (well, simple in principle, but perhaps harder in practice!) by increasing the area of the gate. These considerations will become clearer as we proceed.

When faced with a new design of casting, the first question asked by the casting engineer is 'How fast should it be filled?'

Sometimes there is no choice. On a fast moulding line making 360 moulds per hour there is only 10 seconds for each complete cycle, of which perhaps only 5 seconds may be the available time for pouring. (Although it is worth keeping in mind that even here, as a last resort, the pour time might be doubled if two pouring facilities were to be installed.)

When there *is* a choice the pour time can often be changed between surprisingly large limits. One factor is sometimes the rate of rise of the metal in some sections. The surface of an Al casting becomes marked with striations due to the passage of transverse unzipping waves at a vertical rise velocity below 60 mm s^{-1} (Evans *et al.* 1997). Considerations that control the choice of rate of metal rise in steel castings (Forslund 1954, Hess 1974) indicate that these factors have yet to be properly researched. In practice, a common rate of rise in a steel foundry making castings several metres tall and weighing several tonnes is 100 mm s^{-1}, although, with an improved design of filling system this rate might be reduced. A further limit to the fill time of a steel casting is the possible collapse of the cope when subjected to radiant heat of the rising metal for too long. This problem is reduced by generous venting of the top of the mould via a top feeder for instance, and is further reduced by the practice of providing a white mould coat based on a material such as alumina or zircon, thus absorbing much less of the incident radiation.

A slow rate of rise in the mould cavity can lead to transverse unzipping waves, but although they can leave their witness on the surface of the oxidized surface of the casting they are usually harmless to its internal structure. However, if the alloy has an extra strong surface oxide, or is partially freezing because of a cool pour, the waves lead to such severe surface horizontal laps that the casting is usually not repairable. Types of geometries where this problem is most often seen are illustrated in Figure 2.60. A hollow cylinder cast on its side is a common casualty (2.60a) because of the sudden increase in area to be filled, reducing the rate of rise as the metal reaches the top of the core. The problem is also found on the upper surfaces of tilt castings if the rate of tilt is too slow (2.60b).

Alternatively another constraint on a choice of pour time is the consideration that it may be necessary to fill the mould before freezing starts in its thinnest section (or, more usefully, its

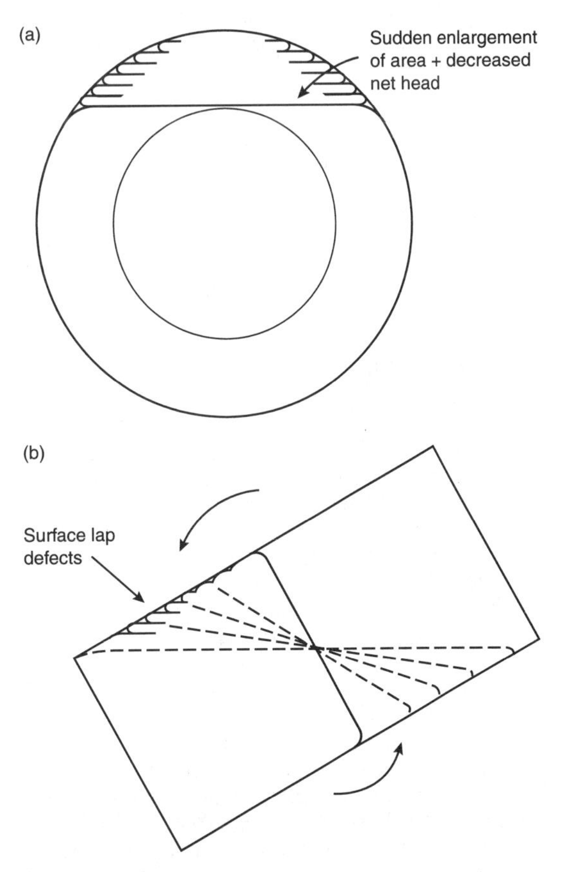

Figure 2.60 *Common lap problems at low rates of rise of liquid surface in the mould (a) in a horizontal pipe or cylinder; (b) on the cope surface of a tilt casting.*

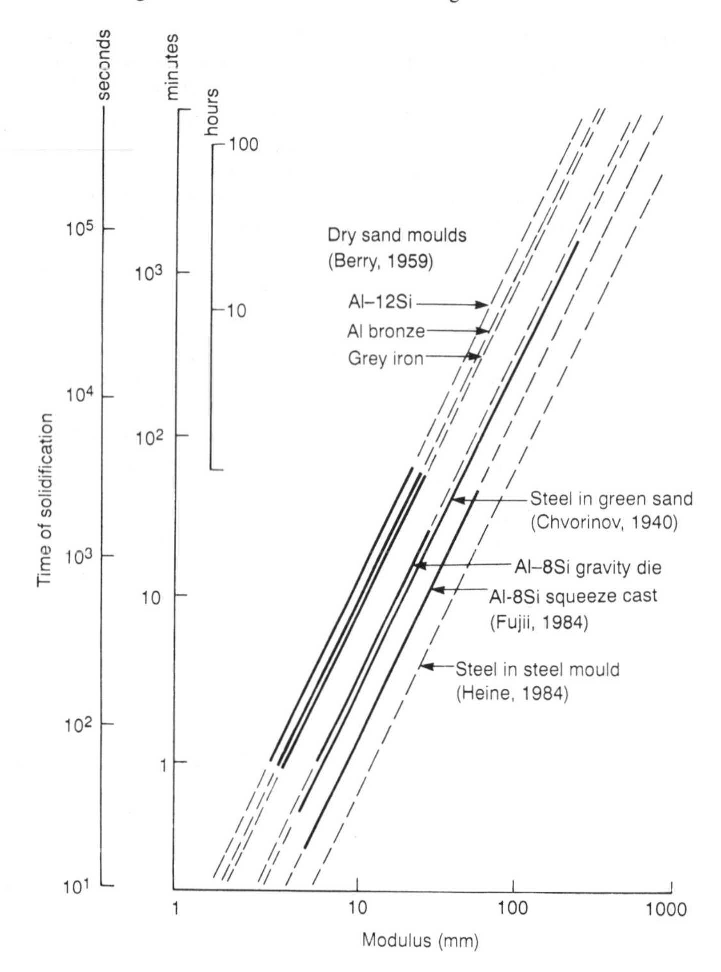

Figure 2.61 *Freezing times for plates in different alloys and moulds.*

smallest modulus). Thus an idea of the time available can be gained from the Figure 2.61. (Readers are recommended to generate such diagrams for themselves for special casting conditions, using embedded thermocouples to determine the freezing times versus modulus relations, e.g. for cast iron in zircon shell moulds, or aluminium-based alloys in investment moulds at 100 °C, etc.)

Clearly filling at too slow a rate does bring its problems. However, many castings are filled very much faster than necessary, and there are benefits to a reduction in this speed. For this reason, having made a choice of an approximate fill time for the casting, it is instructive to consider whether this time could be doubled, or even doubled again. It is surprising how often this is a possibility. Whereas the experienced foundryman will hesitate to extend the pouring time of a familiar casting, his experience will be based usually on a poor filling system. Such systems generate problems such as slopping and surging, and splashes ahead of the main body of melt. The cooling and oxidation of these splashes prior to the arrival of the main flow causes them to be imperfectly assimilated on arrival of the main body of liquid, resulting in the appearance of a 'cold' lap. Thus fill rate or temperature is increased in an effort to avoid this apparent problem, usually resulting in worsening the problem. The provision of a good filling system is not subject to such problems: the advancing liquid front keeps itself together, and so keeps itself warm. The result is that pouring temperature can often be reduced and pouring times extended without penalty.

In general, if there is a wide choice of time, for instance somewhere between 5 and 25 seconds, it is strongly recommended to opt for the maximum time, giving the minimum fill rate. This is because, compared to the faster fill rate, the selection of the slower rate reduces the cross section area of all parts of the filling system, in this case by a factor of 5. This is economically valuable, giving a great boost to yield. The filling channels shrink from appearing 'chunky' to appearing like needles (with the confident but erroneous predictions by all experienced onlookers that such systems will never fill). In addition, there is the benefit that the slimmer filling system actually works better, improving the quality of the casting by giving less room for the metal to jump and splash. Random scrap from pouring defects is thereby reduced. These are important benefits.

On the arrival of a completely new design of casting, the choice of the time to fill the mould can sometimes be impressively arbitrary, with perhaps no-one in the foundry having any clear idea on the time to use. Nevertheless, a value that seems reasonable can be tried, and can always be modified on a subsequent trial.

The important fact to remember is that provided the pouring basin is kept filled to at least its designed level, the filling time is not allowed to vary by chance as in a hand-cut running system, and is not under the control of the pourer, but remains accurately under the control of the casting engineer.

2.3.7.5 Fill rate

Having selected a fill time, the *average fill rate* is, of course, simply the total poured weight divided by the total time in a convenient unit such as $kg\,s^{-1}$. This requires to be converted to the average volume fill rate by dividing by the density of the liquid metal, giving a value in such units as $m^3\,s^{-1}$.

Even this value cannot be used directly. This is because the filling system has to be sized to take the significantly higher rate of flow at the beginning of the pour. The average fill rate is, of course, less than the initial fill rate because the high initial rate is not maintained. The metal slows as the mould fills, the fill rate finally falling to zero if the metal level in the mould finally reaches the same level as that in the pouring basin. To make allowance for this effect, it is convenient to assume that the initial fill rate is a factor of approximately 1.5 times higher than the average fill rate. This factor is actually precisely correct if the casting is a uniform plate with its top level with the pouring basin, as shown in Appendix 1. However, in general, the factor is not particularly sensitive to geometry, as can be demonstrated by such exercises as checking the fill times of extreme examples such as a cone filled via its tip compared to it inverted and filled via its base.

The *initial flow rate, Q*, preferably in units $m^3\,s^{-1}$, is the value to be used for defining the size of all of the remaining features of the filling system that we require to calculate.

Incidentally, for a given volume flow rate, the mould will fill in the same time whether aluminium or iron is poured (Galileo would have known this). Thus the system described below applies to all metals and alloys, perhaps to the surprise of many of us who have unwittingly accepted the dogma that each metal and alloy requires its own special system.

Later, when the first mould is poured, the filling should be timed with a stopwatch as a check of the running system design. The actual time should be within 10 per cent of the predicted time. In fact the agreement is often closer than this (Kotschi and Kleist 1979), to the amazement of doubters of casting science!

After the first casting is produced, it may be clear that it needs a casting rate either slower (allowing some solidification during pouring) or faster (to avoid cold lap-type defects). These modifications to the rate can be easily and quickly carried out by minor adjustment (usually only millimetres of changes to dimensions are required) to the size of the filling channels. Again, it is useful to emphasize that such changes remain under the control of the casting technologist (not the pourer).

2.3.7.6 Sprue (down-runner) design

Now that an initial rate of pouring has been chosen, how can we achieve it accurately, limiting the rate of delivery of metal to precisely this chosen value? Theoretically it can be achieved by tailoring a funnel in the mould of exactly the right size to fit around a freely falling stream of metal, carrying just the right quantity (Figure 2.13). We call this our down-runner, or sprue.

The theoretical dimensions of the sprue can therefore be calculated as follows. If a stream of liquid is allowed to fall freely from a starting velocity of zero, then after falling a height h it will have reached velocity v. The height h always refers to the height to the melt surface in the pouring basin. This zero datum is one of the great benefits of the offset basin compared to the conical basin (the starting velocities can never be known with any accuracy when working with a conical basin). Thus we have

$$v = (2gh)^{1/2}$$

To obtain the sprue sizes it is necessary to realize that the low velocity v_1 at the top of the sprue must be associated with a large cross-sectional area A_1. At the base of the sprue the higher metal speed v_2 is associated with a smaller area

A_2. If the falling stream is continuous it is clear that conservation of matter dictates that

$$Q = v_1 A_1 = v_2 A_2 = v_3 A_3 \; etc.$$

where subscript 3 can refer to any downstream location for the local values of the area of the stream and its velocity (for instance the area and velocity at the gates). Since the velocities are now known from the height that the melt has fallen (neglecting any losses at this stage), and Q has been decided, each of the areas of the filling system can now be calculated.

In nearly all previous treatises on running systems the important dimension of the sprue for controlling the precise rate of flow has been assumed to be the area of the exit. This part of the system has been assumed to act as 'the choke', regulating the rate of flow of metal throughout the whole running system. It is essential to revise this thinking. If the sprue is correctly designed to just touch the surface of the falling liquid at all points, the whole sprue is controlling. There is nothing special about the narrowest part at the sprue exit. We shall continue this concept so far as we can throughout the rest of the filling system. If we achieve the target of fitting the dimensions of the flow channels in the mould just to fit the natural shape of the flowing stream, it follows that no one part is exerting control. The whole system is all just as large as it needs to be; the channels of the filling system just touch the flowing stream at all points.

Even so, after such features as bends and filters and other complications, the energy losses are not known precisely. Thus there is a sense in which the sprue (not just its exit, remember) is doing a good job of controlling, but beyond this point the precision of control may be lost to some extent after those features that introduce imponderables to the flow. (In the fullness of time we hope to understand the features better. Even now, computers are starting to make useful inroads to this problem area.)

Thus, as long as the caster pours as fast as possible, attempting to fill the pouring basin as quickly as possible, and keeping the basin full during the whole of the pour, then he or she will have no influence on the rate of filling inside the mould; the sprue (the whole sprue, remember) will control the rate at which metal fills the mould.

For most accurate results it is best to calculate the sprue dimensions using the formulae given above, and using the alloy density to obtain the initial volume flow rate Q.

However, for many practical purposes we can take a short cut. It is possible to construct a useful nomogram for Al assuming a liquid density of 2500 kg m^{-3} and for the dense alloys based on Fe and Cu assuming a liquid density around 7500 kg m^{-3} (Figure 2.62). Thus areas of sprues at the top and bottom can be read off, and the sprue shape formed simply by joining these areas by a straight taper. Using the diagram it is simple to read off areas of the sprue at any other intermediate level if it is desired to provide a more accurately formed sprue having a curved taper. Recall that the heights are measured in every case from the level of metal in the pouring basin, regarding this as the zero datum.

The nomogram is easy to use. For instance if we wish to pour an aluminium alloy casting at an average rate of 1.0 kg s^{-1}, corresponding, of course, to an initial rate of 1.5 kg s^{-1}, Figure 2.62 is used as follows. The 1.5 kg s^{-1} rate with a depth in the basin (the top level down to the level of the sprue entrance) of 100 mm, and a sprue length of 200 mm (total head height to the top of the melt in the basin of 300 mm), then its entrance and exit areas can be read from the figure as approximately 440 mm^2 and 250 mm^2 respectively. Remember from section 2.3.2.3 that it is advisable to increase the area of the entrance by approximately 20 per cent to compensate for errors, particularly the error introduced if the sprue shape is approximated to a straight taper. Thus the final sprue entrance should be close to 500 mm^2.

As a check on the nomogram read-outs for our aluminium alloy casting, we can now calculate the dimensions numerically using the equations given above. At 1.0 kg s^{-1} average fill rate, corresponding to an initial rate 1.5 kg s^{-1}, assuming a liquid density of 2500 kg m^{-3}, we obtain an initial volume flow rate $Q = 1.5/2500 = 0.6 \times 10^{-3}\,\text{m}^3\text{s}^{-1}$. We can calculate that the falls of 100 mm and 300 mm are seen to cause the melt to accelerate to a velocity of 1.41 and 2.45 m s^{-1}, giving areas of 424 and 245 mm^2 respectively. These values are in reasonable agreement with those taken directly from the nomogram.

The cross-section of the filling system can, of course, be round or square, or even some other shape, provided the area is correct (we are neglecting the small corrections required as a result of increased drag as sections deviate further from a circle). However, in view of making the best junction to the runner, a slot sprue and slot runner are strongly recommended for most purposes. (Multiple sprues might be useful to connect to a number of runners. Several such sprues would be expected to work better than one large sprue as a result of improved constraint of the metal during its fall as shown in Figure 2.63.)

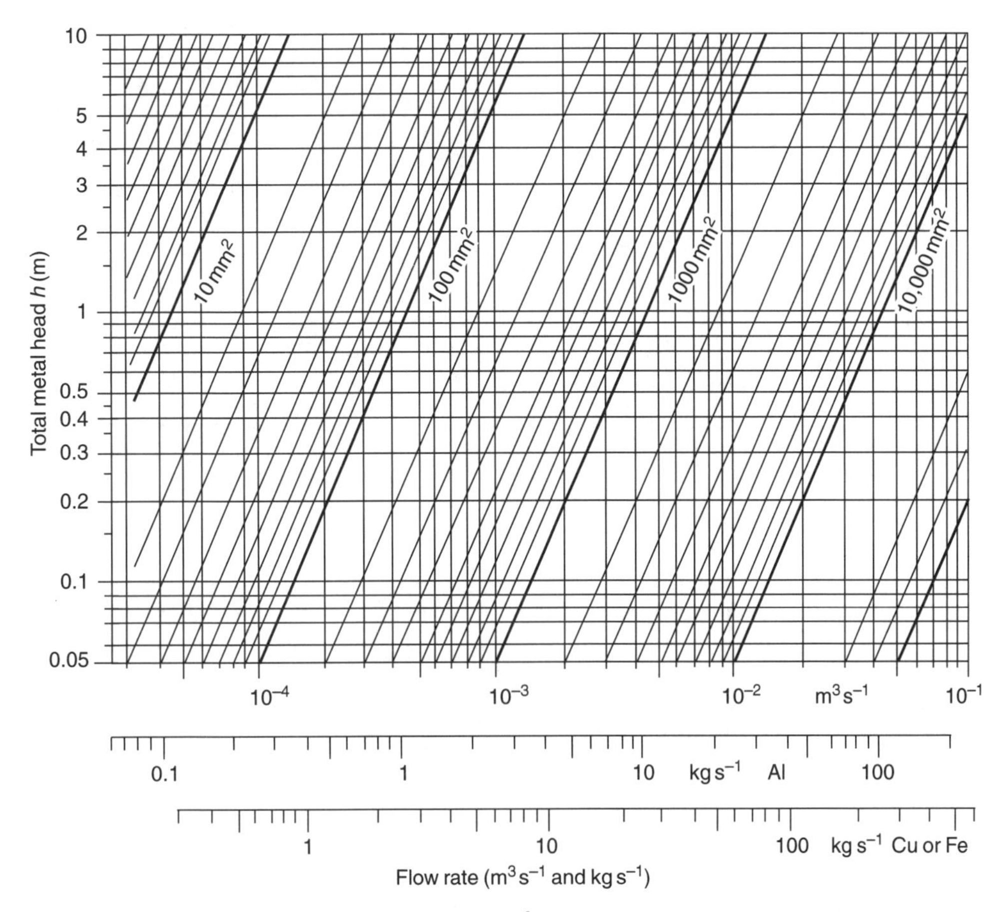

Figure 2.62 *Nomogram giving approximate sprue areas (mm²) for light and dense metals as a function of flow rate and head height.*

If we were to choose a slot-shape for the sprue, convenient sizes might be in the region of $7 \times 70\,mm^2$ entrance and $5 \times 50\,mm^2$ exit. (If two sprues were used as shown in Figure 2.63 these areas would, of course, be halved.)

2.3.7.7 Runner

Taking a simple turn from the sprue exit into the runner, the runner will have dimensions 50 mm wide by 5 mm deep. The inside radius of this turn should be at least approximately 1 or 2 times the thickness of the channel, thus we shall choose 10 mm.

However, the melt will be travelling close to $2.45\,m\,s^{-1}$. The problem remains, 'How to get the speed down from this value, nearly five times too high, to a mere $0.5\,m\,s^{-1}$ at the gate without causing damage to the flow en route?' This is *the* central problem for the design of a good filling system. This central problem seems in the past either to have been overlooked, or to have solutions proposed that do not work. At this point we need to appreciate the possible solutions with some care.

We only have a limited number of strategies for speed reduction. At this stage of the development of the technology the options appear to include

(i) Filtration.
(ii) A number of right-angle bends in succession (Jolly *et al.* 2000).
(iii) A by-pass runner design acting in a surge control mode, calculated to introduce the melt through the gate at the correct initial rate (the rate increases later of course when the surge container is full).

Considering our first option. If the filter reduces the flow rate by a factor of 4 or 5 (computer simulation might assist to provide a better

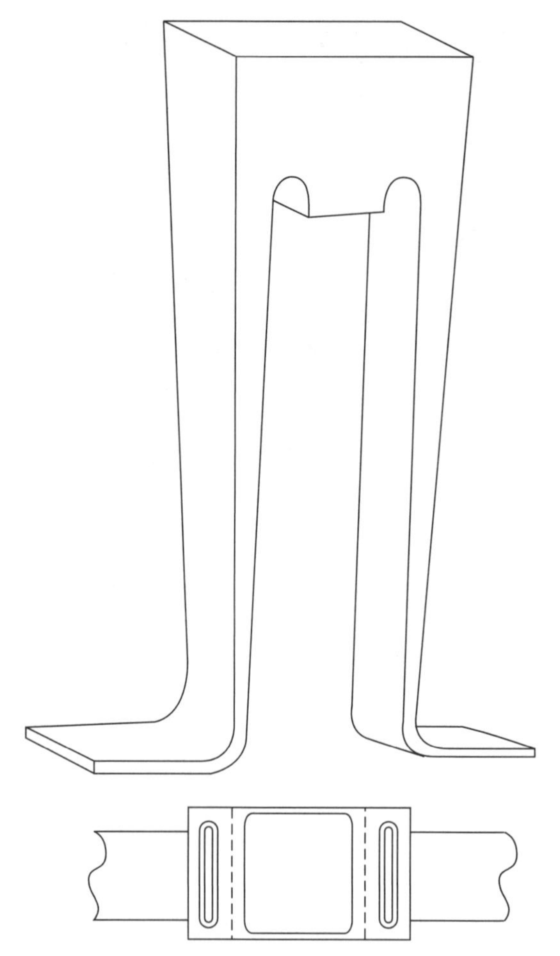

Figure 2.63 *An arrangement for a single basin and two sprues leading to two runners.*

figure) then the runner exit from the filter would require to be increased in area by a factor of four or five. We shall choose a value of 4 to give a margin of safety, helping to ensure that the runner was properly filled and slightly pressurized. Its area would then be 4×245 and so close to $1000\,\mathrm{mm}^2$. The dimensions of the slot runner after the filter might then be $10 \times 100\,\mathrm{mm}$. This is a rather large width that would be in danger of collapse because of mould expansion if a high melting point material such as an iron or steel were to be cast in a silica sand mould. The runner would be expected to survive for an Al alloy.

However, it would perhaps be more convenient if the runner were divided into two runners of $10 \times 50\,\mathrm{mm}$. Much depends on the layout of the mould and the filling system. Two runners might be more conveniently filled from two separate sprues, possibly exiting from the opposite ends of a suitably modified basin, complete with a central pouring well and vertical steps either side as illustrated in Figure 2.63. The problem in this case is the expense of two filters.

The second option, using a succession of right angle bends, is only recommended if a good computer simulation package for flow in narrow filling systems is available to test the integrity of the flow (most simulation packages do not predict flows accurately, most cannot cope with thin sections, and most cannot cope with surface tension). If a proven software package to simulate flow is available, a reasonable solution can be found largely by trial and error along the lines of the development described by Jolly *et al.* (2000). This approach is not described further here.

The third option can be a good solution. An approximate procedure is as follows. The speed of flow into a by-pass is assumed to be constant (this is clearly an overestimate, but therefore errs on the side of safety) at $Q = 0.6 \times 10^{-3}\,\mathrm{m}^3\,\mathrm{s}^{-1}$. If the by-pass volume is positioned above the level of the runner, rising up to height H (Figure 2.37), where H is perhaps at least 20 or 30 mm or more above the height of the bottom of the casting, the gradual filling of the by-pass trap will cause the metal in the gate to experience a gradually higher filling pressure. At the point at which the overflow is filled, the pressure comes on to the gate from the full height of the metal in the pouring basin. At this instant the casting should be filled to some depth at least 20 or 30 mm above the gate, so that any jetting into the mould when the full filling rate comes into effect will be to some extent suppressed. (The precise depth to suppress completely the formation of bifilms remains to be researched.)

If we assume an approximate model (awfully primitive, but better than nothing) that the area A_o of the overflow is sufficiently large to ensure that the head of liquid it contains rises at the critical velocity V_{crit}, then we can assume the liquid in the gate will follow its rise at a roughly similar rate. Thus we can define the area A_o of the overflow required for this to happen Q/V_{crit}.

It remains to work out what height H is required for the overflow.

If negligible metal enters the gate compared to that entering the overflow we have the majority of the flow rate Q entering the overflow of volume A_oH. Thus the time required to fill the overflow is $t = A_oH/Q$. Furthermore, if the velocity V_{crit} through the gate area A_3 remains roughly constant (even though the area A_c of the base of the casting is starting to be filled to a depth h) the statement for volume conservation is $V_{crit} \cdot A_3 = V_c \cdot A_c$. Also, the average velocity of rise in the casting is given by $V_c = h/t$. Thus the time required to fill the casting to a height h above the runner (neglecting the relatively trivial amount contained in the gate) is given by $t = hA_c/V_c \cdot A_3$. Equating these two estimates for times gives

$$A_o H/Q = hA_c/V_c \cdot A_3$$

Rearranging to give the height of the overflow we obtain finally

$$H = (Q/A_o)(h/H)(A_c/A_3)(1/V_{crit})$$

As mentioned earlier, it is sensible to arrange the overflow to be a cylinder and connected tangentially to the runner. In this way, by avoiding unnecessary turbulence and filling more progressively, a better quality of metal is preserved for future recycling within the foundry.

The careful sizing of overflows to suppress the early jetting of melt through the gates is strongly recommended. To the author's knowledge, the technique has been relatively little used so far. More experience with the technique will almost certainly lead to greater sophistication in its use.

2.3.7.8 Gates

In general, it is essential that the liquid metal flows through the gates at a speed lower than the critical velocity so as to enter the mould cavity smoothly. If the rate of entry is too high, causing the metal to fountain or splash in the mould cavity the battle for quality has probably been lost.

The gate should enter at the very base of the casting, if possible at right angles onto a thin section, as has been described earlier. Gating directly across a flat floor of a casting is to be avoided if possible—a thin jet of metal skating across a flat surface is a recipe for mould expansion defects of various sorts that will spoil the surface of the casting. The casting will also be at risk from the formation of an oxide flow tube that may constitute a serious internal discontinuity in the casting. If directed at right angles against a core, higher velocities can be tolerated, since the thin section in the casting effectively acts as an extension of the runner system, helping to spread and thus reduce the velocity before the melt arrives in a section large enough to allow the melt room to damage itself. Thus gating onto a core is often useful providing, of course, that we have succeeded to design the filling system to remain free from entrained air.

For our example casting, the velocity at the base of the sprue is $2.45\,\mathrm{m\,s^{-1}}$. Thus to achieve $0.5\,\mathrm{m\,s^{-1}}$ through the gate(s), and if no friendly core is conveniently sited, we shall require an expansion of the area compared to that of the base of the sprue by a factor $A_2/A_3 = 5$ approximately. In terms of the gating ratio much loved by the traditionalists among us, we are using 1 : 1 : 5 for this casting. The use of this size of gate assumes that we are gaining no advantage from a by-pass runner design. If a good by-pass design could be devised an acceptable ratio might then become 1 : 1 : 1 effectively easing subsequent cut-off, and reducing any possible problems of hot-spots or convection at this location.

Sometimes the by-pass cannot be provided. Even if available, it may be useful to use both the by-pass and the enlarged gate until such a time that our understanding of filling systems makes it clear that such belt-and-braces solutions are not required.

Where the gates form a T-junction with the casting, the maximum modulus of the gates should be half of that of the casting (if, as will be normal, no feeding is planned to be carried out via the gates). Thus, in general, the thickness of the gates needs to be less than half the thickness of the wall. This forces the shape of the gates to be usually of slot form. If made especially thin in alloys of good thermal conductivity, the gate can sometimes be usefully employed to act as a cooling fin soon after the filling of the casting.

The other major consideration that must not be overlooked is the problem of the transverse, or lateral, velocity of the melt in the mould cavity as it spreads away from the ingate. This can easily exceed the critical velocity despite the velocity in the gate itself being correctly controlled. In this case a single gate may have to be divided to give multiple gates as described in section 2.3.2.6.

The area of the gate required to reduce the gate velocity to below the critical velocity, and the limitations of its thickness, sometimes dictates a length of slot significantly longer than the casting. In this situation there is little choice but to revise the design of the filling system, selecting a correspondingly longer fill time so that the gate can be shortened to fit the length available. Alternatively, a by-pass runner design may be the solution. If a solution cannot be found, the conclusion has to be accepted that the casting as designed cannot be made so as to enjoy reliable properties and performance. A serious discussion with the designer will probably be required.

Rule 3

Avoid laminar entrainment of the surface film (the non-stopping, non-reversing condition)

3.1 Continuous expansion of the meniscus

If the liquid metal front continues to advance at all points on its surface, effectively, continuing to expand at all points on its surface, like a progressively inflating balloon, then all will be well. This is the ideal mode of advance of the liquid front.

In fact, we can go further with this interesting concept of the requirement for continuous surface expansion. There is a sense in that if surface is lost (i.e. if any part of the surface experiences contraction) then some entrainment of the surface necessarily occurs. Thus this can be seen to be an all-bracing and powerful definition of the condition for entrainment of defects, simply that surface must not be lost. Clearly, surface is effectively lost by being enfolded (in the sense that the fold now disappears inside the liquid, as has been the central issue described under Rule 2), or by simply shrinking (leading to folding) as described below. Thus in a way this condition 'Avoid loss of surface' can be seen to supersede the conditions of critical velocity or Weber number. It promises to be a useful condition that could be recognized in numerical simulation, and thus be useful for computer prediction of entrainment.

In practice, however, an uphill advance of the liquid front, if it can be arranged in a mould cavity, is usually a great help to keep the liquid front as 'alive' as possible, i.e. keeping the meniscus moving, and so expanding and creating new oxide.

While the surface is being continuously expanded when filling the mould the casting has the benefit that the older, thicker oxide is continuously being displaced to the walls of the mould where it becomes the skin of the casting. Thus very old and very thick oxide does not normally have a chance to form and become entrained. In fact, one of the great benefits of a good filling system is to ensure that the older oxides on the surface of the ladle or pouring basin etc. do not enter the mould cavity. When inside the mould cavity, the continued expansion of the surface ensures that the surface oxide is brought into contact with the mould surface, and so becomes the skin of the casting (and not entrained inside the liquid where it would constitute a defect). For instance, in the tilt pouring of aluminium alloy castings, the filling of a new casting can be checked by dropping a fragment of paper on the metal surface as the tilt commences. This marker should stay in place, indicating that the old skin on the metal was being retained by the runner, so that only clean metal underneath could flow into the mould cavity. If the paper disappears into the runner, the runner is not doing its job. In a way, the use of the tea-pot pouring ladle and bottom-pour ladles common in the steel casting industry are in response to their special problems, in which the high rates of reaction with the environment at such high temperatures encourages the surface oxide to grow from microscopic to macroscopic thickness, to constitute the familiar slag layer.

Problems arise if the front becomes pinned by the rate of advance of the metal front being too slow, or if it stops or reverses. Loss of surface area by enfolding bifilm defects can then occur in two ways:

(i) If the liquid front stops, a thick surface film has chance to form. This may become so thick and strong that the front can no longer re-start its advance if pressure increases to encourage flow a little later. This thick film may be subsequently entrained as the general advance re-starts, so that metal overflows and submerges it. As the new metal rolls over the old film a new fresh oxide is laid down over the old thick oxide, forming our familiar double film. This can constitute a large geometrical defect, sometimes in the form of a vertical tubular crack, and sometimes a large horizontal crack extending across the whole casting, or as a horizontal lap around its complete perimeter. Such bifilm defects are characterized microscopically by asymmetrical components; one being the thick underlying, stationary film, and the other the younger thinner film provided by the meniscus that rolls up against it (Figure 3.1).

(ii) If the liquid front reverses, the shiny, swelling front of the liquid experiences a brief moment as it is flattened, prior to reversing its curvature in the opposite direction. The flat form has slightly less area, so that the small excess of surface may be entrained by random folding. This can form tiny but insidious surface cracks.

Both of these actions occur in various ways during casting. We shall consider them in detail in this section.

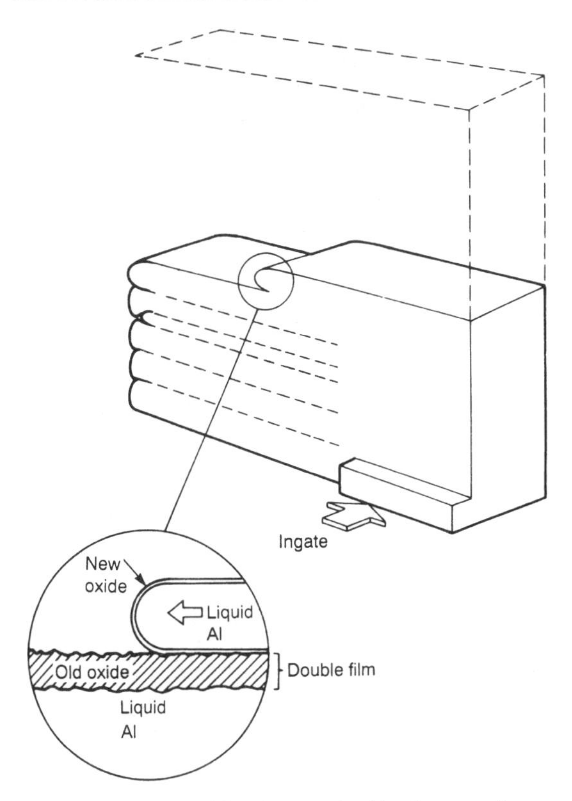

Figure 3.1 *An unstable advance of a film-forming alloy, showing the formation of horizontal laps as the interface intermittently stops and re-starts by bursting through and flooding over the surface film.*

3.2 Arrest of vertical progress

A sudden increase in cross-sectional area of the casting, such as the extensive horizontal areas at *B* and *C* in Figure 3.2 tend to bring the general advance of the liquid front practically to a stop. Such interruptions to the advance of the front are likely to result in lap-type defects at *b* and both at c_1 and c_2.

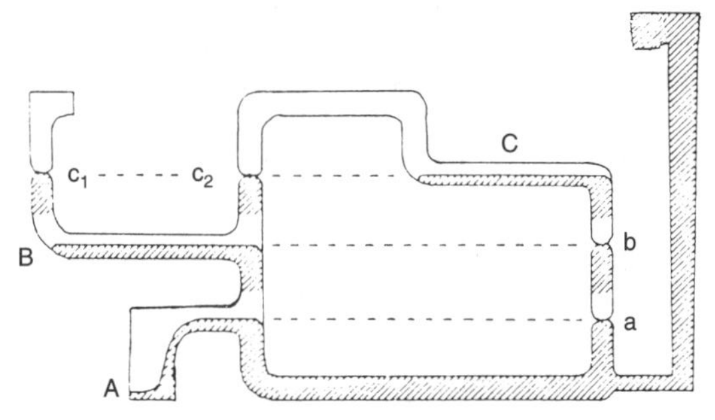

Figure 3.2 *Steady filling via the bottom gate is interrupted because of overflow to the heavy section A, and filling of extensive horizontal surfaces B and C, leading to the danger of lap defects at distant, apparently unrelated regions of the casting a, b, c_1 and c_2.*

The film on the melt thickens while the front is stopped. It can be submerged later if metal breaks through at some point and flows over it. During the process of submerging the film, the newly arriving metal rolls over the thicker oxide, so rolling in place its own, newer oxide film. In this way an asymmetric double oxide layer is created, with dry sides facing dry sides so that the double layer forms an unbonded interface as a crack. This process can result in a major defect, often spanning the casting from wall to wall.

Another good name for this defect is an *oxide lap*. (It is to be distinguished from the solidification defect that often has a superficially similar appearance, called here a *cold lap*.

The method of treating these defects is quite different. A cold lap can be cured by increasing the casting temperature, whereas increasing the temperature is likely to make an oxide lap worse.)

3.3 Waterfall flow

Instead of a large horizontal defect, a tubular or even cylindrical defect that I call an oxide flow tube can form in several ways. If the liquid falls vertically, as a plunging jet, the falling stream is surrounded by a tube of oxide (Figure 3.3). Despite the high velocity of the falling metal inside, the oxide tube remains stationary, thickening with time, until finally surrounded by the rising level of the metal in the mould cavity. This rising metal rolls up against the oxide tube, forming a double oxide crack. Notice the curious cylindrical form of this crack and its largely vertical orientation. The arrest of the advance of the front in this case occurred by the curious phenomenon that although the metal was travelling at a high speed parallel to the jet, its transverse velocity, i.e. its velocity at right angles to its surface, was zero. It is the zero velocity component of a front that allows the opportunity for a thick oxide skin to develop.

These oxide flow tubes are often seen around the falling streams of many liquid metals and alloys as they are poured. The defects are also commonly seen in castings. Although occasionally located deep inside the casting where they are not easily found, they are often clearly visible if formed against the casting surface, especially in Al alloys in permanent moulds.

Even with the best design of gravity-poured system, the rate of fill of the mould may be far from optimum at certain stages during the fill. For instance, Figure 3.2 shows a number of common geometrical features in castings that cause the advance of the liquid metal to come to a stop. The heavy section filled downhill at *A* will cause the metal front to stop at point *a*, possibly causing a lap-type defect at a point on the casting well away from the real cause of the problem. In an uncharacteristic lapse of rigour, the author often refers to this problem as the 'waterfall' effect. This always occurs if the liquid falls into a recess. Until the recess is filled, the remainder of the liquid front cannot advance. There are several reasons for avoiding any 'waterfall' action of the metal during the filling of the mould.

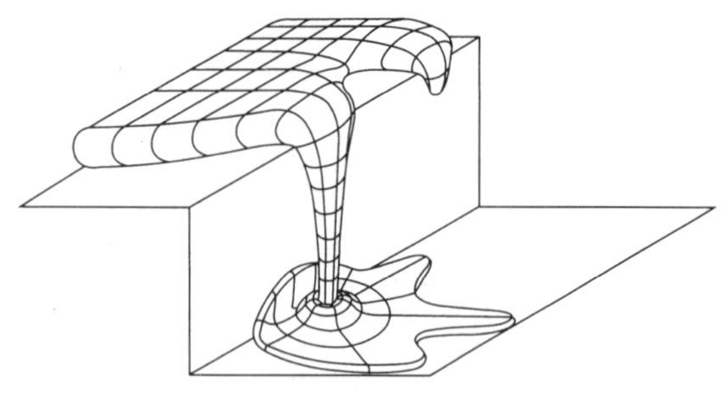

Figure 3.3 *Waterfall effect leading to a vertical oxide flow tube (among other defects).*

(i) A cylindrical oxide flow tube forms around the falling stream itself. If the fall is from a reasonable height, the tube is shed from time to time, and plunges into the melt where it will certainly contribute to severe random defects. The periodic shedding of oxide flow tubes into the melt is a common sight during the pouring of castings. Several square metres of oxide area can be seen to be introduced in this way within a minute or so.
(ii) The plunging jet is likely to exceed the critical velocity. Thus the metal that has suffered the fall is likely to be impaired by the addition of randomly entrained bifilms.
(iii) As the melt rises around the tube, supporting it to some degree and reducing the height of the fall, the flow tube remains in place and simply thickens. As the general level of the melt rises around the tube, the new oxide rolls up against the surface of the cylinder, forming the curious cylindrical bifilm that acts as a major cylindrical crack around a substantially vertical axis.
(iv) During the period of the waterfall action, the general rise of the metal in the rest of the mould will be interrupted, causing an oxide to form across the whole of the stationary level surface. Thus a major horizontal lap defect may be created.

Defects 1 and 2 above are the usual fragmented and chaotic type of bifilm. Defects from 3 and 4 are the major geometrical bifilms.

Waterfall problems are usually easily avoided by the provision of a gate into the mould cavity at every low point in the cavity. Occasionally, deep recesses can be linked by channels through the mould or core assembly; the links being removed during the subsequent dressing of the casting.

3.4 Horizontal stream flow

If the melt is allowed to spread without constraint across a horizontal thin-walled plate, gravity can play no part in persuading the flow to propagate on a broad stable front, as would happen naturally in a vertical or sloping plate (Figure 3.4). The front propagates unstably in the form of a river bounded by river banks composed of thickening oxide. This mode of flow occurs because the oxide at the flow tip is thin, and easily broken, allowing the front to advance here, but not elsewhere, where the oxide on the front is allowed to thicken, so restraining any advance. This is a classical instability situation leading to a kind of dendritic advance of a front. The meandering advance leads to a situation where its sinuous oxide flow tube is sealed into the casting as the liquid metal finally arrives to envelop it.

The avoidance of extensive horizontal sections in moulds is therefore essential for reproducible and defect-free castings. Any horizontal sections should be avoided by the designer, or by the caster tilting the mould. The tilting of the mould is more easily said than done with most of our automatic moulding and casting lines, and represents a serious deficiency in much of our standard foundry equipment. This deficiency needs to be addressed in future equipment. In contrast, a tilting facility is easily provided, and, in principle, can be programmed into the filling process by some casting techniques such some tilt casting machines, and in the Cosworth Process, where the mould is held in a rotatable fixture during casting. The flow across such inclined planes is therefore progressive, if slow, but the continuous advance of the front at all points assists the aim of keeping the meniscus 'alive'.

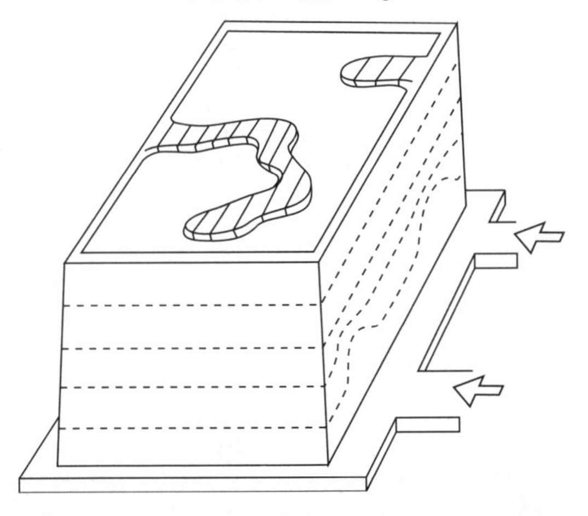

Figure 3.4 *The filling of a rectangular box type casting, illustrating the progressive advance of the front that characterizes the filling of vertical walls. The horizontal top, however, fills unpredictable meanders of river-like flows, leading to horizontal oxide flow tubes.*

A fascinating example of a flow tube can be quoted from observations of uphill flow in an open channel driven by a travelling magnetic field from a linear motor sited under the channel. When used to drive liquid aluminium alloy uphill, out of a furnace and into a higher-level receiver, the travelling melt is seen to flow inside its oxide tube. When the magnetic field is switched off, the melt drains out of the oxide tube and back into the furnace, the tube collapsing flat on the bottom of the channel. However, when the field is switched on again, the same oxide tube magically refills and continues to pass metal as before. Clearly the tube has considerable strength and resilience. It is sobering to think that such features can be built into our castings, but remain unsuspected and almost certainly undetectable. Clearly, the casting methods engineer requires vigilance to ensure that such defects cannot be formed.

The vertical oxide flow tube is probably more common than any of us suspect. The example given below is simply one of many that could be described.

Figure 3.5 illustrates the bronze bell hung outside the railway station in Washington DC. Horizontal weld repairs record for all time the fatal hesitations in the pouring process that led to the horizontal oxide bifilms that would have appeared as horizontal laps. The vertical weld repairs record the passage of the falling streams that created the vertical flow tubes, the oxide laps that led to cracks through most of the thickness of the casting. This is a common source of failure for bells, nearly all of which are top-poured through the crown. The renowned Liberty bell (the only survivor of three attempts, all of which cracked) reveals a magnificent example of a flow tube defect that starts at the crown, curves sinuously around and over the shoulder, and finally falls vertically. Although there are many examples of bells that exhibit these long cracks, it is perhaps all the more surprising that any bells survive the top-pouring process. It seems likely that in the majority of cases of bells of thicker section the oxide flow tube is not trapped between the walls of the mould to create a through-thickness pair of parallel cracks. In such thicker sections the tube is more likely to be detached and carried away, crumpling into a somewhat smaller defect that can be accommodated elsewhere. It is to be hoped that the new resting place of the defect will not pose any serious future threat to the product. Clearly, the top pouring of castings is a risky manufacturing technique.

Figure 3.5 *Nick Green (tall, handsome and young) and the author (less tall, not so good looking, and significantly older) inspect the fine 8000 kg American Legion Freedom Bell outside the railway station in Washington DC, unfortunately spoiled by welds in an attempt to repair cracks caused by horizontal oxide laps and vertical oxide tube defects (photograph courtesy of author's wife).*

Oxide flow tubes are common defects seen in a wide variety of castings that have been filled across horizontal sections or down sloping downhill sections. The deleterious flow tube structures described above that form when filling *downwards* or *horizontally* are usually eliminated when filling *vertically upwards*, i.e. in a *counter-gravity* mode. The requirement that the meniscus only travels uphill is sacred. However, even in this favourable mode of filling, a related oxide lap defect, or even a cold lap defect, can still occur if the advance of the meniscus is stopped at any time, as we have seen.

3.5 Hesitation and reversal

If the meniscus stops at any time, it is common for it to undergo a slight reversal. Minor reversals to the front occur for a variety of reasons. Some of these are discussed below.

(i) A reversal will practically always occur when a waterfall is initiated. This occurs because at the point of overflow, the liquid will be at a level slightly above the overflow, dictated by the curvature of its meniscus, i.e. for a liquid Al alloy it will be about 12.5 mm above the height of the overflow since this is the height of the sessile drop. However, immediately after the overflow starts, the general liquid level drops, no longer supported by the surface tension of the meniscus. In the case of liquid aluminium alloy this fall in general level of the liquid will be perhaps about 6 mm, just enough to flatten the more distant parts of the meniscus against the rest of the mould walls.

(ii) Hesitations to an advancing flow will often be accompanied by slight reversals because of inertial effects of the flow. Momentum perturbations during filling will cause slight gravity waves, the surface therefore experiencing minor slopping and surging motion, oscillating gently up and down.

These minute reversals of flow flatten the oxidized surface of the meniscus. When advancing, the meniscus adopts a rounded form, but when flattened, the oxidized surface now occupies less area. A fold necessarily develops, wrinkling the surface, endangering the melt with the possibility of the entrainment of this excess oxide once the melt is able to continue its advance. The folding in of a small crack attached to the surface of the casting is illustrated in Figure 3.6. Such shallow surface cracks occurring as a result of hesitation and/or reversal of the front are common in aluminium alloys, and are revealed by dye penetrant testing.

It is instructive to estimate the maximum depth that such oxide folds might have. Following Figure 3.6, if the front of the liquid in Figure 3.6a is a cylinder of radius r, the perimeter of the quarter of a cylinder is $\pi r/2$, so that the maximum length of excess surface if the melt level now drops a distance r is $\pi r/2 - r = r((\pi/2) - 1) = r/2$ approximately. The radius of the meniscus r is approximately 6 mm for liquid aluminium (as a result of the total height of a sessile drop being approximately 12 mm), giving the excess length 3 mm. If this is folded just once to create a bifilm, its potential depth is therefore found to be approximately 1.5 mm.

If the melt continues its downward oscillation the defect can be straightened out as shown in

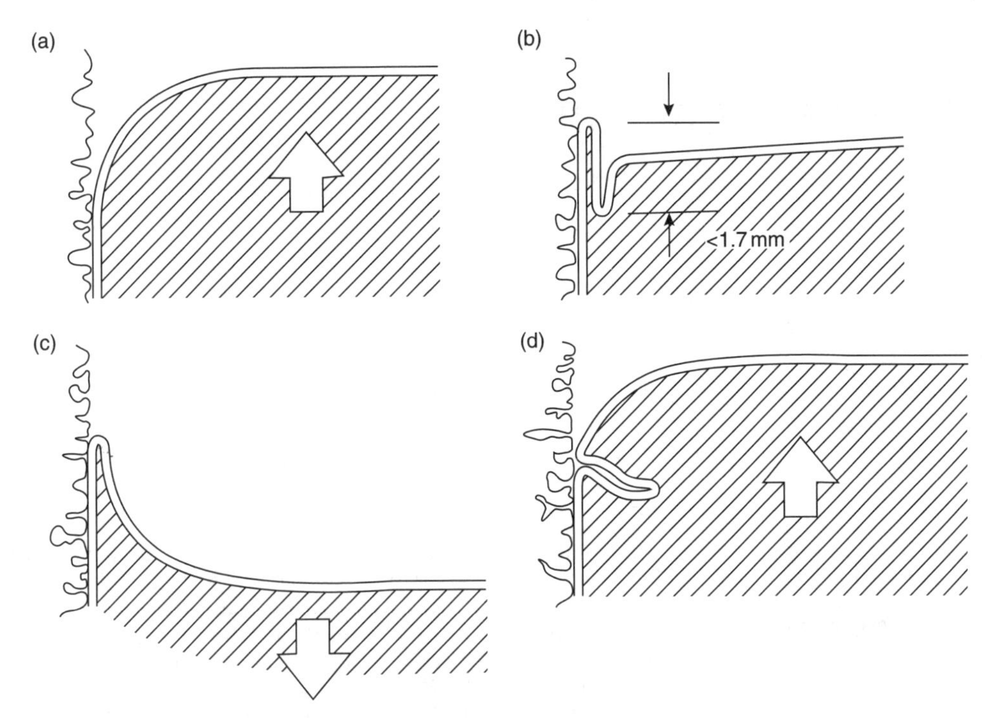

Figure 3.6 *The creation of a bifilm crack by the reversal of the front, causing the meniscus to flatten and enfold in the excess surface area. Surface cracks of the order of a millimetre depth can be formed in this way.*

Figure 3.6c. Alternatively, if the bifilm created in this way holds itself closed, possibly because of viscous adhesion (i.e. the trapped liquid metal takes time to escape from between the films) or possibly as the result of other forces such as Van der Waals forces, then there is the danger that additional folds may be created on each oscillation cycle.

In fact, many of these defects are not as deep as the maximum estimate of 1.5 mm for several reasons: (i) the melt surface may not drop the full distance r; (ii) the film may be folded more than once, creating a greater number of shallower folds; and (iii) the fold-like crack may hinge to lie flat against the surface of the casting. The action of internal forces as a result of flow of the liquid may be helpful in this respect. For these reasons such defects are usually only a fraction of a millimetre deep, so that they can often be removed by grit blasting. Only relatively rarely do they reach the maximum possible depth approaching 1.5 mm. Even so, for castings requiring total integrity, that may be designed for conditions of service involving high stress or fatigue, these minor oscillations of the front are very real threats that are best avoided.

The ultimate solution, as we have emphasized here, is that the melt should be designed to be kept on the move, advancing steadily forwards at all times.

Rule 4

Avoid bubble damage

Entrainment defects are caused by the folding action of the (oxidized) liquid surface. Sometimes only oxides are entrained, as doubled-over film defects, called bifilms. Sometimes the bifilms themselves contain small pockets of accidentally enfolded air, so that the bifilm is decorated by arrays of trapped bubbles. Much, if not all, of the microporosity observed in castings either is or has originated from a bifilm. Sometimes, however, the folded-in packet of air is so large that its buoyancy confers on it a life of its own. This oxide-wrapped bubble is a massive entrainment defect that can become important enough to power its way through the liquid and sometimes through the dendrites. In this way it develops it own distinctive damage pattern in the casting.

The passage of a single bubble through an oxidizable melt is likely to result in the creation of a bubble trail as a double oxide crack, a long bifilm, in the liquid. Thus even though the bubble may be lost by bursting at the liquid surface, the trail remains as permanent damage in the casting.

The bubble trail occurs because the bubble is nearly always attached to the point where it was first entrained in the liquid. The enclosing shroud of oxide film covering the crown of the bubble attempts to hinder its motion. However, if the bubble is sufficiently buoyant, its buoyancy force will split this restraining cover. Immediately, of course, the oxide re-forms on the crown, and splits and re-forms repeatedly. In this way the bubble progresses by its skin sliding around the bubble, gathering together in a mass of longitudinal pleats under the bubble as a trail that leads back to the point at which the bubble was first entrained as a packet of gas.

The structure of the trail is a kind of collapsed tube. In section it is star-like but with a central portion that has resisted complete collapse because of the rigidity of the oxide film (Figure 4.1b). This is expected to form an excellent leak path if it joins opposing surfaces of the casting, or if cut into by machining. In addition, of course, the coming together of the opposite skins of the bubble during the formation of the trail ensure that the films make contact dry side to dry side, and so constitute our familiar classical bifilm crack.

Poor designs of filling systems can result in the entrainment of much air into the liquid stream during its travel through the filling basin, during its fall down the sprue, and during its journey along the runner. In this way dozens or even hundreds of bubbles can be introduced into the mould cavity. When so many bubbles are involved the later bubbles have problems rising through the maze of bubble trails that remain after the passage of the first bubbles. Thus the escape of late-arriving bubbles is hampered by the accumulation of the tangle of residual bifilms. If the density of films is sufficiently great, fragments of bubbles remain entrapped as permanent features of the casting. This messy mixture of bifilm trails and bubbles is collectively christened '*bubble damage*'. In the experience of the author, bubble damage is probably the most common defect in castings, but up to now has been almost universally unrecognized.

Bubble damage is nearly always mistaken for shrinkage porosity as a result of its irregular form, usually with characteristic cusp-like morphology. When seen on polished sections, the cusp forms that characterize bubble damage are often confused with cusps that are

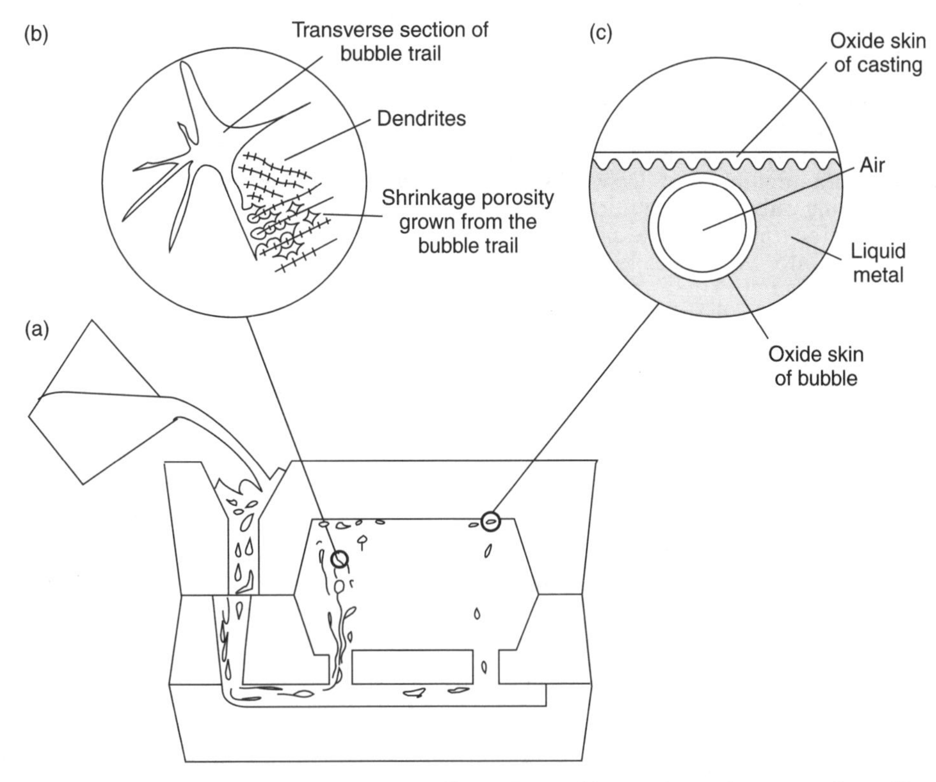

Figure 4.1 *(a) Pattern of bubble damage in a casting; (b) trails invisible in radiography are usually visible on transverse sections; (c) small entrained air bubbles do not have sufficient buoyancy to break the double oxide barrier to escape to the atmosphere.*

associated with interdendritic shrinkage porosity. However, they can nearly always be distinguished with complete certainty by their difference in size. Careful inspection of the dendrite arm spacing will usually reveal that cusps that would have formed around dendrites as the residual interdendritic liquid is sucked into the dendrite mesh are usually up to ten times smaller than cusps that are caused by the folds of oxide in bubble trails (Figure 4.1b). Clearly, the two are quite distinct and totally unrelated.

Bubble damage is commonly observed just inside and above the first (or sometimes the last) ingate from the runner (Figure 4.1a). The large bubbles have sufficient buoyancy to escape up the first ingate, but smaller bubbles can be carried the length of the runner, to appear through the farthest ingate. Alternatively, they can even be carried back once again if there is a back wave. This non-uniform distribution associated sometimes with first and sometimes with some other ingate position is a common but not universal feature of bubble damage. This is because the presence of cores, and sometimes strong flows of metal inside the mould cavity can cause the bubble path to deviate a long way from a direct vertical path to the surface. Highly indirect paths are commonly observed in video radiography studies. Nevertheless, the common feature of bubble damage is its non-uniform distribution.

If the bubbles completely escape the remaining trails can float around, finally settling some distance from their source. Irregular masses of oxides in odd corners of castings have been positively identified as groups of tangled bubble trails. The bubbles have moved on and escaped, but their trails have remained in suspension. They have broken free from their moorings (the point at which the bubble was first entrained) and have travelled, tumbling and ravelling as they go, carried by the sweeping and circulating flow of the liquid during the filling process. Texan founders will recognize an analogy with tumbleweed.

Another common feature of bubble damage is the entrapment of small bubbles just under the cope skin of the casting (Figure 4.1c). They are prevented from escaping only by the thickness of the oxide skin on the casting and their own oxide skin. Both these films require to be broken. (This is achieved by larger bubbles because of their stronger buoyancy forces, but not by smaller bubbles. The dividing line between large and small bubbles seems to be in the region of 5 mm diameter for many light and dense alloys.) Such bubbles, sitting only a double thickness of oxide depth under the top skin of the casting are commonly broken into when shot blasting, or on the first machining cut. These too are commonly observed in video radiographic studies.

Close optical examination of the interiors of bubbles and bifilms in an aluminium alloy casting often reveals some shiny dendrite tips characteristic of shrinkage porosity. This adds to confusion of identification, because shrinkage cavities will often form, expanding an existing bifilm, unfurling and opening it, and finally sucking one or both of its films into the dendrite mesh. Subsequently only fragments of the originating oxides will sometimes be found among the dendrites. This process has been observed in video radiographic studies of castings. An unfed casting has been seen to draw in air bubbles at a hot spot on its surface. The bubbles floated up in succession, but the later bubbles became trapped by dendrites. As solidification progressed, shrinkage caused the air bubbles to gradually convert to shrinkage cavities. The perfectly round and sharp radiographic images were seen to become 'furry' and indistinct as the liquid meniscus was sucked into the surrounding mesh of dendrites. Finally, the defect resembled an extensive shrinkage cavity; its origin as a gas bubble no longer discernible.

Other real-time radiography has shown bubbles entrained in the runner, and swept through the gate and into the casting. The upward progress of one bubble in the region of 5 to 10 mm diameter appeared to be arrested, the bubble circulating in the centre of the casting, behaving like a balloon on a string. The string, of course, being the bubble trail acting as a tether. Other bubbles of various sizes up to about 5 mm diameter in the same casting were observed to float to the top of the casting, coming to rest under the oxide skin of the cope surface. These bubbles had clearly broken free from their tethers, probably as a result of the extreme turbulence during the early part of the filling process. The central bubble was marginally just too small to tear free from its trail. In addition, it may have lost some buoyancy as a result of loss of oxygen during its rise, or perhaps more likely, it ascended as far as it did because of assistance from the force of the flow of the melt. When this abated higher in the mould cavity, its buoyancy alone was insufficient to split its oxide skin, so that its upward progress was halted.

Where many bubbles have passed through an ingate into the mould, a cross-section of the ingate will reveal some central porosity. These are the bubble trails, pushed ahead of the growing dendrites, and so concentrated in the centre of the ingate section. Close examination will confirm that this porosity is not shrinkage porosity, but a mass of double oxide films, the bubble trails. In Al alloys they appear as a series of dark, non-reflective oxidized surfaces interleaved like the flaky, crumpled pages of an old sepia-coloured newspaper.

In some stainless steels the phenomenon is seen under the microscope as a mixture of bubbles and cracks. (A remarkable combination! Without the concept of the bifilm such a combination would be extremely difficult to explain.) In these strong materials the high cooling strain leads to high stresses that open up the double oxide bubble trails.

In grey iron cylinder heads the bubbles and their trails are coated not with oxide but with a lustrous carbon film. The carbon film appears to be somewhat more rigid than most oxide films, and so resists to some extent the complete collapse of the trail, and retains a more open centre. In effect, the bubbles punch holes through the cope surfaces of the casting, so that their trails form highly efficient leak paths.

The bubble trail is usually a collapsed, or nearly closed, tube. However, completely open bubble trails have been observed by Divandari in pressure die castings (Figure 4.2). In this process the very high injection velocities, of the order of 10 to 100 times the critical velocity for entrainment, naturally entrains considerable quantities of air and mould gases. These extraordinary conditions are perhaps better described in terms of atomization and emulsification of the air and the metal. The very high pressure (up to 100 MPa, or 15 000 psi) applied during casting is mainly used to compress these unwanted gases to persuade them to take up the minimum volume in the casting. If, however, the die is opened before the casting is fully solidified, as is usual to maximize productivity, the entrained bubbles may experience a reduction in their surrounding mechanical support, allowing the bubbles to expand under their immense internal pressure. At the same time, of course, their bubble trails will also be re-inflated. Such open bubble trails in pressure die-cast components are expected to be serious sources of leakage, particularly when broken into by machining operations. In this case the problem is greatly reduced (although perhaps never quite eliminated) by sacrificing some productivity,

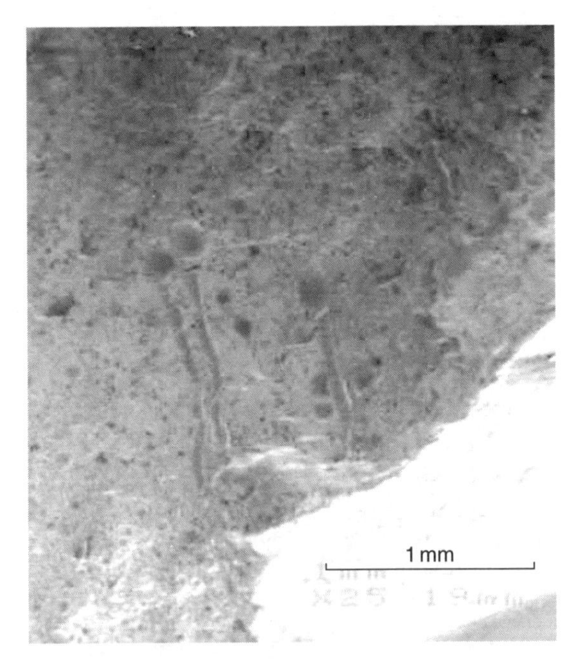

Figure 4.2 *Re-inflated bubbles in a Zn alloy pressure die casting (Divandari and Campbell 2003).*

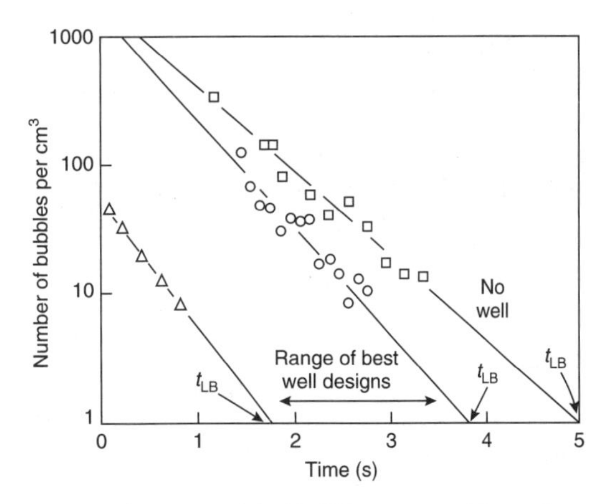

Figure 4.3 *Water model of bubbles entrained by surface turbulence in a well (Isawa and Campbell 1994) showing the decrease of bubbles with time in different well designs, extrapolated to the time for the last bubble, t_{LB}, for runners twice the area of the sprue exit.*

allowing the castings to solidify more completely before opening the die.

4.1 Gravity-filled running systems

In gravity-filled running systems the requirement to reduce bubbles in the liquid stream during the filling of the casting calls for offset stepped basins, or other advanced filling systems. The conventional conical or funnel-shaped pouring basin cannot be permitted. The requirement also demands properly engineered and manufactured sprues. The sprue is required to be tapered, the taper calculated to match, or very slightly compress, the natural form of the falling stream; the stream naturally narrows during its fall because of its acceleration under the action of gravity. By tailoring the shape of the sprue to the natural shape of the stream the melt has the best chance to avoid the entrainment of air. Parallel or reversed taper sprues are not recommended. They may be permitted only if special precautions are adopted such as the provision of a filter and bubble trap combination in the entrance to the runner, as close as possible to the sprue exit.

It is mandatory that the taper of the sprue contains no perturbations to upset the smooth fall of the liquid metal. Thus it must be well-fitting with the pouring basin, and accurately matched in size and alignment at mould or die joints; no steps, ledges, or abrupt changes in direction are permissible (a typical sprue mismatch across a mould joint is shown in Figure 4.1). Also no branching or joining of other ducts, runners, gates or sprues is allowable. All such features (unfortunately especially common in investment castings) have the potential for the introduction of air into the stream, or the uncontrolled premature escape of droplets and dribbles of liquid into other parts of the mould cavity The dividing of sprues might become allowable at some future date when such features have been properly researched by accurate computer simulation and video radiography.

At one time it was mandatory that each sprue had a sprue well at its base. The well was thought to facilitate the turn of the metal through the right angle bend into the runner with minimum turbulence. All the work on well development had been based on water models, and the standard runner had been one of large area, the expanded area specially selected with a view to slow the flow. However, more recent work in the author's laboratory using both water models and liquid metals observed by video radiography has demonstrated that, at best, the well is no better than no well at all for such large area runners, and at worst, causes considerable extra turbulence. The excellent work by Isawa (1993) noted that even the best designs of wells that he was able to optimize introduced hundreds of bubbles that took 2 to 4 seconds to clear (Figure 4.3).

Thus the new designs of filling systems incorporate no well at the base of the sprue. This departure from tradition is possible only because the new filling systems are characterized by runners of approximately the same area as that of the sprue exit (Section 2.3.2.5). It is to be noted that the traditional choice of sprue exit/runner/gate ratios of 1 : 2 : 2 and 1 : 2 : 4 etc. are automatically bad. The runner is too large to fill completely, regrettably ensuring bubble damage problems.

An additional beneficial consequence of the avoidance of a well is the addition of friction to the liquid provided by the additional solid surface of the mould at the point impacted by the metal as it turns the corner, so slowing the velocity of the melt to the greatest extent. If the sprue/runner junction is nicely formed, bubbles are formed for precisely zero seconds. This awkward way of making a simple statement that no bubbles are formed is deliberate. It emphasizes the contrast with filling systems that have been accepted as conventional up to now. Nowadays it is not necessary to accept a design that introduces any bubbles at all.

It is mandatory that no interruption to the pour occurs that leads to the lowering of the melt in the pouring basin below the minimum design level. If the sprue entrance is unpressurized in this way air will enter the running system. In the worst instance of this kind, if the basin level drops to the point that the sprue entrance becomes uncovered this has to be viewed as a disaster. A provision must be made for the foundry to reject automatically any castings that have suffered an interrupted pour, or slow pour that has allowed the basin to empty to a level below the designed minimum level.

To be safe, it is worth ensuring that basins are provided that are at least twice if not four times the required minimum depth to keep the sprue filled, and ensuring that the pourer keeps well above the minimum level. In this way the casting may run a little faster, but air will be excluded and bubble damage avoided.

4.2 Pumped and low-pressure filling systems

Pumped systems such as the Cosworth Process, or low-pressure casting systems into sand moulds or dies, are highly favoured as having the potential to avoid the entrainment of bubbles if, and only if, the processes are carried out under proper control. The reader needs to be aware that good control of a *potentially* good process should not be assumed; it is required to be demonstrated.

For instance, for pumped systems, bubbles can be released erratically from the interior wall of a tube launder system, especially if it is not cleaned with regular maintenance, as well as from the underside of a badly designed distribution plate used in some counter-gravity systems such as the early variants of the low pressure sand (LPS) system (Figure 4.4c).

Although low-pressure filling systems can, in principle, satisfy the requirement for the complete avoidance of bubbles in the metal,

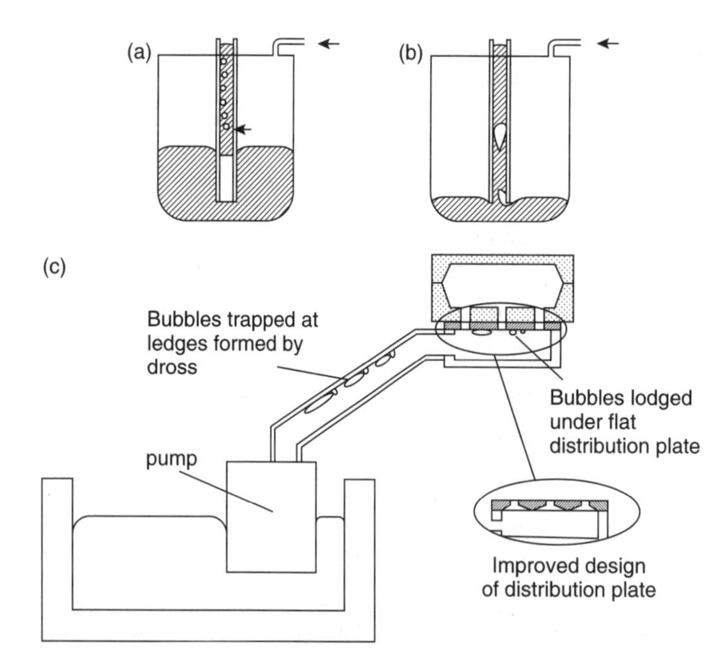

Figure 4.4 *Bubbles introduced by defective counter-gravity systems. (a) Leak in a riser tube of a low pressure die casting machine; (b) the dangerous ingestion of massive bubbles when the melt level is too low; (c) bubbles entrapped by dross and poor design features of some pumped systems, that are released erratically and thus damage castings in a non-reproducible way.*

a leaking riser tube in a low-pressure casting machine can lead to a serious violation (Figure 4.4a). The stream of bubbles from a leak in a defective riser tube will float directly up the tube and enter the casting. Unfortunately, this problem is not rare. Thus regular checks for such leakage, and the rejection of castings subjected to such consequent bubble damage, will be required.

The other major problem with conventional low-pressure delivery systems as used for light alloy casting where the melt is contained within a pressurized vessel is that the topping up of the pressure vessel itself usually damages the quality of the melt. The uncontrolled fall first from the foundry transfer ladle, then down a chute, and finally into the melt is an unsatisfactory transfer process introducing much bubble damage into the metal (Figure 2.49).

The Griffin Process counter-gravity process for the production of steel wheels for rail rolling stock makes an interesting comparison. In this case, of course, the pressurized furnace contains liquid carbon steel. The large density difference between the steel and buoyant defects such as bubbles, bubble trails and other entrained oxides encourages such materials to float out relatively quickly, so that the topping up of the furnace does not necessarily introduce permanent damage; by the time the mould is cast a good quality of steel has developed. In this way a high-integrity safety-critical product can be routinely produced. Even so, one can imagine that the deoxidation practice, leaving different amounts of Si, Mn and Al, plus others such as Ca, could influence the flotation time significantly.

For aluminium alloys, however, the near-neutral buoyancy of the introduced defects means that very few have time to float out, and of the remainder, not all are subsequently removed by the filter, if any, at the entrance to the mould cavity. Even the prior use of rotary degassing units cannot be relied on to effect a complete treatment of the melt.

In fact, in low-pressure casting units (see Figure 2.49) it is difficult to see how enclosed pressure vessels can be made to deliver liquid alloy of good quality. Much emphasis has been placed on the precise control over delivery rates and volumes for such units. The quality of the delivered melt, however, can only remain far from optimum. The use of such technology cannot be recommended at this time.

Rule 5

Avoid core blows

5.1 Background

When sand cores are surrounded by liquid metal, the heating of the sand and its binder causes large volumes of gas to be generated in the core. Normally, the core will be designed so that the gas can escape through the core prints and so be dissipated in the mould. In this way we hope that the pressure inside the core is prevented from rising to high levels. In some circumstances, however, the pressure of gas in the cores may rise to such a level, higher than the pressure in the liquid, with the result that a bubble is forced out into the melt. It is *blown* into existence. *Blow defect* is therefore a good name for this type of gas pore. Bubbles formed in this way are of large size, and so highly buoyant. They rise through the metal leaving oxidised bubble trails in their wakes.

This is, of course, another form of bubble damage as has been discussed under Rule 4. However, it is sufficiently distinct that it benefits from separate consideration.

For instance bubble damage arising from surface turbulence in the filling system is generated by the high velocities in the front end of the system (in the basin, sprue or runner). The high shear stresses in the melt ensure that the bubbles are chopped mainly into small sizes, in the range 1 to 10 mm diameter. Some of the smaller bubbles have been observed in video radiographic studies to coalesce in the gate. These coalesced bubbles float quickly, before any significant solidification has taken place, and so burst at the liquid surface and escape. Bubbles smaller than about 5 mm diameter have only a tenth of the buoyancy of the 10 mm bubbles, and cannot split the oxides that bar their escape (Figure 4.1c). If they succeed to reach the top of the casting they therefore remain trapped at a distance only a double oxide skin depth beneath the surface of the casting.

Turning now to the quite different type of bubble given off by the outgassing of a core, these bubbles are large. In irons and steels the single core blow bubble is about 13 mm diameter. In light alloys the effective bubble diameter is approximately 20 mm (Figure 6.22, *Castings 2003*). Although these large bubbles have high buoyancy, they are not produced immediately. The timing of their eruption into the melt determines the kind of defect that is formed in the casting. If, in relatively thick sections, the bubble detaches prior to any freezing, the repeated arrival of bubbles at the surface of the casting can result in repeated build-up of bubble skins, forming a puff-pastry of the multiple leaves of oxide, known as an *exfoliation defect* (Figure 5.1b). More usually, the core takes time to warm up, and takes further time to build up its internal pressure, thus allowing time for some freezing to occur. Thus by the time the bubble is finally forced into existence, it rises to sit under the frozen layer of solid (Figure 5.1a).

Once a core has blown its first bubble, additional bubbles are easily formed, since the bubble trail seems usually to remain intact, and keeps re-inflating to pass an additional bubble along its length. (The effect is interestingly similar to the re-inflating of the oxide tube with metal as described in Section 3.4.) The bubbles contain a variety of gases, including water vapour, that are aggressively oxidizing to metals such as aluminium and higher melting point metals. Bubble trails from core blows are usually particularly noteworthy for their characteristically thick and leathery double

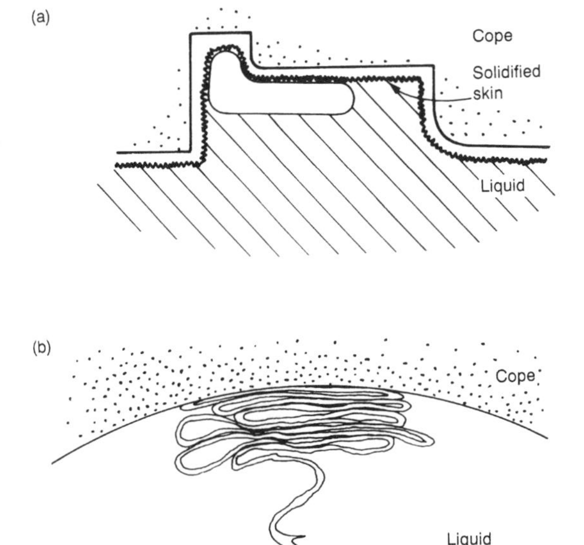

Figure 5.1 *(a) A core blow—a trapped bubble containing core gases evolved after some solidification; (b) an exfoliated dross defect produced by copious gas from a core blow prior to any solidification.*

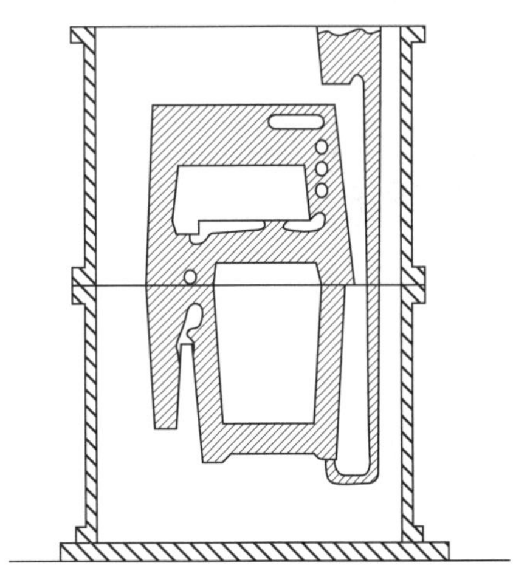

Figure 5.2 *Casting in a close-fitting steel box on an unvented flat steel plate, showing blows from an upwardly oriented feature on the lower part of the mould.*

oxide skin, built up from the passage of many bubbles. This thick skin is part of the reason why core blows result in such efficient leak defects through the upper sections of castings. The reason is that they are, of course, automatically connected from a cored volume of the casting, and often penetrate to the adjacent core (since little solidification will usually have occurred between cores to stop it). Alternatively, in thicker sections, they travel to the very top of the casting.

After the emergence of the first blow into the melt, the passage of additional bubbles contributes to the huge growth of some blow defects. Often the whole of the top of a casting can be hollow. The size of the defect can sometimes be measured in fractions of metres.

Blows can form from moulds. Whereas founders are familiar with the problem of blows from cores, blows from moulds are rarely considered. In fact this is a relatively common problem (even though this section remains entitled 'core blows' as a result of common usage). The huge volumes of gas that are generated inside the mould have to be considered. They need room to expand and flow. Any visitor to an iron or steel foundry will be impressed with the jets of flame issuing from the joints of moulding boxes. Effectively the gases and volatiles will be fighting to get out. It is prudent therefore to provide them with escape routes, since escape via the liquid metal in the mould cavity can spell disaster for the casting.

The build-up of back-pressure inside the mould cavity, leading to incompletely filled moulds, is most easily dealt with by the provision of one or more whistlers. These are narrow, pencil-shaped vents through the cope.

The escape of gases entrapped in the mould cavity is made more difficult by the application of mould coats, so that pressures can be doubled (Ohnaka 2003), making the provision of whistler vents more necessary.

The build-up of pressure can be even more severe in moulds that are enclosed in steel boxes, and which are sat on a steel plate or on a concrete floor. The gases are relatively free to escape from the cope, but gases attempting to escape from the drag are sealed in by the overlying liquid in the mould cavity (Figure 5.3). The problem is enhanced if the casting is a tight fit in the moulding box, as is usually the case of course, since the casting engineer is always trying to get as much value as possible out of each box (Figure 5.2). In fact the build up of pressure inside sand moulds crammed into tight-fitting steel boxes has, in the author's experience, contributed to a number of spectacularly defective castings, and in one instance, to a casting that persistently refused to fill its mould because the back-pressure of gases rose so high.

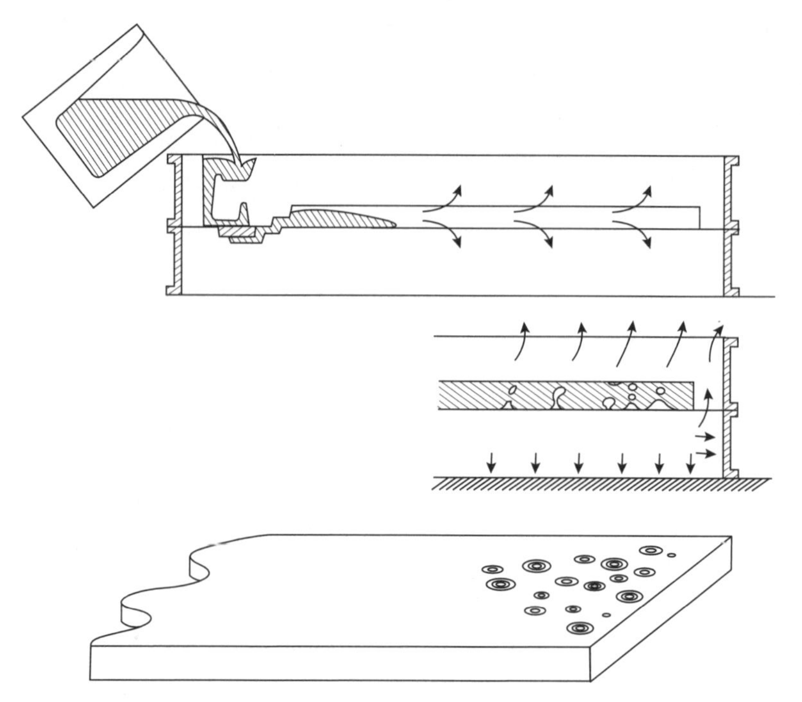

Figure 5.3 *A large flat plate casting with an enclosed drag. Volatiles are driven ahead and condense in the cooler distant mould, exacerbating defects at the far end.*

The build-up of pressure in the drag has been observed by the author to lead to severe blow defects in metre square flat plates of a bronze alloy, particularly towards the far end of the plate where the condensation of volatiles driven ahead of the melt adds to the amount already available from the sand binder (Figure 5.3). The provision of woven nylon vent tubes through the drag was quite inadequate; the enormous quantities of gas simply overwhelmed this painstaking but useless provision. The drag needed to vent from the whole of its lower surface area by standing the mould clear of the ground, or standing it on a deeply ridged base plate.

It is worth commenting on the curious but common provision of whistler vents through the top of the mould in an effort to eliminate 'gas' porosity in the casting; when the founder sees a core blow he will often apply a mould vent. Regrettably, this action is totally misguided. A moment's reflection reveals the self-evident fact that the porosity (i.e. the entrained air bubbles) is already in the metal, and the metal itself would have to rise up the vent to eliminate the porosity from the casting. The error in thinking arises because of the confusion between gas entrained in the melt, and gas entrapped in the mould cavity. When these are separated into their logically separate categories, confusion disappears, and the correct remedial action can be identified.

Although blows can be formed above flat plates as described above, it is to be noted that they form much more easily from upward pointing features of cores or moulds (Figure 5.2). The effect is the upside-down equivalent of the droplet of water detaching from the tip of a stalactite. Thus the removal of upwardly pointing features, or the inverting of the whole casting, is often a useful tactic.

Considerable volumes of water vapour are given off from clay-based core repair and mould repair pastes. This is because the clay contains water of crystallization, so that even after thoroughly drying the core repair at 100 or 200°C, the water bound in the structure of the clay remains unchanged, only being released at a high temperature, in the region of perhaps 600°C. Thus the water is released only when the clay contacts the liquid metal. This is particularly unfortunate, because the clay is composed of such fine particles that it is substantially impermeable, preventing the escape of the water into the core or mould, so that the water is forced to boil off through the metal. Repair of cores with clay-based pastes therefore generally leads automatically to blow defects. The wide use of core repair pastes illustrates that this danger is little known. The use of such materials is to be avoided unless followed by baking at a temperature that can be demonstrated to avoid the generation of blows in the melt.

The generation of blows off chills is the result of an almost identical process. When a block metal chill is placed in a bonded aggregate mould, the pouring of the metal causes a rapid outgassing of the volatiles in the aggregate/binder mixture. The volatiles, particularly water vapour, are driven ahead of the spreading liquid metal, and condense on any cold surface, such as a metal chill. When the liquid metal finally arrives and overruns the chill the condensates boil off. Since the chill is impermeable, the vapour is forced to bubble through the melt.

To demonstrate that a chill, a core, or assembly of cores, does not produce blows may require a procedure such as the removal of all or part of the cope or overlying cores, and taking a video recording of the filling of the mould. If there are any such problems, the eruption of core gases will be clearly observable, and will be seen to result in a boiling action, creating a froth of surface dross that would of course normally be entrapped inside the upper walls of the casting. A series of video recordings might be found to be necessary, showing the steady development of solutions to a core-blowing problem, and recording how individual remedies resulted in progressive elimination of the problem. The video recording requires to be retained by the foundry for inspection by the customer for the life of the component. Any change to the filling rate of the casting, or core design, or the core repair procedure, would necessitate a repeat of this exercise.

For castings with a vertical joint where a cope cannot be conveniently lifted clear to provide such a view, a special sand mould may be required to carry out the demonstration that the core assembly does not cause blows from the cores at any point. This will have to be constructed as part of the tooling to commission the casting. This will have to be seen as an investment in quality assurance.

5.2 Prevention

By far the best solution to the evolution of gases from cores is the use of a sand binder for the core that has little or no evolution of gas as the core becomes hot. This would represent a perfect solution. The best hopes here are the inorganic binders that contain no water of crystallization. However, the few binders that have so far been developed to meet this criterion are usually not satisfactory in other ways. The perfect core binder has yet to be developed!

In the meantime, one of the best actions to avoid blows from cores (or more occasionally from moulds) is to increase the permeability of the core by the use of a coarser aggregate or by the use of venting. Since the core print is usually the area where all the escaping gas has to concentrate, a simple hole through the length of the print makes a huge impact on the problem, as has been shown previously by the author (*Castings* 2003). For some aluminium alloy castings this can be a complete solution. However, of course, if the vent hole can be continued to the centre of the core this is even better. The further provision of easy escape for gases through the mould and out to the atmosphere is necessary for copper-based and iron and steel castings where the outgassing problem becomes severe because of the higher temperatures. Many readers will have seen the impressive jets of burning carbon monoxide issuing from vents in moulds of iron and steel castings for many minutes after pouring. Figure 5.4 illustrates a succession of improved venting techniques.

For low-volume production involving the making of cores by hand, a vent can be provided along a curved path through a core by laying a

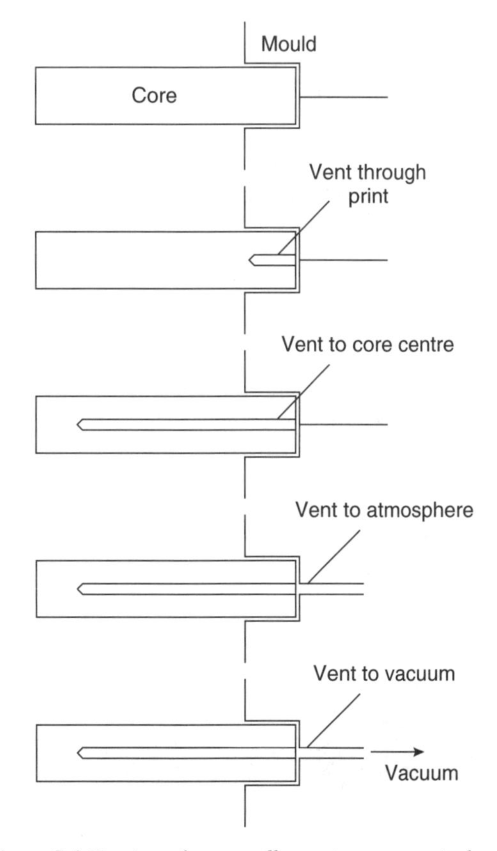

Figure 5.4 *Venting of a core, illustrating progressively improved techniques.*

waxed rope inside while it is being made. The core is subsequently heated to melt the wax so that the rope can be pulled out. The concerns that the wax itself, now percolated into the core, would add to the volatiles and so counter any benefit is, thankfully, unfounded. The provision of the vent is an overwhelming benefit.

For more delicate low-volume work, the author has witnessed long curved cores for aerospace castings being drilled by hand, using a drill bit fashioned from a length of piano wire held in a three-jaw rotary chuck driven by a small motor. The tip of the high-carbon-steel wire is hammered flat, and ground to a sharp point shaped like an arrow head. The core is drilled by hand in a series of straight lengths, the piano wire drill buzzing quickly through the core. Each hole is targeted to intersect the previous hole, the straight holes emerging on the bends, where the openings are subsequently plugged by a minute wipe of refractory cement (Figure 5.5). The complete vent is checked to ensure that it is continuous, and free from leaks, by blowing smoke in to one of the vent openings, and watching for the smoke to emerge from the far opening. Only when the smoke emerges freely at the far end, and from no other location, is the core accepted for use. It is then stored in readiness for mould assembly.

Occasionally, instead of an opening to the atmosphere, it is necessary to link the outside opening to a vacuum line. This is relatively common practice in gravity die (permanent mould) casting, increasing the efficiency of the extraction of gases from a resin-bonded sand core. However, the evolution of volatiles from the binder creates problems by condensing as sticky resins and tars in the vacuum line, so that, for long production runs, regular attention is required to avoid blockage, often dictating the timing of the withdrawal of the tooling for maintenance.

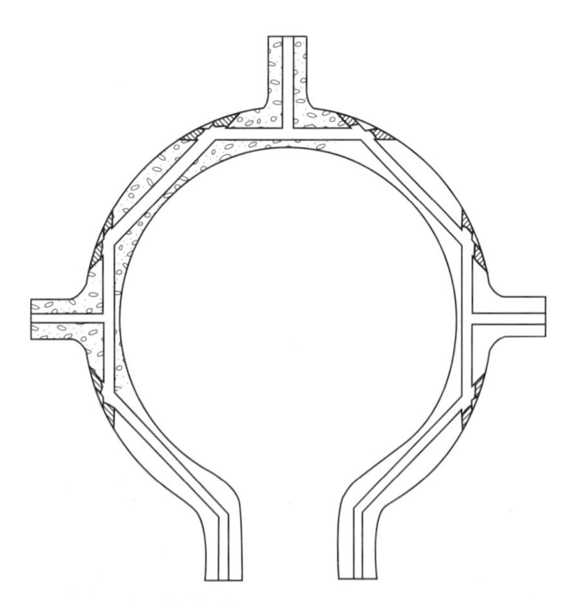

Figure 5.5 *Drilled holes to vent a narrow circular core.*

The reader is advised caution with regard to the application of a vacuum line to aid venting. The author once tried this on an extensive thin-section core with a single small area print around which was poured liquid stainless steel at 1600°C. The resulting rapid build-up of pressure was so dramatic that it blew off the vacuum connection with a bang! However, the discerning reader will notice the extreme circumstances described here, and rightly conclude that in this case the author was testing the patience of Providence.

The prevention of blows from condensation on chills is widely known and generally well applied. The chill should normally be coated with a ceramic wash or spray that is afterwards thoroughly dried to give an inert, permeable and non-wetted surface layer. The effective permeability of the surface can be further enhanced by providing deep V-grooves in a criss-cross pattern. The grooves are bridged to some degree by the action of the surface tension of the melt, so that the bottoms of the grooves act as surface vents, tunnelling the expanding vapours to freedom ahead of the advancing melt. Additionally, the V-grooves are thought to enhance the effectiveness of the chilling action by increasing the contact between the casting and the chill.

If there is an option, it is far better to arrange that the core vents through prints that are directed vertically upwards (Figure 5.6). This is because as the melt rises in the mould, the volatiles migrate through the core ahead of the metal, concentrating in the last part of the core. If the core is vented at its base this is a potential disaster. The volatiles are too far from the print, and will continue to be pushed ahead, finally being pushed into the form of an eruption of bubbles from the top of the core. This problem can be reduced by covering the core with liquid metal as quickly as possible. Venting from the base is then given its best chance.

Even so, a print allowing outgassing from the top of the core is ideal. If a vent cannot be provided up the centre of the top print, a top print is still valuable, even though it may contain no central vent, because the volatiles will travel up the core surface. This can be seen on core prints that emerge from the tops of aluminium alloy castings. The melt is seen to

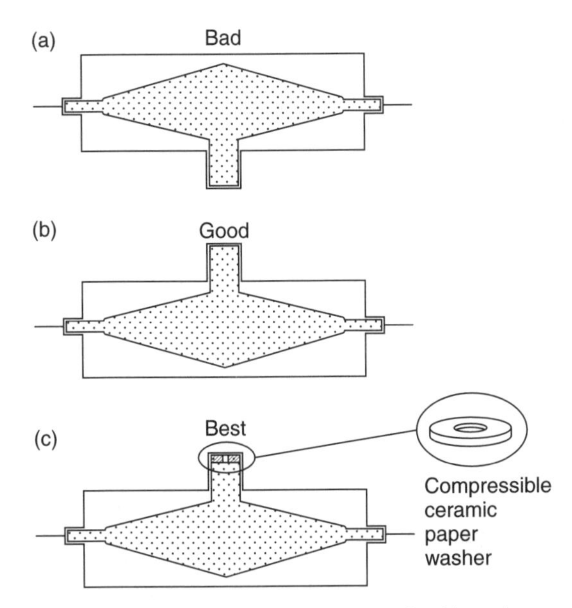

Figure 5.6 *Vertical upwards venting, preferably with a soft print, is ideal. However, the addition of a central vent hole through the core print, or even down into the centre of the core, would be even better.*

flutter, trembling against the side of the core print as gas rushes up in the form of mini waves, causing ripples to radiate out across the surface of the melt.

The provision of soft ceramic paper gaskets, preferably with a central hole, shaped like a washer, placed on the end of core prints is an excellent provision for the escape of gases. This simple remedy prevents the melt from flashing over the end of the print to block the vent (Figure 5.6). The compressible washer allows for the sealing of the core print against ingress by liquid metal, but allows closure of the mould without danger of the crushing of the core.

Finally, if the core can be covered quickly with liquid metal, and the pressure in the metal quickly raised to be at all times greater than the internal pressure generated inside the core, then bubble formation will be suppressed. Thus simply filling the mould faster is often a quick and complete solution. The provision of an additional top feeder to increase hydrostatic pressure needs care, since if the feeder has large volume the delay in the rise of pressure to fill it may be counter-productive. If feeding of the casting is not really required, the sprue and pouring basin can provide the early pressurisation that is needed, it would be better to leave well alone and not be tempted to provide a top feeder.

For those interested in quantifying some of the problems of core outgassing and the effect of sizes of vents and temperatures, etc., the previous volume '*Castings* (2003)' derives an approximate analytical formula to describe the physics of core blows. Eventually, it is hoped, we can look forward to the day when computer simulation will provide an accurate description of each core and mould, allowing in detail for the effect of intricate geometries and the complicated effect of rate of filling that are sometimes encountered. A welcome start has been made by Maeda and colleagues (2002) who demonstrate a computer simulation of the flow of gas through an aggregate core. Perhaps we can now look forward to such studies becoming a commonplace feature of the design of a new casting.

Rule 6

Avoid shrinkage damage

6.1 Feeding systems design background

Before getting launched into this section, we need to define some terms.

There is widespread confusion in parts of the casting industry, particularly in investment casting, between the concepts of *filling* and *feeding* of castings. It is essential to separate these two concepts.

> *Filling* is self-evidently the short period during the pour, and refers to the filling of the filling channels themselves and the filling of the mould cavity. This may only last seconds or minutes.
>
> *Feeding* is the long, slow process that is required during the contraction of the liquid that takes place on freezing. This process takes minutes or hours depending on the size of the casting. It is made necessary as a result of the solid occupying less volume than the liquid, so the difference has to be provided from somewhere. This contraction on solidification is a necessary consequence of the liquid being a structure resembling a random close-packed array of atoms, compared to the solid, which has denser regular close packing in a structure known as a crystal lattice.

Figure 6.1 illustrates the three separate shrinkage problems that occur during the cooling of a metal.

1. The liquid grows in density as it cools. However, this simple thermal contraction in the liquid state is not usually a significant problem because most of the superheat (the temperature above the liquidus) of a melt is usually lost during or quickly after pouring.
2. The main problem is the contraction on solidification. This is around 3% for many steels, but over 6% for Al alloys (Table 6.1). This is the contraction requiring to be fed by a feeder. Its principal action is simply that of a reservoir (there are other important functions of feeders that we shall consider later).
3. The subsequent contraction in the solid state remains a problem for the patternmaker. We

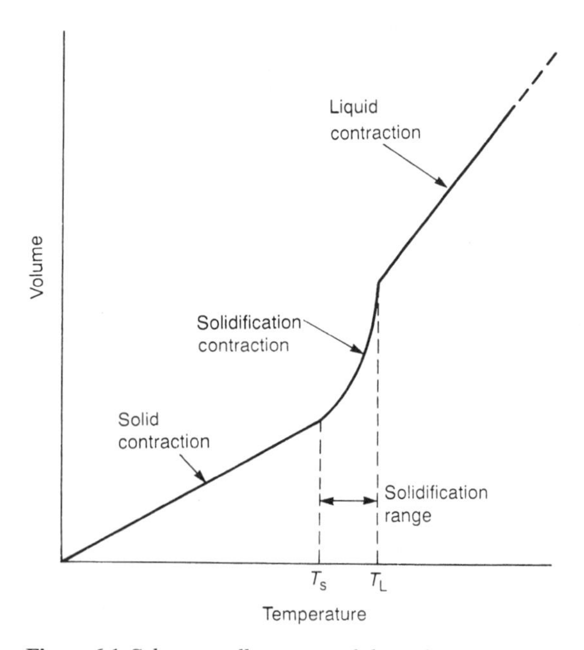

Figure 6.1 *Schematic illustration of three shrinkage regimes: in the liquid; during freezing; and in the solid.*

shall not concern ourselves with the pattern-maker's problems in this section.

We shall concentrate our attention on the main problem, the contraction on freezing as listed in 2 above. To allow ourselves the luxury of some repeated emphasis and further definitions: to provide for the fact that extra metal needs to be fed to the solidifying casting to compensate for the contraction on freezing, it is normal to provide a separate reservoir of metal. We shall call this reservoir a *feeder*, since its action is to *feed* the casting, i.e. to compensate for the solidification shrinkage (obvious really!). In much casting literature the reservoir is known, non-obviously, as a *riser*, and worse still, may be confused with other channels that communicate with the top of the mould, such as vents, or whistlers, since metal *rises* up these openings too. The author reserves the name *riser* for the special kind of feeder described in Section 2.3.2.7, that is connected to the side of the casting via a slot gate, and in which metal rises up at the same time as it rises in the mould cavity.

It is most important to be clear that the filling (sometimes called the running) system is not normally required to provide any significant feeding. The filling system and the feeding system have two quite distinct roles: one fills the casting, and the other feeds the shrinkage during solidification. (On occasions it is possible and valuable to carry out some feeding via the filling system, but this requires the special precautions that are described later.)

The main question relating to the provision of a feeder on a casting is 'Should we have a feeder at all?' This constitutes Rule 1 for feeding. This is a question well worth asking, and we shall return to it later. Just for the moment we shall assume that the answer is 'Yes'. The next question is 'How large should it be?'

There is of course an optimum size. Figure 6.2a illustrates a section of a feeder on a plate casting in which the required shrinkage volume is just nicely concentrated in the feeder. This is the success we all hope for. However, success is not always easily achieved, and Figure 6.2 b, c and d show the complication posed by the different shrinkage behaviour of different alloys. The pure Al and the Al–12Si alloy are both short freezing range, and contrast with the Al–5Mg alloy which is a long freezing range material.

Some additional points of complexity in the operation of feeders in real life need to be emphasized.

(i) The Mg-containing alloy in Figure 6.2d will almost certainly contain some fine, scattered microporosity that will have acted to reduce the apparent shrinkage cavity.
(ii) The complicated form of the pipe in Al–12Si alloy almost certainly reflects the presence of large oxide films that were introduced by the pouring of the castings. These large planar defects fragment both the heat flow and the mass flow in the feeder, and the short freezing range and surface tension conspire to round off the cavities in the separated volumes of liquid. In addition, the oxide, together with the solidifying crust on the top surface of the feeder also has some strength and rigidity, again complicating the collapse of the feeder top, and influencing the shape of the shrinkage pipe as it, and its associated oxide skin, gradually expands downwards. These effects are additional reasons for the 20 per cent safety factor often used for the calculation of feeder sizes. Feeders often do not have the simple carrot-shaped shrinkage pipe predicted by the computer. Figure 6.2e gives a further excellent example of the action of flow from a feeder diverted and fractured by the presence of large bifilms.

Table 6.1 Solidification shrinkage for some metals

Metal	*Crystal structure*	*Melting point* °*C*	*Liquid density* (kg/m^3)	*Solid density* (kg/m^3)	*Volume change* (%)	*Ref.*
Al	fcc	660	2368	2550	7.14	1
Au	fcc	1063	17 380	18 280	5.47	1
Co	fcc	1495	7750	8180	5.26	1
Cu	fcc	1083	7938	8382	5.30	1
Ni	fcc	1453	7790	8210	5.11	1
Pb	fcc	327	10 665	11 020	3.22	1
Fe	bcc	1536	7035	7265	3.16	1
Li	bcc	181	528	–	2.74	4,5
Na	bcc	97	927	–	2.6	4,5
K	bcc	64	827	–	2.54	4,5
Rb	bcc	39	1437	–	2.3	4,5
Cs	bcc	29	1854	–	2.6	4,5
Tl	bcc	303	11 200	–	2.2	2
Cd	hcp	321	7998	–	4.00	2
Mg	hcp	651	1590	1655	4.10	3
Zn	hcp	420	6577	–	4.08	2
Ce	hcp	787	6668	6646	−0.33	1
In	fct	156	7017	–	1.98	2
Sn	tetrag	232	6986	7166	2.51	1
Bi	rhomb	271	10 034	9701	−3.32	1
Sb	rhomb	631	6493	6535	0.64	1
Si	diam	1410	2525	–	−2.9	2

References: 1, Wray (1976); 2, Lucas (quoted by Wray, 1976); 3, This book; 4, Iida and Guthrie (1988); 5, Brandes (1983).

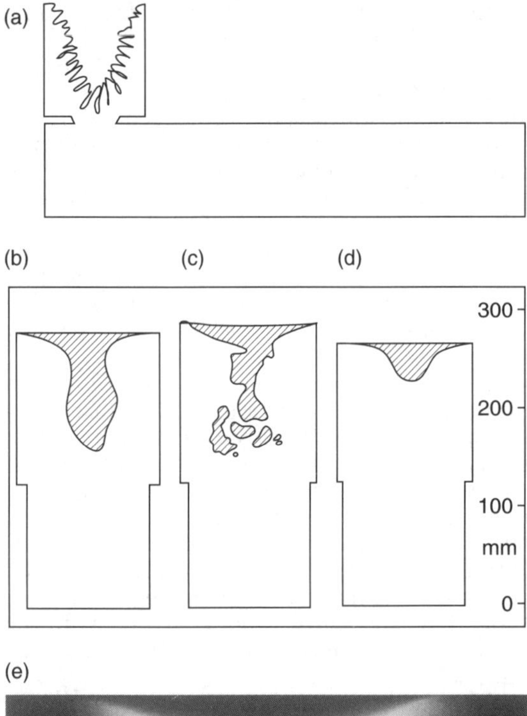

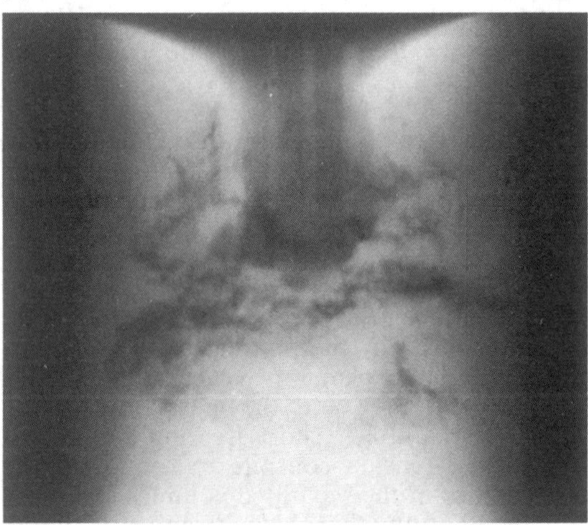

Figure 6.2 *Cross-section of (a) simple plate casting, nicely fed, with all of its shrinkage porosity concentrated in the feeder; (b) 99.5Al; (c) Al–12Si; (d) Al–5Mg; (e) radiograph of Al–12Si alloy feeder (courtesy Foseco 1988).*

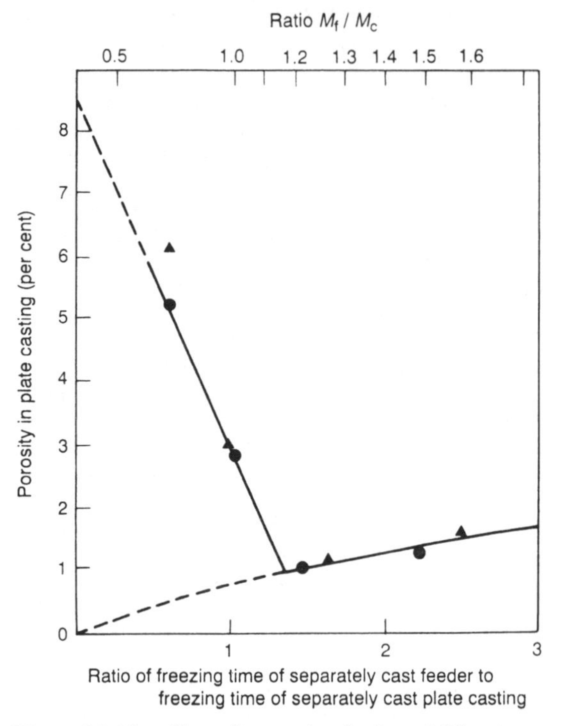

Figure 6.3 *The effect of increasing feeder solidification time on the soundness of a plate casting in Al–12Si alloy. Data from Rao* et al. (1975).

Figure 6.3 shows the results of Rao *et al.* (1975), who investigated the feeding of a simple plate casting in Al–12Si alloy by planting on successively larger feeders. Interestingly, when the data are extrapolated backwards to zero feeder size the porosity is indicated to be approximately 8 per cent, which is close to the theoretical 7.14 per cent solidification shrinkage for pure aluminium (Table 6.1) and may indicate that 1 per cent or so of thermal contraction due to superheat may have contributed to the total shrinkage contraction. At a feeder modulus of around 1.2 times the modulus of the casting, the casting is at its most sound. The residual 1 per cent porosity is probably dispersed gas porosity (i.e. gas precipitated into dispersed microscopic bifilms so as to open them). As the feeder size is increased further the solidification of the casting is now progressively delayed by the nearby mass of metal in the feeder. Thus while this excessive feeder is no disadvantage in itself, the delay to solidification of the whole casting increases the time available for further precipitation of hydrogen as gas porosity. However, it is clear from this work that an undersized feeder will result in very serious porosity. In contrast, an oversize feeder causes less of a problem, increasing porosity slightly by the opening of bifilms and thereby reducing mechanical properties. In addition, of course, the oversize feeder does adversely influence the freezing time (important for cycle time in permanent moulds) and reduces the metallic yield thus adversely influencing the economics!

Figure 6.4 generalizes the finding of Rao and colleagues, to show the expected relation between gas and shrinkage porosity. Clearly, as the feeder size is increased, the optimum feeder

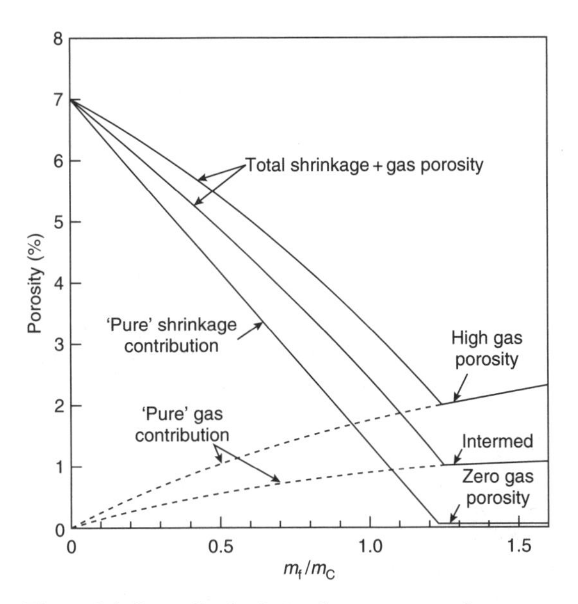

Figure 6.4 *Generalized relation between gas and shrinkage as feeder size is increased in terms of the modulus ratio.*

size is hardly changed by the amount of gas in solution. However, as the gas content increases, the minimum level of porosity that can be achieved steadily rises, although never usually exceeds 1 or 2 per cent, compared to the 7 or 8 per cent contribution of shrinkage. Clearly, it is more important to deal with the shrinkage problems than with gas problems in castings. (This conclusion might raise the eyebrows of practised foundry people. It needs to be kept in mind that most of what previously has been generally described as *'gas'* in castings, has actually been entrained *air bubbles* as a result of our poor filling systems.)

Where computer modelling is not carried out, the following of the *seven feeding rules* by the author is strongly recommended. Even when computer simulation is available, the seven rules will be found to be good guidelines. For the computer itself, following Tiryakioglu's reduced rules constitutes the most powerful logic and is recommended, although the same rules also constitute a useful check for those who determine feeder sizes by pen and paper.

In addition to observing all the requirements of the Rules for feeding, the use of all *five mechanisms* for feeding (as opposed to only liquid feeding) should also be used to advantage. This will be found to be especially useful when attempting to achieve soundness in an isolated boss or heavy section where the provision of feed metal by conventional techniques may be impossible. However, a reminder of the attendant dangers of the use of solid feeding are presented later.

6.1.1 Gravity feeding

As opposed to *filling uphill* (which is of course quite correct) *feeding* should only be carried out *downhill* (using the assistance of gravity).

Attempts to feed uphill, although possible in principle, can be unreliable in practice, and may lead to randomly occurring defects that have all the appearance of shrinkage porosity. In castings of modest size feeding uphill appears to be successful as will be discussed in Section 6.4 'Active feeding'. In many castings, particularly larger castings, problems occur when attempting to feed uphill because of the difficulties caused by two main effects: (i) adverse pressure gradient as discussed below; and (ii) adverse density gradient leading to convection as dealt with in Chapter 7.

The atmosphere is capable of holding up several metres of head of metal. For liquid mercury the height is approximately 760 mm, being the height of the old-fashioned atmospheric barometer of course. Equivalent heights for other liquid metals are easily estimated allowing for the density difference. Thus for liquid aluminium of specific gravity 2.4 compared to liquid mercury of 13.9, the atmosphere will hold up about $(13.9/2.4) \times 0.76 = 4$ m of liquid aluminium.

While no pore exists, the tensile strength of the liquid will in fact allow the metal, in principle, to feed to heights of kilometres, since in the absence of defects the liquid can withstand tensile stresses of thousands of atmospheres. The liquid can, in principle, hang up in a tube, its great weight stretching its length somewhat. However, the random initiation of a single minute pore will instantly cause the liquid to 'fracture', causing such feeding to stop and go into reverse. The liquid in the tube will fall, finally stabilizing at the level at which atmospheric pressure can support the liquid. Thus any height above that supportable by one atmosphere is clearly at high risk.

Moreover, there is even worse risk of attempting to feed against gravity. If there is a leak path to atmosphere, allowing atmospheric pressure to be applied in the liquid metal inside the solidifying casting, the melt will then fall further, the action of gravity tending to equalize levels in the mould and feeder. Thus, if the feeder is sited below the casting, the casting will completely empty of residual liquid. Regrettably, this is an efficient way to cast porous castings and sound feeders.

Clearly, the initiation of a leak path to atmosphere (via a double oxide film, or via a liquid region in contact with the surface at a hot spot) is rather easy in many castings, making the whole principle of uphill feeding so risky that it should not be attempted in circumstances where porosity cannot be tolerated. It is a pity that the comforting theories of pushing liquid uphill by atmospheric pressure or even hanging it from vast heights using the huge tensile strength of the liquid cannot be relied upon in practice.

For most purposes, the only really reliable way to feed is *downhill*, using gravity.

6.1.2 Computer modelling of feeding

Good computer models have demonstrated their usefulness in being able to predict shrinkage porosity with accuracy. A simulation using a reliable modelling software package should now be specified as a prior requirement to be carried out before work is started on making the tooling for a new casting. This minor delay will have considerable benefits in shortening the overall development time of a new casting, and will greatly increase the chance of being 'right first time'.

However, at this time many computer simulations are inadequate for other reasons that require to be recognized. For instance these include:

(i) no allowance for the effect of thermal conduction in the cast metal (rare);
(ii) no allowance for the important effects due to convection in the liquid (common);
(iii) neglect of, or only crude allowance for, the effect of the heating of the mould by the flow of metal during filling (rare);
(iv) no capability of any design input. Thus gating and feeding designs will be required as inputs (universal at this time).

For the future, it is to be expected that software packages will evolve to provide intelligent solutions to all these requirements. Examples of a good start in this direction are shown by Dantzig and co-workers (Morthland *et al.* 1995 and McDavid and Dantzig 1998). In the meantime, it remains necessary to use computer models with some discretion. For instance in the work by the Morthland team they warn that the results are specific to the feeding criterion used. If a more stringent temperature gradient criterion were used (for instance $2\,K\,cm^{-1}$ instead of $1\,K\,cm^{-1}$) the feeder would have been larger.

The above approaches to the optimization of the feeding requirements of castings have involved the use of numerical techniques such as finite element and finite difference methods. Ransing *et al.* (2003) propose a geometrical method based on an elegant extension of the Heuvers circle technique. This technique is described later in the section describing the feed path requirements for feeding (Section 6.2 Feeding Rule 5).

6.1.3 Random perturbations to feeding patterns

In aluminium castings, flash of approximately 1 mm thickness and only 10 mm wide has been demonstrated to have a powerful effect on the cooling of local thin sections up to 10 mm thick, speeding up local solidification rates by up to ten times (see Section 6.5.3). The effect is much less in ferrous castings because of their much lower thermal conductivity.

Thus for high conductivity alloys flash has to be controlled, or used deliberately, since, in moderately thick sections, it has the potential to cut off feeding to more distant sections. The erratic appearance of flash in a production run may therefore introduce uncertainty in the reproducibility of feeding, and the consequent variability of the soundness of the casting. Flash on thick sections is usually less serious because convection in the liquid in thick sections conveys the local cooled metal away, effectively spreading the cooling effect over other parts of the casting, giving an averaging effect over large areas of the casting. In general however, it is desirable that these uncertainties are reduced by good control over mould and core dimensions.

The other known major variable affecting casting soundness in sand and investment castings is the ability of the mould to resist deformation. This effect is well established in the case of cast irons, where high mould rigidity is a condition for soundness. However, there is evidence that such a problem exists in castings of copper-based alloys and steels. A standard system such as statistical process control (SPC), or other techniques, should be seen to be in place to monitor and facilitate control of such changes. Permanent moulds such as metal or graphite dies are relatively free from such problems. Similarly many other aggregate moulding materials are available that possess much lower thermal expansion rates, and so produce castings of greater accuracy and reproducibility. Many of these are little, if any, more expensive than silica sand. A move away from silica sand is already under way in the industry, and is strongly recommended.

The solidification pattern of castings produced from permanent moulds such as gravity dies and low pressure dies may be considerably affected by the thickness and type of the die coat which is applied. A system to monitor and control such thickness on an SPC system should be seen to be in place.

For some permanent moulds, pressure die-casting and some types of squeeze casting the feeding pattern is particularly sensitive to mould cooling. After the development and acceptance of the casting, any further changes to cooling channels in the die, or to the cooling spray during die opening, will have to be checked to ensure that corresponding deleterious changes have not been imposed on the casting. The quality of the water used for cooling also requires to be seen to be under good control if deposits inside the system are not to be allowed to build up and so cause changes in the effectiveness of the cooling system with time.

6.1.4 Dangers of solid feeding

It is often possible to make a casting without feeders despite a large feeding demand. Because, in favourable conditions, the casting can collapse plastically, the shrinkage volume is merely transferred from the inside to the outside of the casting. Here, if the volume is distributed nicely, the shrinkage will cause only a negligible and probably undetectable reduction in the size or shape of the casting.

If the outside shrinkage is not distributed so favourably, but remains concentrated in a local region, a surface sink is the result.

When operating without feeders, a second possibility is the formation of shrinkage pores, grown from initiation sites (almost certainly bifilms), so that solid feeding immediately fails. It seems that such events tend to be triggered by rather large, rather open bifilms, whose size might be measured in centimetres.

A further possibility of much smaller bifilms will be common, but not easily perceived. If the melt has a distribution of small, possibly microscopic, bifilms, these will be unfurled to some extent by the reduced pressure in the unfed region thus being converted from crumpled compact features of negligible size to flat thin extensive cracks. Thus although the casting may continue to appear perfectly sound in the unfed region, and solid feeding declared to be a complete success, the mechanical properties of this part of the casting will be reduced. In particular, although the yield strength of the region will be hardly affected, that part of the casting will exhibit reduced strength and ductility.

If the localized shrinkage problems are even more severe, the distribution of small bifilms will develop further. After unfurling to become flat cracks, additional reduction of pressure in the liquid will open them further to become visible microporosity. The pores may even grow to such a size that they become visible on radiographs.

Thus in view of the action of a feeder to pressurize the melt and so help to resist the unfurling and even the inflating of bifilms, the bifilms, are still present, but simply remain out of sight. Using a domestic analogy from home decoration, there is a very real sense in which adding feeders to castings is almost literally 'papering over the cracks'.

6.1.5 The non-feeding roles of feeders

Feeders are sometimes important in other ways than merely providing a reservoir to feed the solidification shrinkage during freezing.

We have already touched on the effect that feeders can have on the metallurgical quality of cast metal by helping to restrain the unfurling and opening of bifilms by maintaining a pressure on the melt. This action of the feeder to pressurize the casting therefore helps to maintain mechanical properties, particularly ductility.

A further key role of many feeders, however, is merely as a flow-off or kind of dump. Many filling system designs are so poor that the first metal entering the mould arrives in a highly damaged condition. The presence of a generous feeder allows some of this metal to be floated out of the casting. This role is expected to be hindered, however, in highly cored castings where the bifilms will tend to attach to cores in their journey through the mould.

In general, experience with the elimination of feeders from Al alloy castings has resulted in the casting 'tearing itself apart'. This is a clear sign of the poor quality of metal probably resulting mainly from the action of the poor running system. The inference is that the casting is full of serious bifilm cracks. These remain closed, and so invisible, while the feeder acts to pressurize the metal. If the pressurization from the feeder is removed the bifilms will be allowed to open, becoming visible as cracks. This phenomenon has been seen repeatedly in X-ray video radiography of freezing castings. It is observed that good filling systems do not lead to the casting tearing itself apart, even though the absence of a feeder has created severe shrinkage conditions. In this situation the casting shrinks a little more (under the action of solid feeding) to accommodate the volume difference.

The action of the feeder to pressurize the melt during solidification is useful in further ways. Both summarizing and thinking further we have:

(i) As we have seen, pressurization raises mechanical properties, particularly ductility.

(ii) Pressurization together with some feeding helps to maintain the dimensions of the casting. Although the changes in dimensions by solid feeding are usually small, and can often be neglected, on occasions the changes may be outside the dimensional tolerance. A feeder to ensure the provision of liquid metal under some modest pressure is then required.

(iii) Pressurization can delay or completely prevent blow defects from cores.

In summary, providing the filling system design is good so as to avoid creating large bifilms, and provided the solidification rate is sufficiently fast to retain the inherited population of bifilms compact, castings that do not require feeders for feeding should not be provided with feeders.

6.2 The seven feeding rules

Although the conditions for feeding were originally listed as six rules (*Castings 1991*), at that time the basic first rule was implicitly assumed. Only since has it been recognized as having sufficient importance to be listed as a separate Rule, bringing the total to seven. The originally overlooked first rule is '*Do not feed (unless necessary)*'.

The great literature on the feeding of castings is mainly concerned with two feeding rules: The first is *The feeder must solidify at the same time as, or later than, the casting*. This is Chvorinov's heat-transfer criterion.

The second and widely understood and well-used Rule, usually known as the volume criterion is as follows: *The feeder must contain sufficient liquid to meet the volume contraction requirements of the casting.*

However, there are additional rules that are also often overlooked, but which define additional thermal, geometrical and pressure criteria that are absolutely necessary conditions for the casting to freeze soundly.

The junction between the casting and the feeder should not create a hot spot, i.e. have a freezing time greater than either the feeder or the casting. This is a problem, which, if not avoided, leads to 'underfeeder shrinkage porosity'. The junction problem is a widely overlooked requirement. It often overrides the Chvorinov requirement, making the feeder size calculated by the condition stipulated by the Great Master to be insufficient.

There must be a feed path to allow feed metal to reach those regions that require it. This communication criterion appears so self-evident it is understandable why this criterion has been often overlooked as part of the overall logic. Nevertheless it does have a number of geometrical implications which are not self-evident, and which will be discussed.

There must be sufficient pressure differential to cause the feed material to flow, and the flow needs to be in the correct direction (obvious when spelled out!).

There must be sufficient pressure at all points in the casting to maintain the dimensional accuracy of the casting and to suppress the formation and growth of cavities. The reduction of the rate of unfurling of bifilms is also an important and largely unrecognized role of the feeder, being a largely invisible contribution to the mechanical properties of the casting alloy.

It is essential to understand that all of the above criteria must be fulfilled if castings are to be produced that require soundness, accuracy and high mechanical properties. The reader must not underestimate the scale of this problem. The breaking of only one of the rules may result in ineffective feeding, and a defective casting. The wide prevalence of porosity in castings is a sobering reminder that solutions are often not straightforward. Because the calculation of the optimum feeder size is therefore so fraught with complications, is dangerous if calculated wrongly, costs money to cast on, and more money to cut off, the casting engineer is strongly recommended to consider whether a feeder is really necessary at all. This is our first question. You can see how valuable it is to ask this. We shall start with this rule.

Feeding Rule 1: *Do not feed (unless necessary)*

Rule 1 is perfectly applicable to most thin-walled castings. In fact the addition of a feeder to a thin-walled casting will often impair the casting, causing misruns as a result of the feeder filling preferentially to the casting or simply delaying the filling and pressurization of the casting itself. This rule was mentioned in *Castings 1991* but was not given the status of a Rule. This was an oversight. It is probably the most important rule of all. For instance if a feeder is incorrectly sized, violating any one of the subsequent Rules, the consequences are so serious in some cases that it is likely that the casting would have been better with no feeder at all.

Probably 50 per cent of small and medium-sized castings do not need to be fed. This is especially true as modern castings are being designed with progressively thinner walls. In fact, as we have already mentioned, the siting of heavy feeders on the top of thin-walled castings is positively unhelpful for the filling of the casting, since the slow filling of the feeder delays the filling of the thin sections at the top of the casting, with consequent misruns.

As a general rule, therefore, it is best to avoid the placing of feeders on thin-walled castings. The low feed requirement of thin walls can be partly understood by assuming that of the total 7 per cent solidification shrinkage in an aluminium casting, 6 per cent is easily provided along the relatively open pathway through the growing dendrites. Only about the last 1 per cent of the volume deficit is difficult to provide. Thus if this final percentage of contraction on freezing has to be provided by solid feeding, moving the walls of the casting inwards, this becomes, at worst, 0.5 per cent per face, which on a 4 mm thick wall is only 20 μm. This small movement is effectively unmeasurable since it is less than the surface roughness. If this deficit does appear as internal porosity then it is in any case rather limited, and normally of little consequence in commercial castings. (It may require some attention in castings for safety-critical and aerospace applications.)

The other feature of thin-walled castings is that considerable solidification will often take place during pouring. Thus the casting is effectively being fed via the filling system. The extent to which this occurs will, of course, vary considerably with section thickness and pouring rate. If the section thickness (or rather, modulus; see below) of the filling system is similar to that of the casting, then feeding via the filling system might be a valuable simplification and cost saving. This important and welcome benefit to cost reduction is strongly recommended.

Of the remainder of castings that do suffer some feeding demand, many could avoid the use of a feeder by the judicious application of chills or cooling fins. The general faster freezing of the casting might then allow the provision of sufficient residual feeding via the filling system as indicated above. Minor revisions, opening up restrictions to the feed path along the length of the filling system may provide valuable (and effectively 'free') feeding from the pouring basin.

However, this still leaves a reasonable number of castings that have heavy sections, isolated heavy bosses, or other features which cannot easily be chilled and thus need to be fed. The remainder of this section is devoted to getting these castings right.

Feeding Rule 2:
The heat-transfer requirement

The heat-transfer requirement for successful feeding can be stated as follows: the freezing time of the feeder must be at least as long as the freezing time of the casting.

Nowadays this problem can be solved by computer simulation of solidification of the casting. Nevertheless it is useful for the reader to have a good understanding of the physics of feeding, so that computer predictions can be checked, since many computer simulations are not especially accurate at the present time, and much of the basic input data are not well known. Also, of course, computer time could be usefully avoided in sufficiently simple cases. In this chapter we shall concentrate on approaches which do not require a computer.

We have seen in Chapter 4 that the freezing time of any solidifying body is approximately controlled by its ratio (volume)/(cooling surface area), known as its modulus, m. Thus the problem of ensuring that the feeder has a longer solidification time than that of the casting is simply to ensure that the feeder modulus m_f is larger than the casting modulus m_c. To allow a factor of safety, particularly in view of the potential for errors of nearly 20 per cent when converting from modulus to freezing time, it is normal to increase the freezing time of the feeder by 20 per cent, i.e. by a factor of 1.2. Thus the heat-transfer condition becomes simply

$$m_f > 1.2\, m_c \qquad \textbf{(6.1)}$$

It is important to notice that the modulus has dimensions of length. Using SI units it is appropriate to use millimetres. (Take care to note that in French literature the normal units are centimetres, and in the USA at the present time, a confusing mixture of millimetres, centimetres and inches, to the despair of all those promoters of the welcome logic of the units of the *Systeme International*. It is essential therefore to quote the length units in which you are working.)

The modulus of a feeder can be artificially increased by the use of an insulating or exothermic sleeve. It can be further increased by an insulating or exothermic powder applied to its open top surface after casting. Recent developments in such exothermic additions have attempted to ensure that after the exothermic reaction is over, the spent exothermic material continues in place as a reasonable thermal insulator. These products are constantly being further developed, so the manufacturer's catalogue should be consulted when working out minimum feeder sizes when using such aids. However, as a guide as to what can be achieved at the present time, a cylindrical feeder in an insulating material is only $0.63D$ in diameter compared with the diameter D in sand. This particular insulated feeder therefore has only 40 per cent of the volume of the sand feeder. Useful savings can therefore be made, but have, of course, to be weighed against the cost of the insulating sleeve and the organizational effort to purchase, store, and schedule it, etc. However, a further benefit that is easily overlooked from the use of a more compact feeder is the faster pressurization of thin sections that may aid filling, and so reduce losses due to occasional incomplete filling of mould cavities, and the faster pressurization of cores to reduce the chance of blows.

When working out the modulus of the casting it is necessary to consider which parts are in good thermal communication. These regions should then be treated as a whole, characterized by a single modulus value. Parts of the casting that are not in good thermal communication can be treated as separate castings. For instance, castings of high thermal conductivity such as those of aluminium- and copper-based alloys can nearly always be treated as a whole, since when extensive thin sections cool attached thicker sections and bosses, the thin sections act as cooling fins for the thicker sections. Conversely, of course, the thick sections help to maintain the temperature of thinner sections. The effect of thin sections acting as cooling fins extends for up to approximately ten times the thickness of the thin section.

However, for castings of low thermal conductivity materials such as steel and nickel-based alloys (and surprisingly, the copper-based Al-bronze), practically every part of the casting can be treated as separate from every other. Thus a complex product can be dealt with as an assembly of primitive shapes: plates, cubes, cylinders etc. (making allowance, of course, for their common mating faces, which do not count as cooling area in the modulus estimate).

Table 6.2 lists some common primitive shapes. Familiarization with these will greatly assist the estimation of appropriate feeder modulus requirements.

Feeder Rule 3: *Mass-transfer (volume) requirement*

At first sight it may seem surprising that when Requirement 2 is satisfied then the volume requirement is not automatically satisfied also. However, this is definitely not the case. Although we may have provided a feeder of such a size that it would theoretically contain liquid until after the casting is solid, in fact it may still be too small to deliver the volume of feed liquid that the casting demands. Thus it will be prematurely sucked dry, and the resulting shrinkage cavity will extend into the casting.

Figure 6.5 illustrates that normal feeders are relatively inefficient in the amount of feed metal that they are able to provide. This is because they are themselves freezing at the same time as the casting, depleting the liquid reserves of the reservoir. Effectively, the feeder has to feed both itself and the casting. We can allow for this in the following way. If we denote the efficiency e of the feeder as the ratio (volume of available feed metal)/(volume of feeder, V_f,) then the volume of feed metal is, of course, eV_f. Since the liquid contracts by an amount α during freezing, then the feed demand from both the feeder and casting together is $\alpha(Vf + Vc)$, and hence:

$$eV_f = \alpha(V_f + V_c) \qquad \textbf{(6.2)}$$

or:

$$V_f = (e - \alpha)V_c \qquad \textbf{(6.3)}$$

For aluminium where $\alpha = 7$ per cent approximately (see Table 6.1 for values of α for other metals), and for a normal cylindrical feeder of $H = 1.5D$ where $e = 14$ per cent, we find:

$$V_f = V_c \qquad \textbf{(6.4)}$$

i.e. there is as much metal in the feeder as in the casting! This is partly why the yield (measured as the weight of metal going into a foundry divided by the weight of good castings delivered) in most aluminium foundries is rarely above 50 per cent. In fact, yields of 45 per cent are common. Metal in the running system, and scrap allowance will reduce the overall yield of good castings even further. The economic benefits of higher-yield casting processes such as counter-gravity casting, in which metallic yields of 80 to 90 per cent are common, appear compellingly attractive, especially for high-volume foundries.

For steels the value of α lies between 3 and 4 per cent, depending on whether solidification is to the body-centred-cubic or face-centred-cubic structures. For pure Fe–C steels the fcc structure applies above 0.1 per cent carbon where the melt solidifies to austenite. For $\alpha = 4$ and $e = 14$ per cent, Equation 6.3 gives:

$$V_f = 0.40V_c$$

Table 6.2 Moduli of some common shapes

Shape	*H/D*	*Modulus*			
		100% Cooled area		*Base uncooled*	
Sphere		$\frac{D}{6}$	$0.167D$	–	–
Cube		$\frac{D}{6}$	$0.167D$	$\frac{D}{5}$	$0.200D$
Cylinder	1.0	$\frac{D}{6}$	$0.167D$	$\frac{D}{5}$	$0.200D$
	1.5	$\frac{3D}{16}$	$0.188D$	$\frac{3D}{14}$	$0.214D$
	2.0	$\frac{D}{5}$	$0.200D$	$\frac{2D}{9}$	$0.222D$
Infinite cylinder	∞	$\frac{D}{4}$	$0.250D$	–	–
Infinite plate		$\frac{D}{2}$	$0.500D$	–	–

and for steel that freezes to the bcc structure (delta ferrite) with $\alpha = 3$, and using a feeder of 14 per cent efficiency we have:

$$V_f = 0.27V_c$$

Thus, compared to Al alloys, the smaller solidification shrinkage of ferrous metals reduces the volume requirement of the feeder considerably. For graphitic cast irons the value reduces even further of course, becoming approximately zero in the region of 3.6 to 4.0 per cent carbon equivalent. Curiously, a feeder may still be required because of the difference in timing between feed demand and graphite expansion, as will be described later.

The interesting reverse tapered feeder (Figure 6.5c) has been promoted for many years (Heine 1982, Creese and Xia 1991) and is currently widely used for ductile iron castings.

Even so, the reader needs to be aware that in the opinion of the author, Figure 6.5 may not be as accurate as we would like. At this time, the extent of the uncertainties is not known following the recent work of Sun and Campbell (2003). This investigation of the effect of positive and negative tapers on the efficiencies of feeders, found that the reverse tapered feeder

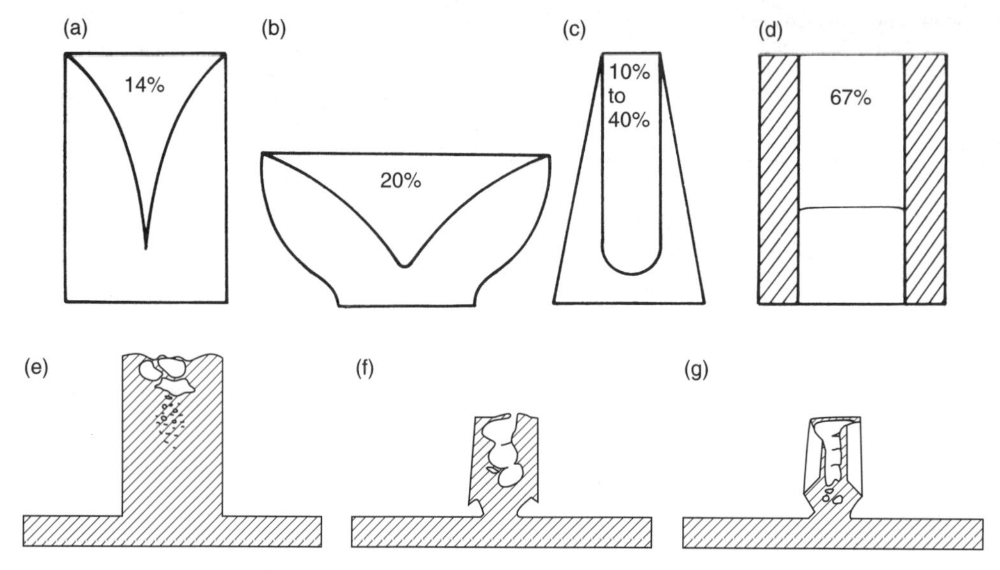

Figure 6.5 *Metal utilization of feeders of various forms moulded in sand. The (a) cylindrical and (b) hemispherical heads have been treated with normal feeding compounds; (c) efficiency of the reverse taper heads depends on detailed geometry (Heine, 1982, 1983); (d) exothermic sleeve (Beeley, 1972). Metal utilization for ductile iron plates with (e) cylindrical sand feeder; (f) insulating feeder; and (g) cruciform exothermic feeder (after Foseco 1988).*

(Figure 6.5c) appeared to be less efficient than parallel sided cylindrical feeders, or even feeders with a slight positive taper. These doubts are an unwelcome sign of the extent of our ignorance of the best feeder designs at this time.

Whether the size of the feeder is dictated by the thermal or volume requirement is related to the geometry of the casting. Figure 6.6 shows a theoretical example, calculated neglecting non-cooling interfaces for simplicity. Curve A is the minimum feeder volume needed to satisfy the thermal condition $m_f = 1.2m_c$; and curve B is the minimum feeder volume needed to satisfy the feed demand criterion based on 4 per cent volume shrinkage and 14 per cent metal utilization from the feeder. Figure 6.6 reveals that chunky steel plates up to an aspect ratio of about 6 or 7 length to thickness are properly fed by a feeder dictated by freezing time requirements. However, thin section steel plates above this critical ratio always freeze first, and so require a feeder size dictated by volume requirements.

In fact the shape of the shrinkage pipe in the feeder is likely to be different for each of these conditions. For instance, the feeder efficiencies shown in Figure 6.5 are appropriate for the feeding of chunky castings because the continuing demand of feed metal from the casting until the feeder itself is almost solid naturally creates a long, tapering shrinkage pipe, resembling a carrot.

In the case of the more rapid solidification of thin castings, the relatively large diameter feeder needed to provide the volume requirement will give a shallowly dished shape in the top of the feeder, since the feed metal is provided early, before the feeder itself has solidified to any great extent. The efficiency of utilization of the feeder will therefore be expected to be significantly higher, as confirmed by Figure 6.6.

Research is needed to clarify this point. In the meantime, the casting engineer needs to treat the present data with caution, and conclusions from Figures 6.5 and 6.6, for instance, have to be viewed as illustrative of general principles rather than numerically accurate. Clearly, it is desirable to achieve smaller, more cost-effective feeders. The change of feeder efficiency depending on whether freezing or volume requirements are operating requires more work to clarify this uncertainty. In the meantime, this problem illustrates the power of a good computer simulation to avoid the necessity for simplifying assumptions.

A further use of feeders where the casting engineer requires care is the use of blind feeders sited low down on the casting. The problems are compounded if such low-sited blind feeders are used together with open feeders placed higher. It must be remembered that during the early stages of freezing the top feeder is supplying metal to the blind feeders as well as the casting. The blind feeders have to be treated as though they are an

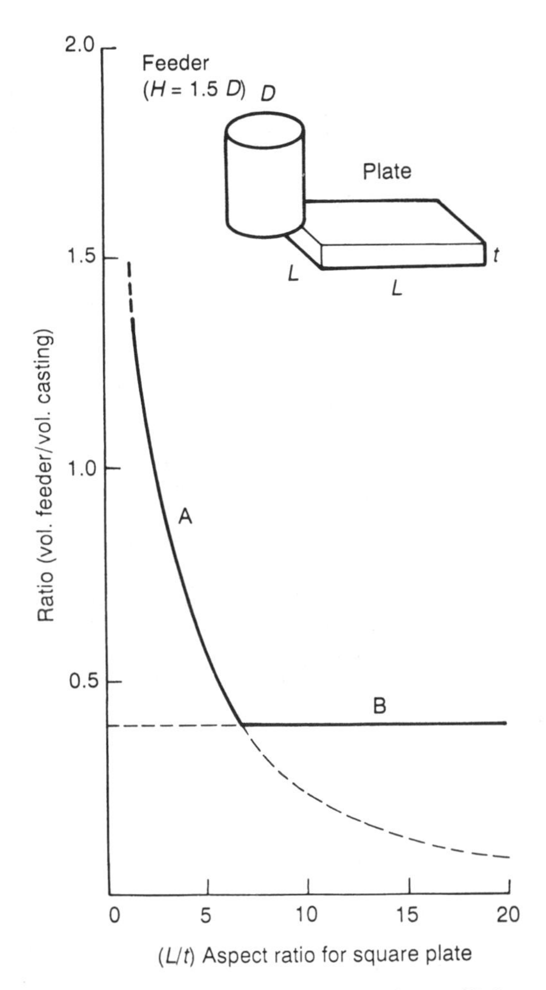

Figure 6.6 *Feeder volume based on a feeder moulded in sand, and calculated neglecting non-cooling interfaces for simplicity. Curve A is the minimum feeder volume to satisfy* $m_f = 1.2\, m_c$*; and curve B is the minimum volume to satisfy the feed demand of 4 per cent volume shrinkage and 14 per cent utilization of the feeder.*

integral part of the casting. The size of the top feeder needs to be enlarged accordingly. The blind feeders only start to operate independently when the feed path from the top feeder freezes off. This point occurs when the solidification front has progressed a distance $d/2$, where d is the thickness of the thickest casting section between the top and the blind feeders. Thus the volume of the blind feeder is now reduced by the $d/2$ thickness layer of solid that has already frozen around its inner walls.

If this caution were not already enough, a further pitfall is that the thickness of solid shell inside the blind feeder may now exceed the length of the atmospheric vent core, creeping over its end and sealing it from the atmosphere. It is therefore prevented from breathing, and is unable to provide any feed metal.

There are therefore subtleties in the operation of blind feeders that make success illusive. It is easy to make a mistake in their application, and the correct operation of the atmospheric vent is not always guaranteed, so it is difficult to recommend their use on smaller castings. For larger castings, where the feeder size is large, the collapse of the top of blind feeders is more predictable, and they become more reliable.

Whereas the size of feeders for alloys such as those based on alloys that shrink in a conventional fashion on freezing are straightforward to understand and work with, graphitic cast irons are considerably more complicated in their behaviour. They are therefore more complicated to feed, and the estimate of feeder sizes subject to more uncertainty.

The amount of graphite that is precipitated depends strongly on factors that are not easy to control, particularly the efficiency of inoculation. In addition, the expansion of the graphite can lead to an expansion of the casting if the mould and its container are not rigid. This leads to a larger volume of casting that requires to be fed, with the danger that the feeder is now insufficient to provide this additional volume. Shrinkage porosity as a result of mould dilation is a common feature of iron castings. One of the ways to reduce this problem is to use very dense, rigid moulds in rigid, well-engineered boxes. Furthermore, the expansion of the graphite can be accommodated without swelling the casting by allowing the residual melt to exude out of the casting and into the feeder. The provision of a small feeder is therefore essential to the production of many geometries of small iron castings, even though subsequent examination of the feeder indicates, mysteriously, that no liquid has been provided by the feeder.

Ductile cast irons commonly use a reverse-tapered feeder such as that shown in Figure 6.5c. If the feeder remained full for too long, its top would freeze over, preventing the delivery of any liquid (recall the coffee-cup experiment Figure 6.10). It is logical therefore to encourage the feeder to start feeding almost immediately, concentrating the action of shrinkage throughout the casting all on the small area at the top of the feeder. In this way the level of liquid in the top of the feeder falls quickly, becoming surrounded by hot metal, so very soon there is no danger of it freezing over. For this to happen it is essential that no feeding continues to be provided via the running system. This would keep the feeder full for too long. Thus it is necessary to design the ingate to freeze quickly. The feeder then works well.

This feeding technique, although at this time used exclusively in the ductile iron industry so far as the author is aware, would be expected to be applicable to a wide variety of metals and alloys.

Feeding Rule 4:
The junction requirement

The junction problem is a pitfall awaiting the unwary. It occurs because the simple act of placing a feeder on a casting creates a junction. As we have seen from Section 2.3.2.6, a junction with an inappropriate geometry will lead to a hot spot. The hot spot may cause a shrinkage cavity that extends into the casting.

The range of simple T-junctions was shown in Figure 2.32. Clearly the problem junctions are those with 1 : 1 ratios of the thickness of the upright to that of the horizontal. If we assume that the feeder when planted on the casting is a kind of T-junction, and if we further assume that the ratios of thickness discussed so far for simple plates are also valid for ratios of modulus of shaped castings, we can, with some justification, extend the T-junction findings to identify the 1 : 1 ratio of feeder modulus to casting modulus is a problem.

The simplest example illustrating the problem clearly is that of the feeding of a cube. The cube casting has the reputation of being notoriously difficult to feed. This is because the casting technologist, carefully following Rule 2, calculates a feeder of 1.0 or perhaps 1.2 times the modulus of the cube. If the cube has side length D, then the feeder of 1 : 1 height to diameter ratio works out to have a diameter of $1.2D$. Thus the cube appears to require a feeder of rather similar volume sitting on top. However, the cube and its feeder are now a single compact shape that solidifies as a whole, with its thermal centre in the centre of the new total cast shape, i.e. approximately in the centre of the junction. The combination therefore develops a shrinkage cavity at the junction, the hot spot between the casting and feeder. When the casting is cut off from the feeder, the porosity that is found is generally called 'under-feeder shrinkage porosity.' This rather pompous pseudo-technical jargon clouds the clear conclusion that the feeder is too small.

Returning to our junction rules; to avoid creating a hot spot we need to ensure that the feeder actually has twice the modulus of the casting. Thus the cube should have had a feeder of side length $2D$. The shrinkage cavity would then have been concentrated only in the feeder.

However, in some cases the junction problem can be avoided. The simplest solution is not to place the feeder directly on the casting so as to create a junction. It happens that this rule is not easily applied to a cube because there is no alternative site for it.

However, in the case of a plate casting, there are options. The feeder should not be placed directly on the plate, but should be placed on an extension of the plate.

The general rules to solve the junction problem are therefore as follows:

1. Appendages such as feeders and ingates should not be planted on the casting so as to create a T- (or an L-) junction (although the L-junction is rather less detrimental than the T-junction). They are best added as extensions to a section, as an elongation to a wall or plate, effectively moving the junction off the casting.
2. If there is no alternative to the placing of the feeder directly on the casting, then to avoid the hot spot in the middle of the junction, the additional requirement that the feeder must meet is, if a T-junction,

$$m_f > 2m_c \qquad \textbf{(6.5)}$$

and if an L-junction

$$m_f > 1.33m_c \qquad \textbf{(6.6)}$$

The value of the constants is taken from Sciama (1974).

Note that no safety factor of 1.2 has been applied to these feeder sizes. This is because the shrinkage cavity does not occur exactly at the geometric centre of the freezing volume trapped at the thermal centre; the cavity 'floats' to the top, and the feed liquid finds its level at the base of the isolated region. Thus the final shrinkage cavity is naturally displaced above the junction interface, giving a natural 'built-in' safety factor.

Note that we have assumed that the feeder is above the casting, so as to feed downwards under gravity. This is the recommended safe way to use feeders. If the feeder (or large gate) were placed *below* the casting, gravity would now act in reverse, so that any shrinkage cavity caused by the junction would float into the casting (effectively, the residual liquid metal in the casting drains into the feeder). This action illustrates one of the dangers of attempting to feed uphill. Conditions in which feeders might be used to feed uphill are discussed later in Section 6.4.

Although it is not a good idea to make the feeder any larger than is really required, if it is only marginally adequate, the tail of porosity seen in Figure 6.2 may on occasions just enter the casting, and may therefore be unacceptable. This necessitates the application of a safety factor, giving a feeder of larger size on average, but still just acceptable even when all the variables are loaded against it. It is common to use the factor 1.2.

Feeding Rule 5: *Feed path requirement*

There must be a feed path. It is clearly no use having feed metal available at one point on the casting, unable to reach a more distant point where it is needed. Clearly there has to be a way through.

In a valuable insight, Heine (1968) has drawn attention to the fact that the highest-modulus regions in a casting are either potential regions for shrinkage porosity if left unfed, or may be feeding paths. He recommends the identification of feed paths that will transport feed metal through castings of complex geometry, such as the hot spots at the T-junctions between plates. (He also draws attention to the fact that certain locations are never feed paths. These include corners or edges of plates, or the ends of bars and cylinders.)

The various ways to help to ensure that feed paths remain open are considered in this section.

Directional solidification towards the feeder

If the feeder can be placed on the thickest section of the casting, with progressively thinner sections extending away, then the condition of progressive solidification towards the feeder can usually be achieved.

A classical method of checking this due to Heuvers can be used in which circles are inscribed inside the casting sections. If the circle diameters increase progressively towards the feeder then the condition is met (Figure 6.7). Lewis and Ransing (1998) draw attention to the fact that Heuvers' technique is only two-dimensional, and that the condition would be more accurately represented in three dimensions by a progressive change in the radius of a sphere, effectively equivalent to the progressive increase in casting modulus towards the feeder. The fundamental reason for tapering the casting in this way is to achieve taper in the liquid flow path (Sullivan *et al.* 1957). For convenience we shall call this the modulus gradient technique.

Failure to provide sufficient modulus gradient towards the feeder can be countered in

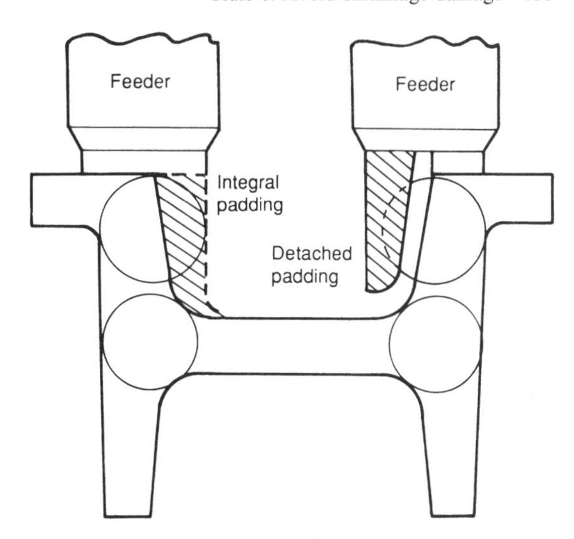

Figure 6.7 *Use of Heuvers circles to determine the amount of attached padding (Beeley 1972) and the use of detached (or indirect) padding described by Daybell (1953).*

various ways by (i) either re-siting the feeder, or (ii) providing additional feeder(s), or (iii) modifying the modulus of the casting. Ransing describes a further option, (iv) in which he proposes a change in heat transfer coefficient. The latter technique is a valuable insight because it is easily and economically computed by a geometrical technique, and so contrasts with the considerable computing reserves and effort required by finite element and finite difference methods. If R is the radius of the inscribed sphere, the local solidification time t is proportional to R^2/h where h is the heat transfer coefficient at the metal/mould interface. Thus at locations 1 and 2 we have $h_2 = h_1(R_1/R_2)^2$. This relation allows an estimate of the change of h that is required to ensure that the freezing time increases steadily towards the feeder. Ransing uses a change of 10 per cent increase of solidification time for each different geometrical section of the casting (i.e. for feeding via a thin section to a thick distant section he increases the freezing time of the thin section over that of the thick by 10 per cent). The technique can be usefully employed in reverse, in the sense that known values of h produced by a chill can quickly be checked for their effectiveness in dealing with an isolated heavy section. In every case the target is to eliminate the local hot spot, and ensure a continuous feed path back to the feeder. This simple technique has elegance, economy and power and is strongly recommended.

The casting modulus can be modified by providing either a chill or cooling fin to speed solidification locally, or by providing extra metal to thicken the section and delay solidification locally. The provision of extra metal on the casting is known as padding. The addition of padding is most usefully carried out with the customer's consent, so that it can be left in position as a permanent feature of the casting design. If consent cannot be obtained then the caster has to accept the cost penalty of dressing off the padding as an additional operation after casting.

Occasionally this problem can be avoided by the provision of detached, or indirect, padding as shown in Figure 6.7. Daybell (1953) was probably the first to describe the use of this technique. The author has found it useful in the placement of feeders close to thin adjacent sections of casting, with a view to feeding through the thin section into a remote thicker section.

The principle of progressive increase of modulus towards the feeder, although generally accurate and useful, is occasionally seen to be not quite true. Depending on the conditions, this failure of the principle can be either a problem or a benefit, as shown below. (Even so, the Ransing technique described above is unusual since it successfully takes this problem into account.)

In a re-entrant section of a casting the confluence of heat flow into the mould can cause a hot spot, leading to delayed solidification at this point, and the danger of local shrinkage porosity in an alloy that shrinks during freezing, such as an Al alloy. Alternatively, in an alloy that expands such as a high carbon-equivalent cast iron, the exudation of residual liquid into the mould as a result of the high internal pressure created during the precipitation of graphite can cause penetration of the aggregate mould material, with unwelcome so-called *burned-on* sand. Such a hot spot can occur despite an apparent unbroken increase in modulus through that region towards the feeder. This is because the simple estimation of modulus takes no account of the geometry of heat flow away from cooling surfaces; all surfaces are assumed to be equally effective in cooling the casting. Such a hot spot requires the normal attention such as extra local cooling by chill or fin, or additional feed via extra padding or feeders.

The failure of the modulus gradient technique can be used to advantage in the case of feeder necks to reduce the subsequent cut-off problem. Feeders are commonly joined to the main casting via a feeder neck, with the modulus of the neck commonly controlled to be intermediate between that of the casting and the feeder; the moduli of casting, neck and feeder are in the ratio 1.0 : 1.1 : 1.2 (Beeley 1972). However the neck can be reduced considerably

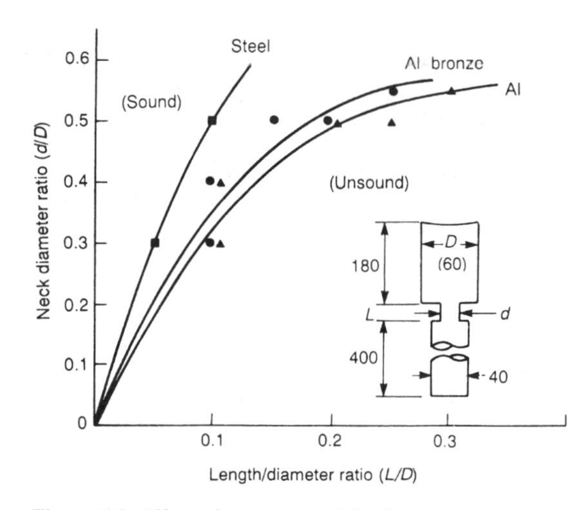

Figure 6.8 *Effect of a constricted feeder neck on soundness of steel, aluminium bronze, and 99.5Al castings. The experimental points by Sciama (1975) denote marginal conditions.*

below this apparently logical lower limit, because of the hot spot effect, and because of the conduction of heat from the neighbouring casting and feeder that helps to keep the neck molten for a longer period than its modulus alone would suggest. This point is well illustrated by Sciama (1975), and his results are summarized in Figure 6.8. The results clearly demonstrate that for steel, feeder necks can be reduced to half of the diameter D of the feeder, providing that they are not longer than $0.1D$. The higher thermal conductivity copper- and aluminium-based materials can have necks almost twice as long without problems.

By extrapolation of these results towards smaller neck sizes, it seems that a feeder neck in steel can be only $0.25D$ in diameter, providing it is no more than approximately $0.03D$ in length. Similarly, for copper and aluminium alloys the $0.25D$ diameter neck can be up to $0.06D$ long. These results explain the action of the Washburn core, or breaker core, which is a wafer-thin core with a narrow central hole, and which is placed at the base of a feeder, allowing it to be removed after casting by simply breaking it off. In separate work the dimensions of typical Washburn cores is recommended to be a thickness of $0.1D$ and a central hole diameter of $0.4D$ (work by Wlodawer summarized by Beeley 1972). The hole size and thickness appear to be very conservative in relation to Sciama's work. However, Sciama may predict optimistic results because he uses a feeder of nearly 1.5 times the modulus of casting, which would tend to keep the junction rather hotter than a feeder with a modulus of only 1.2 times that of the casting (it would be valuable to repeat this work using a more economical feeder). Also, of course, conservatism may be justified where feeding conditions are less than optimum for other reasons in the real foundry environment.

The aspect of conservatism because of the real foundry environment is an interesting issue. For instance, the feeder neck could, in the extreme theoretical case, be of zero diameter when the thickness of the feeder neck core was zero. The *reductio ad absurdum* argument illustrates that an extreme is not worth targeting, especially when sundry debris in suspension, such as metal dendrites or rigid bifilms, could close off a narrow aperture.

Minimum temperature gradient requirement

Experiments on cast steels have found that when the temperature gradient at the solidus (i.e. the temperature at which the final residual liquid freezes) falls to below approximately 0.1–1 K mm^{-1} then porosity is observed even in well-degassed material. Although there is much scatter in other experimental determinations, it seems in general that the corresponding gradients for copper alloys are around 1 K mm^{-1} and those for aluminium alloys around 2 K mm^{-1} (Pillai *et al.* 1976). It seems therefore that the temperature gradient defines a critical threshold of a non-feeding condition. As the flow channel nears its furthest extent, and becomes vanishingly narrow, it will become subject to small random fluctuations in temperature along its length. This kind of temperature 'noise' will occur as a result of small variations in casting thickness, or of density of the mould, thickness of mould coating, blockage or diversion of heat flow direction by random entrained films etc. Thus the channel will not reduce steadily to infinite thinness, but will terminate when its diameter becomes close to the size of the random perturbations.

There has been some discussion about the absolute value of the critical gradient for feeding on the grounds that the degree of degassing, or the standard of soundness, to which the casting was judged, will affect the result. These are certainly very real problems, and do help to explain some of the wide scatter in the results.

Hansen and Sahm (1988) draw attention to a more fundamental objection to the use of temperature gradients as a parameter that might correlate with feeding problems. They indicate that the critical gradient required to avoid shrinkage porosity in a steel bar is five to ten times higher than that required for a plate, and point to other work in which the critical gradient in a cylindrical steel casting is a function of its diameter. Thus the concept of a single gradient which applies in all conditions seems to be at fault. If this can be confirmed, which seems likely, then its use will require to be re-thought.

Feeding distance

It is easy to appreciate that in normal conditions it is to be expected that there will be a limit to how far feed liquid can be provided along a flow path. Up to this distance from the feeder the casting will be sound. Beyond this distance the casting will be expected to exhibit porosity.

This arises because along the length of a flow channel, the pressure will fall progressively because of the viscous resistance to flow. (This effect was covered in more detail in *Castings 2003*.) When the pressure falls to a critical level, which might actually be negative, then porosity may form. Such porosity may occur from an internal initiation event (such as the opening of a bifilm), or from the drawing inwards of feed metal from the surface of the casting, since this may now represent a shorter and easier flow path than supply from the more distant feeder.

There has been much experimental effort to determine feeding distances. The early work by Pellini and his co-workers (summarized by Beeley (1972)) at the US Naval Materials Laboratory is a classic investigation that has influenced the thinking on the concept of feeding distance ever since. They discovered that the feeding distance L_d of plates of carbon steels cast into greensand moulds depended on the section thickness T of the casting: castings could be made sound for a distance from the feeder edge of $4.5T$. Of this total distance, $2.5T$ resulted from the chilling effect of the casting edge; the remaining $2.0T$ was made sound by the feeder. The addition of a chill was found to increase the feeding distance by a fixed 50 mm (Figure 6.9). They found that increasing the feeder size above the optimum required to obtain this feeding distance had no beneficial effects in promoting soundness. The feeding distance rule for their findings is simply:

$$L_d = 4.5T \quad \textbf{(6.7)}$$

Pellini and colleagues went on to speculate that it should be possible to ensure the soundness of a large plate casting by taking care that every point on the casting is within a distance of $2.5T$ from an edge, or $2.0T$ from a feeder.

Note that all the semi-empirical computer programs written since have used this and the

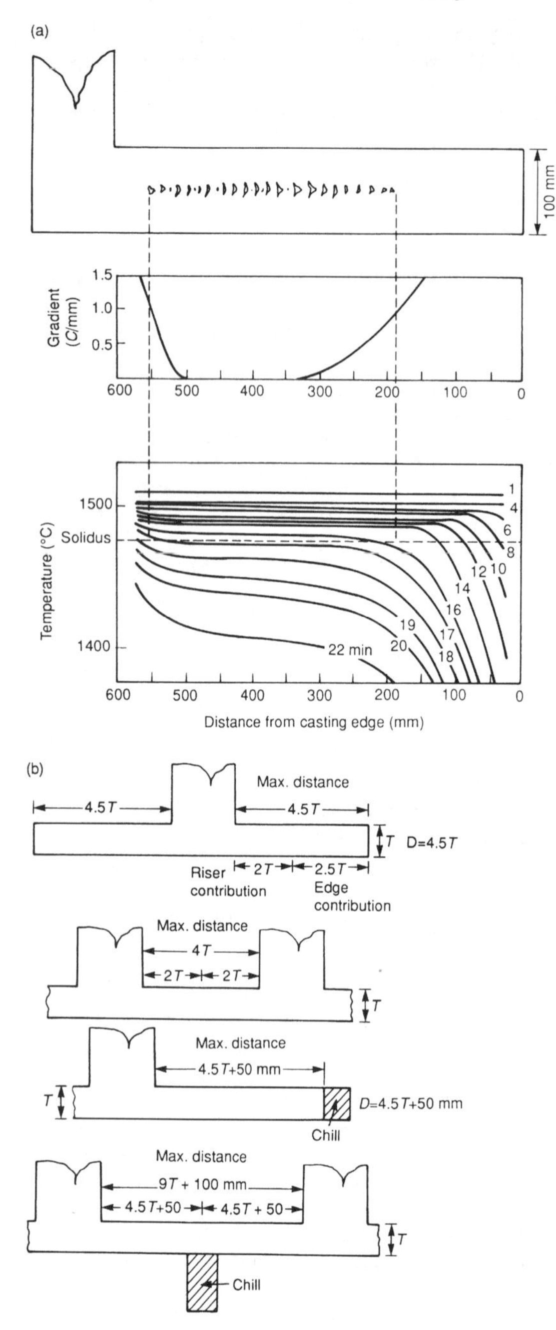

Figure 6.9 *The famous results by Pellini (1953) for (a) the temperature distribution in a solidifying steel bar; and (b) the feeding distances for steel plates cast in greensand.*

associated family of rules as illustrated in Figure 6.9 to define the spacing of feeders and chills on castings. However, the original data relate only to steel in greensand moulds, and only to rather heavy sections ranging from 50–200 mm. Johnson and Loper (1969) have extended the range of the experiments down to sections a thickness of 12.5 mm and have re-analysed all the data. They found that for plates, the data, all in units of millimetres, appeared to be more accurately described by the equation:

$$L_d = 72m^{1/2} - 140 \quad \textbf{(6.8)}$$

and for bars:

$$L_d = 80m^{1/2} - 84 \quad \textbf{(6.9)}$$

where m is the modulus of the cast section in millimetres. The revised equations by Johnson and Loper have usually been overlooked in much subsequent work. What is also overlooked is that all the relations apply to cast mild steels in greensand moulds, not necessarily to any other casting alloys in any other kinds of mould.

In their nice theoretical model, Kubo and Pehlke (1985) find support for Pellini's feeding distance rules for steel castings, but it is a concern that no equivalent rule emerged for Al–4.5Cu alloy that they also investigated.

In fact, colleagues of Flinn (1964) found that whereas the short-freezing-range alloys manganese bronze, aluminium bronze, and 70/30 cupro-nickel all had feeding distances that increased with section thickness, the long-freezing-range alloy tin bronze appeared to react in the opposite sense, giving a reduced feeding distance as section thickness increased. (The nominal composition of this classical long-freezing-range material is 85Cu, 5Sn, 5Zn, 5Pb. It was known among traditional foundrymen as 'ounce metal' since to make this alloy they needed to take one pound of copper to which one ounce of tin, one ounce of zinc, and one ounce of lead was added. This gives, allowing for small losses on addition, the ratios 85 : 5 : 5 : 5.)

Kuyucak (2002) reviews the relations for estimating feeding distance in steel castings, and finds considerable variation in their predictions. This makes sobering reading.

Jacob and Drouzy (1974) found long feeding distances, greater than 15T, for the relatively long-freezing-range aluminium alloys Al–4Cu and Al–75i–0.SMg, providing the feeder is correctly sized.

All this confusion regarding feeding distances remains a source of concern. We can surmise that the opposite behaviour of short- and long-freezing-range materials might be understood in terms of the ratio pasty zone/casting section. For short-freezing-range alloys this ratio is less than 1, so the solidified skin of the alloy is complete, dictating feeding from the feeder, and thus normal feeding distance concepts apply.

For the case of long-freezing-range materials where the pasty zone/casting section ratio is greater than 1, and in fact might be 10 or more, the outer solid portions of the casting are far from solid for much of the period of solidification. The connections of liquid through to the outer surface will allow flow of liquid from the surface to feed solidification shrinkage. In addition, the higher temperature and lower strength of the liquid/solid mass will allow general collapse of the walls of the casting inwards, making an important contribution to the feeding of the inner regions of the casting by the 'solid feeding' mechanism. It is for this reason that the higher conductivity, and lower strength alloys of Al and Cu can be characterized by practically infinite feeding distances, particularly if the alloys are relatively free from bifilms. Internal porosity simply does not nucleate, no matter how distant the casting happens to be from the feeder; the outer walls of the casting simply move inwards very slightly.

Thus although the general concept of feeding distance is probably substantially correct, at least for short-freezing-range alloys, and particularly for stronger materials such as steels, it should be used, if at all, with great caution for non-ferrous metals until it is better understood and quantified. In summary it is worth noting the following:

1. The data on feeding distances have been derived from extensive work on carbon steels cast in greensand moulds. Relatively little work has been carried out on other metals in other moulds.
2. The definition of feeding distance is sensitive to the level of porosity that can be detected and/or tolerated.
3. It is curious that the feeding distance is defined from the edge of a feeder (not its centreline).
4. The quality of the cast metal in terms of its gas and oxide content would be expected to be crucial. For instance, good quality metal achieved by the use of filters and good degassing and casting technique (i.e. with a low bifilm content) would be expected to yield massive improvements in feeding distance. This has been demonstrated by Romero *et al.* (1991) for Al-bronze. Berry and Taylor (1999) report a related effect, while reviewing the benefit to the feeding distance of pressurizing the feeder. This work is straightforwardly understood in terms of the pressure on the liquid acting to suppress the opening of bifilms.

A final note of caution relates to the situation where the concept of feeding distance applies to an alloy, but has been exceeded. When this happens it is reported that the sound length is considerably less than it would have been if the feeding distance criterion had just been satisfied. If true, this behaviour may result from the spread of porosity, once initiated, into adjacent regions. The lengths of sound casting in Figure 6.9a are considerably shorter than the maximum lengths given by Equations 6.5 to 6.7, possibly because the feeding distance predicted by these equations has been exceeded and the porosity has spread. Mikkola and Heine (1970) confirm this unwelcome effect in white iron castings.

Other parameters (criteria functions)

In a theoretical study of the formation of porosity in steel plates of thickness 5 to 50 mm, with and without end chills, Minakawa *et al.* (1985) investigated various parameters that might be useful in assessing the conditions for the onset of porosity in their castings. They looked at G, the temperature gradient along the centreline of the casting at the solidification front, and the fraction solid f_s along the centreline. Neither of these was satisfactory. However, they did find that the parameter $G/V^{1/2}$ suggested by Niyama *et al.* (1982) correctly assessed the difficulty of providing feed liquid under the various conditions of their work, where V is the velocity of advance of the freezing front. In plate-like castings the value of G drops to low levels in the centre of the plate, and at the same time V increases because the front accelerates along the centre of the plate, reaching its highest velocity, requiring feed metal at the highest rate. It thus creates the largest pressure drop to drive this flow. To obtain sound castings, therefore, they found that the value of $G/V^{1/2}$ has to be at least $1.0\,\mathrm{K\,s^{1/2}\,mm^{-3/2}}$.

It would be valuable to know whether this parameter is similarly discriminating for other casting alloys, particularly the high-thermal-conductivity alloys of aluminium and copper.

In another theoretical study Hansen and Sahm (1988) support the usefulness of $G/V^{1/2}$ for steel castings. However, in addition they go on to argue the case for the use of a more complex function $G/V^{1/4}\,V_L^{1/2}$ where V_L is the velocity of flow of the residual liquid.

They proposed this relation because they noticed that the velocity of flow in bars was five to ten times the velocity in plates of the same thickness, which, they suggest, contributes to the additional feeding difficulty of bars compared to plates. (A further contributor will be the comparatively high resistance to collapse that is shown by bars, compared to the efficiency of solid feeding in plates as will be discussed later.) They found that $G/V^{1/4}V_L^{1/2}$ < a critical value, which for steel plates and bars is approximately 1 K $s^{3/4}$ $mm^{-7/4}$. Their parameter is, of course, less easy to use than that due to Niyama, because it needs flow velocities. The Niyama approach only requires data obtainable from temperature measurements in the casting.

Feeding Rule 6: *Pressure gradient requirement*

Although all of the previous feeding rules may be met, including the provision of feed liquid and a suitable flow path, if the pressure gradient needed to cause the liquid to flow along the path is not available, then feed liquid will not flow to where it is needed. Internal porosity may therefore occur.

A positive pressure gradient from the outside to the inside of the casting will help to ensure that the feed material (either solid or liquid) travels along the flow path into those parts of the casting experiencing shrinkage. The various feeding mechanisms (to be discussed later) are seen to be driven by the positive pressure such as atmospheric pressure and/or the pressure due to the hydrostatic head of metal in the feeder. The other contributor to the pressure gradient, the driving force for flow, is the reduced or even negative pressure generated within poorly fed regions of some castings. All of these driving forces happen to be additive; the flow of feed metal is caused by being pushed from the outside and pulled from the inside.

Figure 6.10 illustrates the feeding problems in a complicated casting. The casting divides effectively into two parts either side of the broken line. The left-hand side has been designed to be fed by an open feeder Fl and a blind feeder F2. The right-hand side was intended to be fed by blind feeder F3.

Feeder Fl successfully feeds the heavy section S1. This feeder is seen to be comparatively large. This is because it is required to provide feed metal to the whole casting during the early stages of freezing, while the connecting sections remain open. At this early stage of interconnection of the whole casting, the top feeder is also feeding both blind feeders, of course.

Feeder F2 feeds S5 because it is provided with an atmospheric vent, allowing the liquid to be pressurized by the atmosphere as in the coffee cup experiment illustrated below, so forcing the metal through into the casting. (The reader is encouraged to try the coffee cup experiment.)

The identical heavy sections S3 and S4 show the unreliability of attempting to feed uphill. In S4 a chance initiation of a pore has created a free liquid surface, and the internal gas pressure within the casting happens to be close to 1 atmosphere. Thus the liquid level in S4 falls, finding its level equal to that in the feeder F2. (If the internal gas pressure within the casting

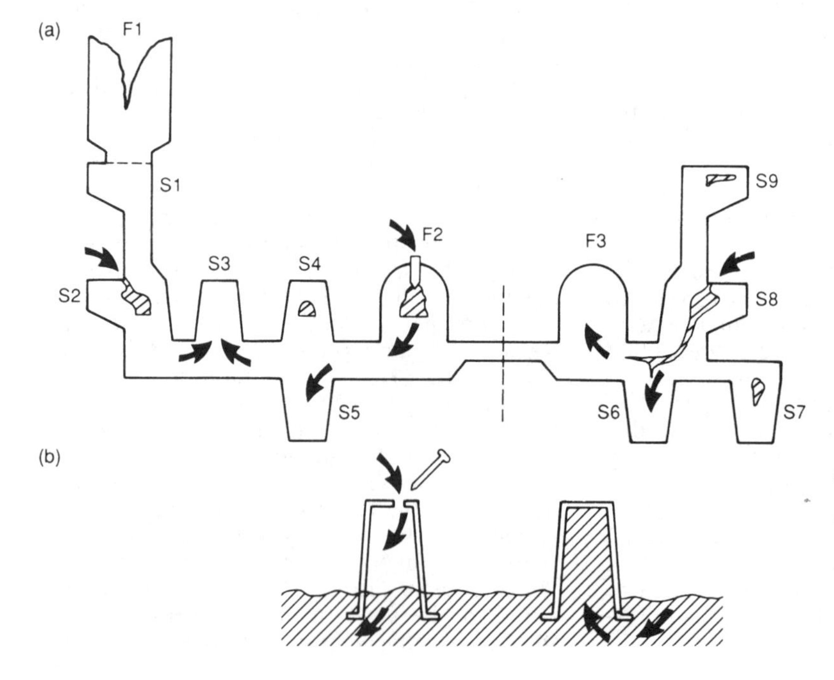

Figure 6.10 *(a) Castings with blind feeders, F2 is correctly vented but has mixed results on sections S3 and S4. Feeder F3 is not vented and therefore does not feed at all. The unfavourable pressure gradient draws liquid from a fortuitous skin puncture in section S8. See text for further explanation. (b) The plastic coffee cup analogue: the water is held up in the upturned cup and cannot be released until air is admitted via a puncture. The liquid it contains is then immediately released.*

had been much less than an atmosphere, then the level in S4 would have been correspondingly higher.) The surface-initiated pore in S2 has grown similarly, equalizing its level exactly with that in the feeder F2, since both surfaces are subject to the same atmospheric pressure. In Section S3, by good fortune, no pore initiation site is present, so no pore has occurred, with the result that atmospheric pressure via F2 (and unfortunately also via the puncture by the atmosphere at the hot spot in the re-entrant section S2) will feed solidification shrinkage here, causing the section to be perfectly sound.

Turning now to the right-hand part of the casting, although feeder F3 is of adequate size to feed the heavy sections S6, S7 and S8, its atmospheric vent has been forgotten. This is a serious mistake. The plastic coffee cup experiment (Figure 6.10) shows that such an inverted air-tight container cannot deliver its liquid contents. The pressure gradient is now reversed, causing the flow to be in the wrong direction, from the casting to the feeder! The detailed reasoning for this is as follows. The pressure in the casting and feeder continues to fall as freezing occurs until a pore initiates, either under the hydrostatic tension, or because of a build-up of gas in solution, or because of the inward rupture of the surface at a weak point such as the re-entrant angle in section S8. The pressure in section S8 is now raised to atmospheric pressure while the pressure in the feeder is still low, or even negative. Thus feed liquid is now forced to flow from the casting into the feeder as freezing progresses. A massive pore then develops because feeder F3 has a large feed requirement, and drains section S8 and the surrounding casting. The defect size is worse than that which would have occurred if no feeder had been used at all!

Section S6 remains reasonably sound because it has the advantage of natural drainage of residual liquid into it. Effectively it has been fed from the heavy section S8. The pressure gradient due to the combined actions of gravity, shrinkage and the atmosphere from S8 to S6 is positive. The only reason why S6 may display any residual porosity may be that S8 is a rather inadequate feeder in terms of either its thermal requirement or its volume, or because the feed path may be interrupted at a late stage.

Section S7 cannot be fed because there is no continuous feed path to it. S9 is similarly disadvantaged. This has been an oversight in the design of the feeding of this casting. In a sand casting it is likely that S7 and S9 will therefore suffer porosity. This will be almost certainly true for a steel casting, but less certain if the casting is a medium-freezing-range aluminium alloy. The reason becomes clear when we consider an investment casting, which, if a high mould temperature is chosen, and if the metal is clean, will allow solid feeding to operate, allowing the sections the opportunity to collapse plastically, and so become internally sound, provided that no pore-initiation event interrupts this action. Solid feeding is often seen in aluminium alloy sand castings, but rarely in steel sand castings because of the greater rigidity of the solidified steel, which successfully resists plastic collapse in cold moulds.

The exercise with the plastic coffee cup shows that the water will hold up indefinitely in the upturned cup until released by the pin causing a hole. The cup will then deliver its contents immediately (but not before!). Blind feeders are therefore often unreliable in practice because the atmospheric vent may not open reliably. Such feeders then act to suck feed metal from the casting, making any porosity worse.

If a blind feeder is provided with an effective atmospheric vent, then the available atmospheric pressure may help it to feed uphill. The maximum heights supportable by one atmosphere for various pure liquids near their melting points are:

Mercury	0.760 m (barometric height)
Steel	1.48 m
Zinc	1.58 m
Aluminium	4.36 m
Magnesium	6.54 m
Water	10.40 m

However, as we have seen, feeding uphill is not altogether reliable, and cannot be recommended as a general technique. To restate the reasons briefly, this is because any initiation of a shrinkage or gas pore, or any inward rupture of the casting surface, will release the internal stress of the casting, removing the pressure difference between the casting and feeder. With the pressures in casting and feeder equalized, the metal level in the casting will fall, and that in the feeder will rise so as to equalize the levels if possible. The result is a porous casting. Blind feeders that are placed low on the casting can be unreliable in practice for this reason.

This loss of pressure difference cannot occur if the feeder is placed above the general level of the casting so that feeding always takes place with the assistance of gravity. Feeder F2 in Figure 6.10 would have successfully fed sections S2 and S4 either if it was taller, or if it had been placed at a higher location, for instance on the top of S4.

It is clear that F3 may not have fed section S8 if the corner puncture occurred, even if it had been provided with an effective vent, because the pressure gradient for flow would have been removed. A provision of an effective vent, and the re-siting of the base of the feeder F3 to the side of S8, would have maintained the soundness of both S6 and S8 and would have prevented the surface puncture at S8. S7 and S9 would still have required separate treatment.

The conclusion to these considerations is: *place feeders high to feed downhill.* This is a general principle of great importance. It is of similar weight to the general principle discussed previously, *place ingates low to fill uphill.* These are fundamental concepts in the production of good castings.

Feeding Rule 7: *Pressure requirement*

The final rule for effective feeding is a necessary requirement like all the others. Sufficient pressure in the residual liquid within the casting is required to suppress both the initiation and the growth of cavities both internally and externally.

This is a *hydrostatic* requirement relating to the suppression of porosity, and contrasts with the previous pressure gradient requirement that relates to the *hydrodynamic* requirements for flow (especially flow in the *correct* direction!)

A fall in internal pressure may cause a variety of problems:

1. Liquid may be sucked from the cast surface. This is particularly likely in long-freezing-range alloys, or from re-entrant angles in shorter-freezing-range alloys, resulting in internal porosity initiated from, and connected to, the outside.
2. The internal pressure may fall just sufficiently to unfurl, but not fully open the population of bifilms. The result will be an apparently sound casting but poor mechanical properties, particularly a poor elongation to failure. (It is possible that some so-called '*diffraction mottle*' may be noted on X-ray radiographs.)
3. The internal pressure may fall sufficiently to open bifilms, so that a distribution of fine and dispersed microporosity will appear. The mechanical properties will be even lower.
4. The internal shrinkage may cause macroshrinkage porosity to occur, especially if there are large bifilms present as a result of poor filling of the casting. Properties may now be in disaster mode and/or large holes may appear in the casting. Figure 6.11 illustrates pressure loss situations in castings that can result in shrinkage porosity. Figure 6.12 illustrates the common observation in Al alloy castings in which a glass cloth is placed under the feeder to assist the break-off of the feeder after solidification. Bafflingly, it sometimes appears that the cloth prevents the feeder from supplying liquid, so that a large cavity appears under the feeder. The truth is that double oxide films plaster themselves against the underside of the cloth as the feeder fills. The half of the film against the cloth tends to weld to the cloth, possibly by a chemical fusing action, or possibly by mechanical wrapping around the fibres of the cloth. Whatever the mechanism, the action is to hold back the liquid above, while the lower half of the (unbonded) double film is easily pulled away by the contracting liquid, opening the void that was originally the microscopically thin interface of air inside the bifilm.
5. If there is insufficient opportunity to open internal defects, the external surface of the casting may sink to accommodate the internal shrinkage. (The occurrence of *surface sinks* is occasionally referred to elsewhere in the literature as '*cavitation*'; a misuse of language to be deplored. Cavitation properly refers to the creation and collapse of minute bubbles, and the consequent erosion of solid surfaces such as those of ships' propellers.)

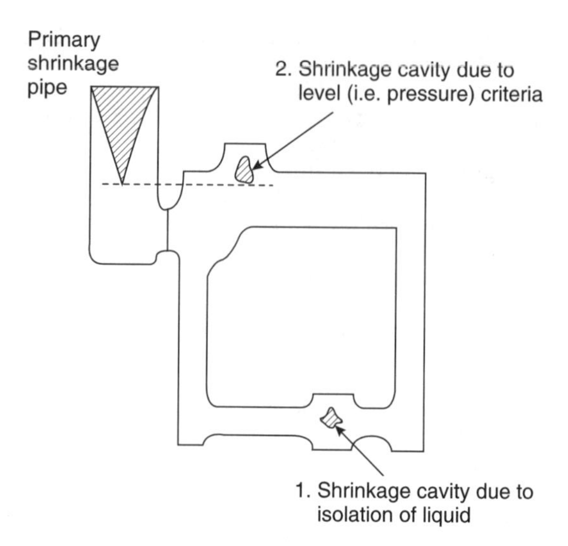

Figure 6.11 *Pressure loss situations in castings leading to the possibility of shrinkage porosity.*

Often, of course, the distribution of defects observed in practice is a mixture of the above list. The internal pressure needs to be

Top half of bifilm mechanically wrapped/welded to fibres of glass cloth, presenting barrier to feed metal

Feeder

Glass cloth

Casting

Unbonded lower half of bifilm sucked into the casting, leading to 'under-feeder' porosity

Fragments of lower half of bifilm sucked into dendrite mesh, and thereby dispersed

Figure 6.12 *The apparent blocking of feed metal by a glass cloth strainer in an Al alloy casting by the action of bifilms collected on the underside of the cloth.*

maintained sufficiently high to avoid all of these defects.

Finally, however, it is worth pointing out that over-zealous application of pressure to reduce the above problems can result in a new crop of different problems.

For instance, in the case of long-freezing-range materials cast in a sand mould, a high internal pressure, applied for instance to the feeder, will force liquid out of the surface-linked capillaries, making a casting having a 'furry' appearance. Overpressures are not easy to control in low-pressure sand casting processes, and are the reason why these processes often struggle to meet surface-finish requirements. In cast iron castings the generation of excess internal pressure by graphite precipitation has been shown by work at the University of Alabama (Stefanescu *et al.* 1996) to lead to exudation of the residual liquid via hot spots at the surface of the casting, leading to penetration of the sand mould. Later work on steels has shown analogous effects (Hayes *et al.* 1998).

In short-freezing-range materials the inward flow of solid can be reversed with sufficient internal pressure. Too great a pressure will expand the casting, blowing it up like a balloon, producing unsightly swells on flat surfaces (Figure 6.13).

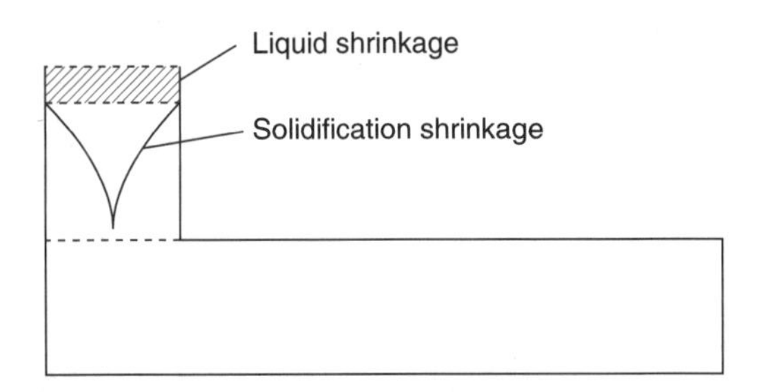

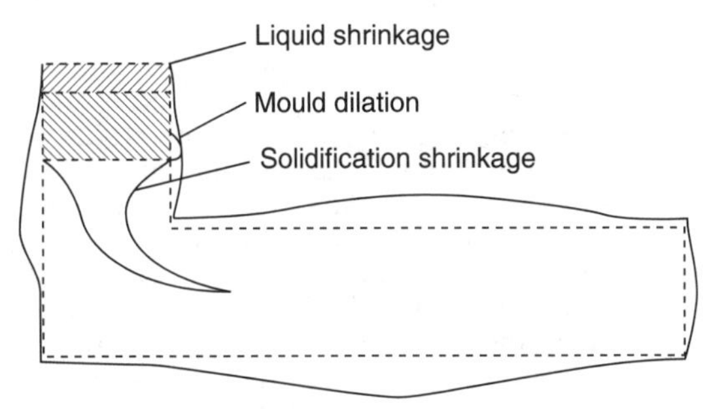

Figure 6.13 *A comparison between the external size and internal shrinkage porosity in a casting as a result of (a) moderate pressure in the liquid, and adequately rigid mould, and (b) too much pressure and/or a weak mould.*

Pressurization of cast iron castings

The successful feeding of cast irons is perhaps the most complex and challenging feeding task compared to all other casting alloys as a result of the curious and complicating effect of pressure. The effects are most dramatically seen for ductile irons.

The great prophet of the scientific feeding of ductile irons was Stephen Karsay. In a succession of chattily written books he outlined the principles that applied to this difficult metal (see, for instance, Karsay 1992 and 2000). He drew attention to the problem of the swelling of the casting in a weak mould as shown in Figure 6.13, in which the valuable expansion of the graphite was lost by enlarging the casting, causing the feeder to be inadequate to fill the increased volume. He promoted the approach of making the mould more rigid, and so better withstanding stress, and at the same time reducing the internal pressure by providing feeders that acted as pressure relief valves. The feeder, after some initial provision of feed metal during the solidification of austenite, would back-fill with residual liquid during the expansion of the solidification of the eutectic graphite. The final state was a feeder that was substantially sound. (Occasionally, one hears stories that such sound feeders would be declared to be evidently useless, having apparently provided no feed metal. However, their removal would immediately cause all subsequent castings to become porous!).

The reproducibility of the achievement of soundness in ductile iron castings is, of course, highly sensitive to the efficiency of the inoculation treatment, because the degree of expansion of graphite is directly affected. This is notoriously difficult to keep under good control, and makes for one of the greatest challenges to the iron founder.

Roedter (1986) introduced a refinement of Karsay's pressure relief technique in which the pressure relief was limited in extent. Some relief was allowed, but total relief was prevented by the premature freezing of the feeder neck. In this way the casting was slightly pressurized, elastically deforming the very hard sand mould, and the surrounding steel moulding box (if any). The elastic deformation of the mould and its box would store the strain energy. The subsequent relaxation of this deformation would continue to apply pressure to the solidifying casting during the remainder of solidification. Thus soundness of the casting could be achieved, but without the danger of unacceptable swells on extensive flat surfaces.

For somewhat heavier ductile iron castings, however, it has now become common practice to cast completely without feeders. This has been achieved by the use of rigid moulds, now more routinely available from modern green-sand moulding units. Naturally, the swelling of the casting still occurs, since, ultimately, solids are incompressible. However, as before, the expansion is restrained to the minimum by the elastic yielding of the mould and its container, and distributed more uniformly. Thus the whole casting is a few per cent larger. If the total net expansion was 3 volume per cent, this corresponds to 1 linear per cent along the three orthogonal axes, so that from a central datum, each point on the surface of the casting would be approximately 0.5 per cent oversize. This uniform and very reproducible degree of oversize is usually negligible. However, of course, it can be compensated, if necessary, by making the pattern 0.5 per cent undersize.

The use of the elastic strains to re-apply pressure is strictly limited because such strains are usually limited to only 0.1 linear per cent or so. Thus only a total of perhaps 0.3 volume per cent can be compensated by this means. This is, as we have seen above, only a fraction of the total volume change that is usual in a graphitic iron, and which permanently affects the size and dimensions of the casting. The judgement of feeder neck sizes to take advantage of such small margins is not easy.

With the steady accumulation of experience in a well-controlled casting facility, the casting engineer can often achieve such an accuracy of feeding that even such a modest gain is considered a valuable asset. Even so, the reader will appreciate that the feeding of graphitic irons is still not as exact a science and still not as clearly understood as we all might wish.

6.3 The new feeding logic

6.3.1 Background

Much of the formal calculation of feeders has been of poor accuracy because of a number of simplifying assumptions that have been widely used. Tiryakioğlu has pioneered a new way of analysing the physics of feeding, having, in addition, the good fortune to have as a critical test his late father's exemplary experimental data on optimum feeder sizes determined many years earlier (E. Tiryakioğlu 1964). The reader is recommended to the original papers by M. Tiryakioğlu (1997–2002) for a complete description of his admirable logic. We shall summarize his approach only briefly here, following closely his excellent description (Tiryakioğlu *et al.* 2002).

As we have seen in Rules 2 to 4, an efficient feeder should (i) remain molten until the portion of the casting being fed has solidified (i.e. the solidification time of the feeder has to be equal to, or exceed, that of the casting), (ii) contain sufficient volume of molten metal to meet the feeding demand of that same portion of the casting, (iii) not create a hot spot at the junction between feeder and casting. An optimum feeder is then defined as the one with the smallest volume, for its particular shape, to meet these criteria. A feeder that is less compact or that has less volume than the optimum feeder will result in an unsound casting.

The standard approaches to solve these problems have usually been based on the famous rule by Chvorinov (1940) for the solidification time of a casting:

$$t = B\left(\frac{V}{A}\right)^n \quad \textbf{(6.10)}$$

where B is the mould constant, V is the casting volume, A is the surface area through which heat is lost, and n is a constant (2 in Chvorinov's work for simple shaped castings in silica sand moulds). The V/A ratio is known as the modulus m, and has been used as the basis for a number of approaches to determining the size of feeders for the production of sound castings, as described in Section 6.2 Feeding Rule 2.

Despite its wide acceptance, Chvorinov's Rule has limitations because of the underlying assumptions used in deriving the equation. As a result of these limitations, the exponent, n, fluctuates between 1 and 2, depending on the shape and size of the casting, and the mould and pouring conditions. One of the reasons for this anomaly is that Chvorinov's Rule originally did not take the shape of the casting into consideration. A new geometry-based model (Tiryakioğlu *et al.* 1997) proved that the modulus includes the effect of both casting shape and size. These two independent factors were separated from each other by the use of a shape factor k where

$$t = B'k^{1.31}V^{0.67} \quad \textbf{(6.11)}$$

and B' is the mould constant. The shape factor, k, is the ratio (the surface area of a sphere of same volume as the casting)/(the surface area of the casting). In Equation 6.11, V assesses the amount of heat that needs to be dissipated for complete solidification, and k assesses the relative ability of the casting shape to dissipate the heat under the given mould conditions.

$$k = \frac{A_s}{A} = \frac{4.837V^{2/3}}{A} \quad \textbf{(6.12)}$$

where A_s is the surface area of the sphere.

Adams and Taylor (1953) were the first to consider *mass transfer* from feeder to casting. They realized that during solidification, a mass of αV_c needs to be transferred from the feeder to the casting (α is fraction shrinkage of the metal). However, as Tiryakioğlu (2002) explains, their development of the concept unfortunately introduced errors so that the final solution was not accurate.

Moreover, the lack of knowledge about the effect of *heat transfer* between a feeder and a casting has led researchers to the mindset of considering the feeder and casting separately. In other words, almost all feeder models have been based on calculation of solidification times for the feeder and casting independently, and then assuming that the same solidification characteristics will be followed when they are combined. However, it should be remembered that the feeder is also a section of the mould cavity and the solidifying metal does not know (or care) which section is the casting and which is the feeder.

The objective of the foundry engineer when designing a feeding system is to have the thermal centre of the total casting (the feeder–casting combination) in the feeder. In fact, all three requirements for an efficient feeder listed (i) to (iii) above can be summarized as a single requirement: *the thermal centre of the feeder–casting combination will be in the feeder*. This new approach, which treats the casting–feeder combination as a single, total casting, constitutes the foundation of Tiryakioğlu's new approach to characterize heat and mass transfer between feeder and casting.

6.3.2 The new approach

Let us consider a plate casting that is fed effectively by a feeder. Knowing that the solidification contraction of the casting is αV_c, this volume is transferred from the feeder to the casting, resulting in the final volume of the feeder being $(V_f - \alpha V_c)$. Solidification contraction of the feeder is ignored since it does not change the heat content of the feeder. If the feeder has been designed according to the rules for efficient feeding, the last part to solidify in this combination is the feeder. In other words, the thermal centre of the casting–feeder combination (the total casting) is in the feeder. Therefore the solidification time of the total casting is exactly the same as the feeder, and both have the same thermal centre.

This is not true for the casting, however. The thermal centre of the casting is also in the feeder, but its solidification time may or may not be equal to that of the total casting. Hence

$$k_t^{1.31}V_t^{0.67} = k_f^{1.31}(V_f - \alpha V_c)^{0.67} \quad \textbf{(6.13)}$$

where subscript t refers to the total casting. So far we have ignored the heat transfer between the casting and the feeder. Using optimum feeder data obtained by systematic changes in feeder size for an Al–12wt%Si alloy (Tiryakioğlu 1964) the solidification times of casting and feeder are compared in Figure 6.14a and total casting and feeder in Figure 6.14b. Figure 6.14a shows the $(t_c - t_f)$ relationship when mass transfer is taken into account and heat transfer is ignored. It should be kept in mind that the scatter in Figure 6.14a is not due to experimental error since all values were calculated. Figure 6.14b shows the relationship between the solidification times of feeder and total casting (feeder+casting combination). Although the agreement in Figure 6.14b is encouraging, at low values the error is up to 30%. However solidification time should be identical for the feeder and the total casting. The error is due to neglect of the heat exchange between feeder and casting. Mass is transferred from the feeder to the casting throughout the solidification process. Since solidification takes place over a temperature range, subtracting αV_c from V_f adjusts for mass exchange completely. For heat exchange however, this treatment assumes isothermal conditions, and therefore is not sufficient. Hence the feeder solidification time needs to be adjusted for the heat exchange with the casting. We can treat this heat exchange as if it were superheat extracted from/given to the feeder. For the superheat model, we will use the model by E. Tiryakioğlu (1964) for its simplicity and its independence from actual pouring temperature. Equation 6.13 can now be rewritten as:

$$k_t^{1.31} V_t^{0.67} = k_f^{1.31}(V_f - \alpha V_c)^{0.67} e^{\xi \Delta T f} \qquad \textbf{(6.14)}$$

where ξ is a constant dependent on the alloy (0.0028°C^{-1} for Al–Si eutectic alloy (Tiryakioğlu 1964) and 0.0033°C^{-1} for Al–7%Si (Tiryakioğlu *et al.* (1997b)), ΔT_f is the temperature change (rise or fall) in feeder because of the heat exchange, and can be easily calculated using Equation 6.14. The sum of change in heat content of the feeder and casting is zero (heat lost by one is gained by the other). Therefore:

$$C(V_f - \alpha V_c)\Delta T_f + C(V_f + \alpha V_c)\Delta T_c = 0 \qquad \textbf{(6.15)}$$

where C is the specific heat of the metal. Hence

$$\Delta T_c = (V_f - \alpha V_c)\Delta T_f/(V_f + \alpha V_c) \qquad \textbf{(6.16)}$$

The total solidification time of the casting can now be written as

$$t_c = k_c^{1.31}(V_f + \alpha V_c)^{0.67} e^{\xi \Delta T c} \qquad \textbf{(6.17)}$$

The solidification times of feeder and casting can now be compared. This comparison is

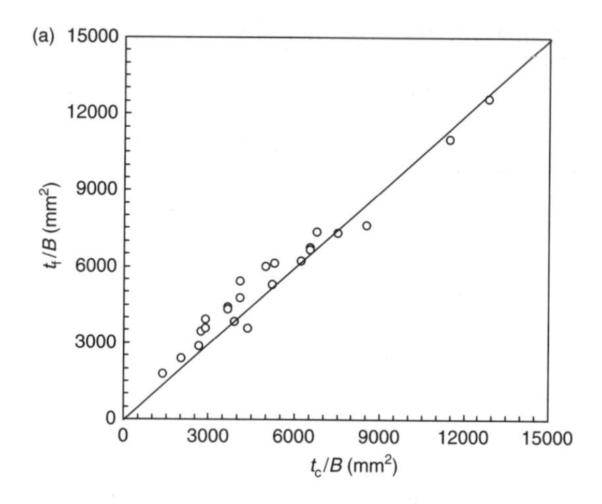

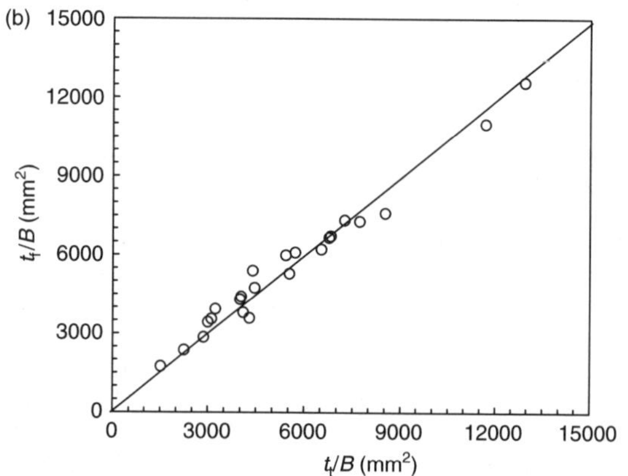

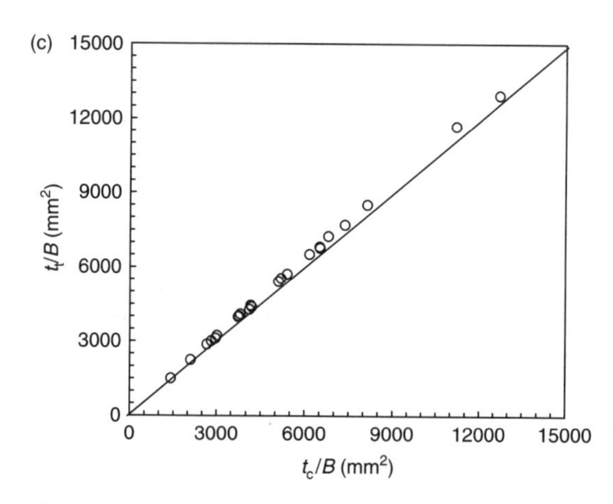

Figure 6.14 *(a) Comparison of calculated solidification times of (a) casting and feeder; (b) total casting and feeder; (c) total casting (or feeder) versus casting after adjustment to account for heat transfer between feeder and casting (Tiryakioğlu* et al. *2002).*

presented in Figure 6.14c, which shows a practically perfect fit and a relationship that can be expressed as:

$$t_t = t_f = \alpha t_c \qquad \textbf{(6.18)}$$

The data for Al–Si alloy shown in Figure 6.14c gives $\alpha = 1.046$.

In a separate exercise, using the data for steel by Bishop and co-workers (1955) assuming ξ of $0.0036°C^{-1}$ for steel (Tiryakioğlu 1964), a similar excellent relationship is obtained where α is found to be 1.005 (Tiryakioğlu 2002). Thus the solidification time of optimum-sized feeders in the feeder–casting combination was found to be only a few per cent longer than that of castings both for Al–Si alloy and steel castings.

We can conclude that for an accurate description of the action of a feeder, both mass and heat transfer from feeder to casting during solidification have to be taken into account simultaneously. Previous feeder models that account for mass transfer assume that the transfer takes place isothermally and at the pouring temperature. This previous assumption overestimates the additional heat brought into the casting from the feeder. The new model incorporates the effect of superheat and is based on the equality of solidification times of feeder and total casting.

The requirements for efficient feeders: (i) solidification time; (ii) feed metal availability; and (iii) prevention of hot-spot at the junction; can be combined into a single requirement when the casting–feeder combination is treated as a single, total casting. The three criteria reduce to the simple requirement: 'The thermal centre of the total casting will be in the feeder'.

The disarming simplicity of this conclusion conceals its powerful logic. It represents the ideal criterion for judging the success of a computer model of a casting and feeder combination.

6.4 Active feeding

Most feeding systems on castings are passive. They work by themselves without outside intervention. (Even those counter-gravity systems to which pressure is applied to enhance feeding are not considered to be 'active' in the sense discussed below.)

There has recently been introduced a novel system of feeding in which a side feeder is pressurized and is thus encouraged to feed uphill (Figure 6.15). This approach to feeding was developed for use with automatic moulding where moulds could not be inverted through 180 degrees after pouring. The concept appears to have proved useful within the context of fast, automatic greensand moulding for aluminium alloy castings in a vertically parted mould, where the application of the pressurizing tubes can be automated, and where the casting sizes and weights are limited. Current problems with repeatability may be the teething problems of the new technique that remain to be completely solved by additional effort.

The counter-gravity-filling of such moulds has been a considerable advance in terms of attaining a high quality of metal in the mould cavity. It has also been useful because the space taken by the down sprue is now released as additional moulding area.

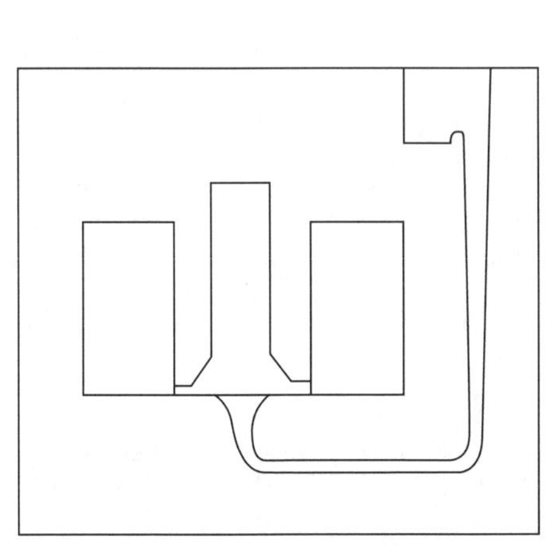

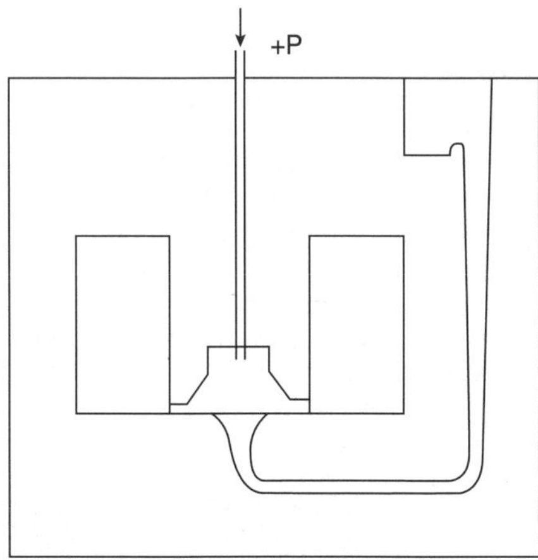

Figure 6.15 *Active feeding in a vertically parted automatic greensand moulding machine.*

The provision of small, compact feeders has been a similar benefit, saving mould space, although this is countered to some extent by the need for a direct line of access for the pressurizing tube to the top of the feeder. Greensand moulds have been shown to develop useful temperature gradients in thin-walled aluminium alloy castings (Rasmussen 1995). This natural gradient, the consequence of a good bottom-gating technique, is exploited by the bottom feeder.

It seems that the danger of convection reversing these advantages is small if the castings are of limited wall thickness and weight, which is the case for most casting produced on vertically-parted automatic moulding machines. At the time of writing, the limits are not yet known. Clearly, at some point of increasing wall thickness and casting weight an extended solidification time will encourage the development of convection, and feeding uphill in such circumstances will become problematic, if not actually impossible.

Thus although active feeding has been investigated by computer simulation and shown to have attractive advantages, its application to sections of 15 mm thickness in Al alloys (Hansen and Rasmussen 1994) seems likely to be close to, if not actually over, a limit at which convection will start to undo the benefits.

6.5 Freezing systems design

In this section on feeding, we are of course mainly concerned with the action of chills and fins to provide localized cooling of the casting. In this way we can assist directional solidification of the casting towards the feeder, thus assisting in the achievement of soundness. This is one of the important actions of these chilling devices. It is, however, not the only action, as discussed below.

Chills also act to increase the ductility and strength of that locality of the casting. From *Castings 2003* the proposal is that this occurs because the faster solidification freezes in the bifilms in their compact form before they have a chance to unfurl. (Recall that the bifilms are compacted by the extreme bulk turbulence during pouring and during their travel through the running system. However, they subsequently unravel, opening up in the mould cavity when conditions in the melt become quiet once again.)

The interesting corollary of this fact is that if chills are seen to increase ductility and strength of a casting, it confirms that the cast material is defective, containing a high percentage of bifilms. Another interesting corollary is that if a casting alloy can be cast without bifilms, chilling should not increase its properties. This rather surprising prediction is fascinating, and, if true, indicates the huge potential for the increase of the properties of cast alloys. All castings without bifilms are therefore predicted to have extraordinary ductility and strength. It also explains our lamentable current condition in which most of us constantly struggle in our foundries to achieve minimum mechanical properties for castings. Some days we win, other days we continue to struggle. The message is clear, we need to focus on technologies for the production of castings with reduced bifilm content, preferably zero bifilm content. The rewards are huge.

Another action of chills is to straighten bifilms. This action occurs because the advancing dendrites cannot grow through the air layer between the double films, and so push the bifilms ahead. Those that are somehow attached to the wall will be partially pushed, straightened and unravelled by the gentle advance of grains. This effect is reported in *Castings* (2003). Thus although a large percentage of bifilms will be pushed ahead of the chilled region, concentrating (and probably reducing the properties) in the region immediately ahead, some bifilms will remain aligned in the dendrite growth direction, and so be largely perpendicular to the mould wall.

The overall effects on mechanical properties of the pushing action are not so easily predicted. The reduction in density of defects by the chill will raise properties, but the presence of occasional bifilms aligned at right angles to the surface of the casting would be expected to be severely detrimental. These complicated effects require to be researched. However, we can speculate that they seem likely to be the cause of troublesome edge cracking in the rolling of cast materials of many types, leading to the expense of machining off the surface of many alloys before rolling can be attempted. The superb formability of electroslag remelted compared to vacuum arc remelted alloys is almost certainly explained in this way. The ESR process produces an extremely clean material because oxide films will be dissolved during remelting under the layer of liquid slag, and will not re-form in the solidifying ingot. In contrast, the relatively poor vacuum of the VAR process ensures that the lapping of the melt over the liquid meniscus at the mould wall will create excellent double oxide films. If considerable depths of the surface are not first removed 'oxide lap defects' will open as surface cracks when subjected to forging or rolling.

Although, as outlined above, the chilling action of chills and fins is perhaps more complicated than we first thought, the chilling action itself on the rate of solidification is well documented and understood. It is this thermal aspect of their behaviour that is the subject of the remainder of this section.

6.5.1 External chills

In a sand mould the placing of a block of metal adjacent to the pattern, and subsequently packing the sand around it to make the rest of the mould in the normal way, is a widely used method of applying localized cooling to that part of the casting. A similar procedure can be adopted in gravity and low-pressure die-casting by removing the die coat locally to enhance the local cooling rate. In addition, in dies of all types, this effect can be enhanced by the insertion of metallic inserts into the die to provide local cooling, especially if the die insert is highly conductive (such as made from copper alloy) and/or artificially cooled, for instance by air, oil or water.

Such chills placed as part of the mould, and that act against the outside surface of the casting are strictly known as external chills, to distinguish them from internal chills that are cast in, and become integral with, the casting.

In general terms, the ability of the mould to absorb heat is assessed by its heat diffusivity. This is defined as $(K\rho C)^{1/2}$ where K is the thermal conductivity, ρ the density, and C the specific heat of the mould. It has complex units $\mathrm{J\,m^{-2}\,K^{-1}\,s^{-1/2}}$. (Take care not to confuse with thermal diffusivity defined as $K/\rho C$, and normally quoted in units of $\mathrm{m^2 s^{-1}}$.) From the room temperature data in Table 6.3 we can obtain some comparative data on the chilling power of various mould and chill materials, shown in Table 6.4.

It is clear that the various refractory mould materials—sand, investment and plaster—are all poor absorbers of heat, and become worse in that order. The various chill materials are all in a league of their own, having chilling powers orders of magnitude higher than the refractory mould materials. They improve marginally,

Table 6.3 Mould and metal constants

Material	*Melting point* (°C)	*Liquid–solid contraction* (%)	*Specific heat* ($\mathrm{J\,kg^{-1}K^{-1}}$)			*Density* ($\mathrm{kg\,m^{-3}}$)			*Thermal conductivity* ($\mathrm{J m\,K^{-1}\,s^{-1}}$)		
			Solid		*Liquid m.p.*	*Solid*		*Liquid m.p.*	*Solid*		*Liquid m.p.*
			20°C	*m.p.*		*20°C*	*m.p.*		*20°C*	*m.p.*	
Pb	327	3.22	130	(138)	152	11680	11020	10678	39.4	(29.4)	15.4
Zn	420	4.08	394	(443)	481	7140	(6843)	6575	119	95	9.5
Mg	650	4.2	1038	(1300)	1360	1740	(1657)	1590	155	(90)?	78
Al	660	7.14	917	(1200)	1080	2700	(2550)	2385	238	–	94
Cu	1084	5.30	386	(480)	495	8960	8382	8000	397	(235)	166
Fe	1536	3.16	456	(1130)	795	7870	7265	7015	73	(14)?	–
Graphite	–	–	1515	–	–	2200	–	–	147		
Silica sand	–	–	1130	–	–	1500	–	–	0.0061	–	
Investment (Mullite)			750	–	–	1600	–	–	0.0038	–	–
Plaster	–	–	840	–	–	1100	–	–	0.0035	–	–

References: Wray (1976); Brandes (1991); Flemings (1974)

Table 6.4 Thermal properties of mould and chill materials at approximately 20°C

Material	*Heat Diffusivity* $(K\rho C)^{1/2}$ ($\mathrm{Jm^{-2}\,K^{-1}s^{-1/2}}$)	*Thermal Diffusivity* $K/\rho C$ ($\mathrm{m^2s^{-1}}$)	*Heat Capacity per unit volume* ρC ($\mathrm{JK^{-1}m^{-3}}$)
Silica sand	3.21×10^3	3.60×10^{-9}	1.70×10^6
Investment	2.12×10^3	3.17×10^{-9}	1.20×10^6
Plaster	1.8×10^3	3.79×10^{-9}	0.92×10^6
Iron (pure Fe)	16.2×10^3	20.3×10^{-6}	3.94×10^6
Graphite	22.1×10^3	44.1×10^{-6}	3.33×10^6
Aluminium	24.3×10^3	96.1×10^{-6}	2.48×10^6
Copper	37.0×10^3	114.8×10^{-6}	3.60×10^6

within a mere factor of 5, in the order steel, graphite, and copper.

The heat diffusivity value indicates the action of the material to absorb heat when it is infinitely thick, i.e. as would be reasonably well approximated by constructing a thick-walled mould from such material. When a relatively small lump of cast iron or graphite is used as an external chill in a sand mould, it does not develop its full potential for chilling as indicated by the heat diffusivity because it has limited capacity for heat.

Thus although the initial rate of freezing of a metal may be in the order given by the above list, for a chill of limited thickness its cooling effect is limited because it becomes saturated with heat; after a time it can absorb no more. The amount of heat that it can absorb is defined as its heat capacity. We can formulate the useful concept of volumetric heat capacity in terms of its volume V, its density ρ and its specific heat C:

$$\text{Volumetric heat capacity} = V\rho C$$

In the SI system its units are $\mathrm{J\,K^{-1}}$. The results by Rao and Panchanathan (1973) on the casting of 50 mm thick plates in Al–5Si–3Cu reveals that the casting is insensitive to whether it is cooled by steel, graphite or copper chills, provided that the volumetric heat capacity of the chill is taken into account.

These authors show that for a steel chill 25 mm thick its heat capacity is $900\,\mathrm{J\,K^{-1}}$. A chill with identical capacity in copper would be required to be 32 mm thick, and in graphite 36 mm. These values led the author to conclude (*Castings 1991*) that copper may therefore not always be the best chill material. However, using somewhat more accurate data, copper is found, after all, to be best. These results are presented in Figure 6.16 showing the relative heat capacities and diffusivities. It is clear that for similar thicknesses of a block chill, copper is always most effective whether limited by heat diffusivity or heat capacity.

Figure 6.17 illustrates that the chills are effective over a considerable distance, the largest chills greatly influencing the solidification time of the casting even up to 200 mm (four times the section thickness of the casting) distant. This large distance is perhaps typical of such a thick-section casting in an alloy of high thermal conductivity, providing excellent heat transfer along the casting. A steel casting would respond less at this distance.

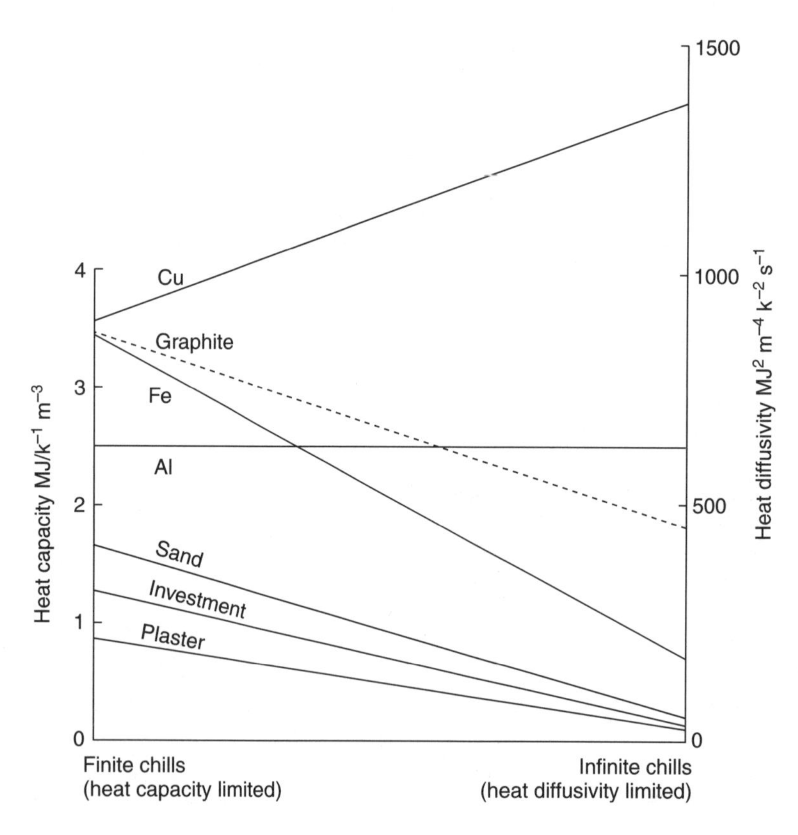

Figure 6.16 *Relative diffusivities (ability to diffuse heat away if a large chill) and heat capacities (ability to absorb heat if relatively small) of chill materials.*

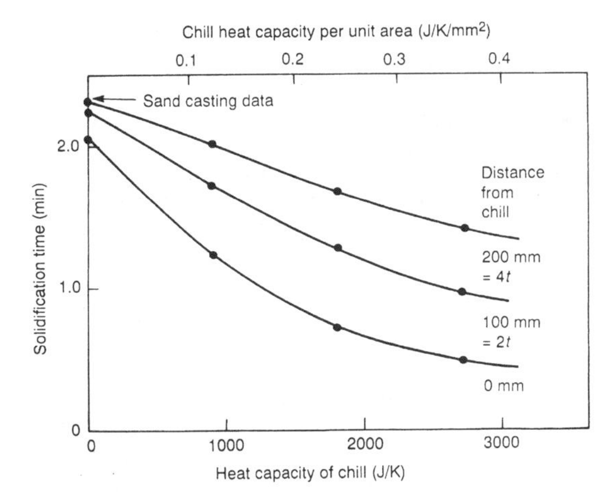

Figure 6.17 *Freezing time of a plate 225 × 150 × 50 mm in Al–5Si–3Cu alloy at various distances from the chilled end is seen to decrease steadily as the chill is approached, and as the chill size is increased (Rao and Panchanathan 1973).*

The work by Rao and Panchanathan reveals the widespread sloppiness of much present practice on the chilling of castings. General experience of the chills generally used in foundrywork nowadays shows that chill size and weight are rarely specified, and that chills are in general too small to be fully effective in any particular job. It clearly matters what size of chill is added.

Computational studies by Lewis and colleagues (2002) have shown that the number, size and location of chills can be optimized by computer. These studies are among the welcome first steps towards the intelligent use of computers in casting technology.

Finally, in detail, the action of the chill is not easy to understand. The surface of the casting against the chill will often contract, distorting away and thus opening up an air gap. The chilled casting surface may then reheat to such an extent that the surface remelts. The exudation of eutectic is often seen between the casting and the chill (Figure 6.18). The new contact between the eutectic and chill probably then starts a new burst of heat transfer and thus rapid solidification of the casting. Thus the history of cooling in the neighbourhood of a chill may be a succession of stop/start, or slow/fast events.

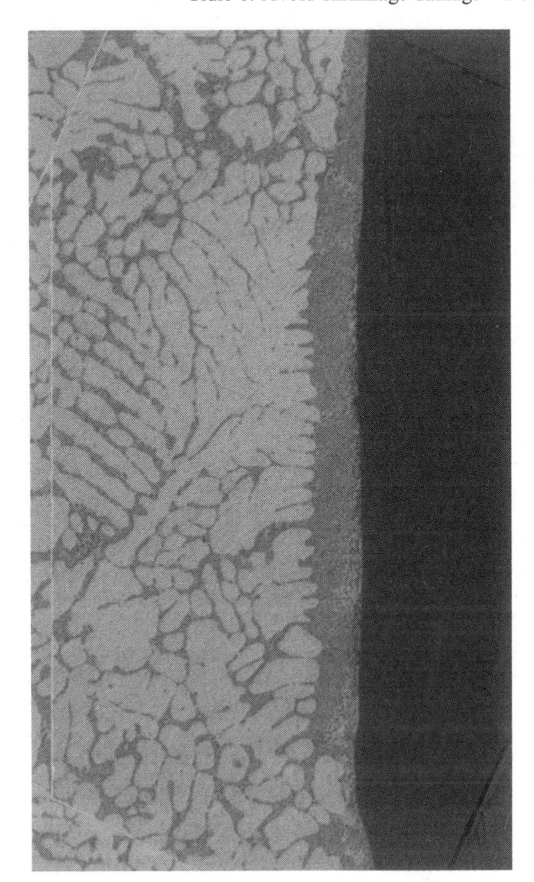

Figure 6.18 *Al–Si eutectic liquid segregation by exudation at a chilled interface of an Al–Si alloy.*

6.5.2 Internal chills

The placing of chills inside the mould cavity with the intention of casting them in place is an effective way of localized cooling. The simple method of mixtures approach (Campbell and Caton 1977) indicates that to cool superheated pure liquid iron to its freezing point, and freezing a proportion of it, will require various levels of addition of cold, solid iron depending on the extent that the added material is allowed to melt (Table 6.5). These calculations take no account of other heat losses from the casting. Thus for normal castings the predictions are likely to be incorrect by up to a factor of 2. This is broadly confirmed by Miles (1956), who top-poured steel into dry sand moulds 75 mm square and 300 mm tall. In the centre of the moulds was positioned a variety of steel bars ranging from 12.5 mm round to 25 mm square, covering a range of chilling from 2 to 11 per cent solid addition. His findings reveal that the 2 per cent solid addition nearly melted, compared to the predicted value for complete melting of 3.5 per cent solid. The 11 per cent solid addition caused extensive (possibly total) freezing of the casting judging by the appearance of the radial grain

Table 6.5 Weight percentage of internal chills in pure cast iron

Calculated addition (%) (Campbell and Caton 1977)	*Observed addition* (%) (Miles, 1956)	*Result*
3.0		Chill completely melted
3.5	≃2	Chill reaches melting point, but does not melt
7.0		50% of melt is solidified
10.5	≃11	100% of melt is solidified

structure in the macrosections. He found 5 per cent addition to be near optimum; it had a reasonable chilling effectiveness but caused relatively few defects.

In the case of the higher additions, where the heat input is not sufficient to melt the chill, the fusing of the surface into the casting has to be the result of a kind of diffusion bonding process. This would emphasize the need for cleanness of the surface, requiring the minimum presence of oxide films or other debris against the chill during the filling of the mould. If Miles had used a better bottom-gated filling technique he may have reduced the observed filling defects further, and found that higher percentages were practical.

The work by Miles does illustrate the problems generally experienced with internal chills. If the chills remain for any length of time in the mould, particularly after it is closed, and more particularly if closed overnight, then condensation is likely to occur on the chill, and blow defects will be caused in the casting. Blows are also common from rust spots or other impurities on the chill such as oil or grease. The matching of the chemical composition of the chill and the casting is also important; mild steel chills will, for instance, usually be unacceptable in an alloy steel casting.

Internal chills in aluminium alloy castings have not generally been used, almost certainly as a consequence of the difficulty introduced by the presence of the oxide film on the chill. This appears to be confirmed by the work of Biswas *et al.* (1985), who found that at 3.5 per cent by volume of chill and at superheats of only 35°C the chill was only partially melted and retained part of its original shape. It seems that over this area it was poorly bonded. At superheats above 75°C, or at only 1.5 per cent by volume, the chill was more extensively melted, and was useful in reducing internal porosity and in raising mechanical properties. The lingering presence of the oxide film from the chill remains a concern however.

The development of a good bond between the internal chill and the casting is a familiar problem with the use of chaplets—the metal devices used to support cores against sagging because of weight, or floating because of buoyancy. A one page review of chaplets is given by Bex (1991). To facilitate the bond for a steel chaplet in an iron or steel casting the chaplet is often plated with tin. The tin serves to prevent the formation of rust, and its low melting point (232°C) and solubility in iron assists the bonding process.

The bond between steel and titanium inserts in Al alloy castings has been investigated in Japan (Noguchi *et al.* 2001) who found only a 10 μm silver coating was effective to achieve a good bond, although even this took up to 5 minutes to develop at the Al–Ag eutectic temperature 566°C. Attempts to achieve a bond with gold plating and Al–Si sprayed alloy were largely unsuccessful.

It seems, therefore, that internal chills in aluminium alloys might be tolerable to tackle porosity problems in castings that are difficult to tackle by other techniques. However, the oxide film remains an ever-present danger. It will persist as a double film (having acquired its second layer during the immersion of the chill) and so pose the risk of leakage or crack formation. Such risks are only acceptable for low duty products.

Brown and Rastall (1986) use the non-bonding of heavier aluminium inserts in aluminium castings to advantage. They use a cast aluminium alloy core inside an aluminium alloy casting to form re-entrant details that could not easily be provided in a pressure die cast product. Also, of course, because the freezing time is shortened, productivity is enhanced. The internal core is subsequently removed by disassembly or part machining, or by mechanical deformation of the core or casting.

6.5.3 Fins

Before we look specifically at fins on castings, it is worth spending some time to consider the concepts involved in junctions of all types between different cast sections. Figure 2.34 shows a complete range of T-junctions between walls of different relative thickness. When the wall forming the upright of the T is thin, it acts as a cooling fin, chilling the junction and the adjacent wall (the top cross of the T) of the casting. We shall return to a more detailed consideration of fins shortly.

When the upright of the T-section has increased to a thickness of half the casting section thickness, then the junction is close to thermal balance, the cooling effect of the fin balancing the hot-spot effect of the concentration of metal in the junction. Kotschi and Loper (1974) were among the first to evaluate junctions and highlighted this special case.

By the time that the upright of the T has become equal to the casting section, the junction is a hot spot. This is common in castings. Foundry engineers are generally aware the 1 : 1 T-junction is a problem. It is curious therefore that castings with even wall thickness are said to be preferred, and that designers are encouraged to design them. Such products necessarily contain 1 : 1 junctions that will be hot spots. However, because the 1 : 1 thickness junction is such an intractable problem, Mertz and Heine (1973) suggest that it should be fed from its end, along its length, and thereby used as a feeding path. In fact, they go further and recommend generous convex radii for the fillets and the planting of a pad on the far side of the T to maximize the feeding distance along the junction, as illustrated in Figure 6.19.

When finally the section thickness of the upright of the T is twice the casting section, then the junction is balanced once again, with the casting now acting as the mild chill to counter the effect of the hot spot at the junction. We have considered these junctions merely in the form of the intersections of plates. However, we can extend the concept to more general shapes, introducing the use of the geometric modulus $m = \text{(volume)/(cooling area)}$. It subsequently follows that an additional requirement when a feeder forms a T-junction on a casting is that the feeder must have a modulus two times the modulus of the casting. The hot spot is then moved out of the junction and into the feeder, with the result that the casting is sound. This is the basis behind Rule 4 for feeding discussed in Section 6.2.

Pellini (1953) was one of the first experimenters to show that the siting of a thin 'parasitic' plate on the end of a larger plate could improve the temperature gradient in the larger plate. However, the parasitic plate that he used was rather thick, and his experiments were carried out only on steel, whose conductivity is poor, reducing useful benefits.

Figure 6.20 shows the results from Kim *et al.* (1985) of pour-out tests carried out on 99.9% pure aluminium cast into sand moulds. The faster advance of the freezing front adjacent to the junction with the fin is clearly shown. (As an aside, this simple result is a good test of some computer simulation packages. The simulation of a brick-shaped casting with a cast-on fin should show the cooling effect by the fin. Some relatively poor computer algorithms do not take into account the conduction of heat in the casting, thus predicting, erroneously, the appearance of the junction as a hot spot.)

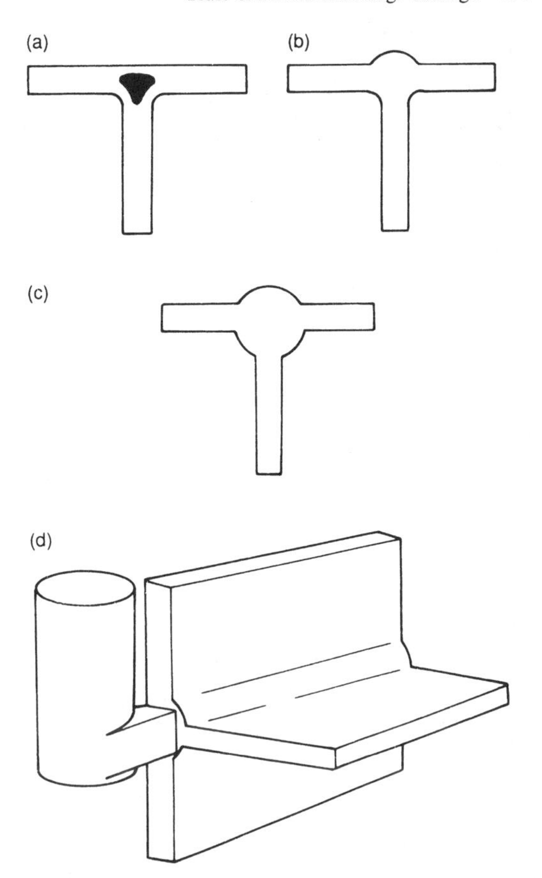

Figure 6.19 *(a) T-junction with normal concave fillet radius; (b) marginal improvement to the feed path along the junction; (c) convex fillets plus pad that doubles feeding distance along the junction; and (d) practical utilization of a T-junction as a feed path (Mertz and Heine 1973).*

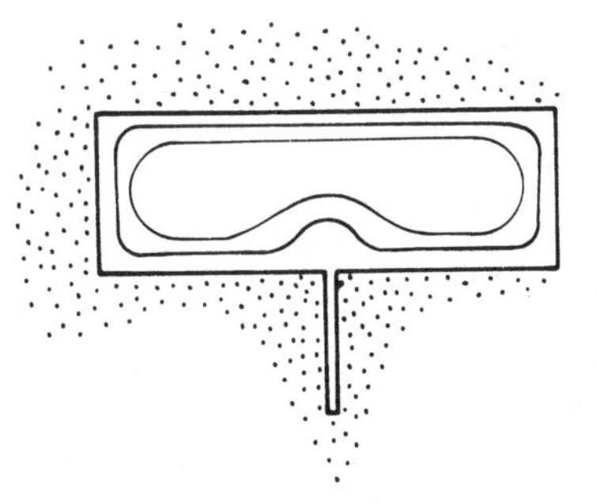

Figure 6.20 *A T-junction casting in 99.9 Al by Kim* et al. *(1985) showing successive positions of the freezing front.*

Creese and Sarfaraz (1987) demonstrate the use of a fin to chill a hot spot in pure Al castings

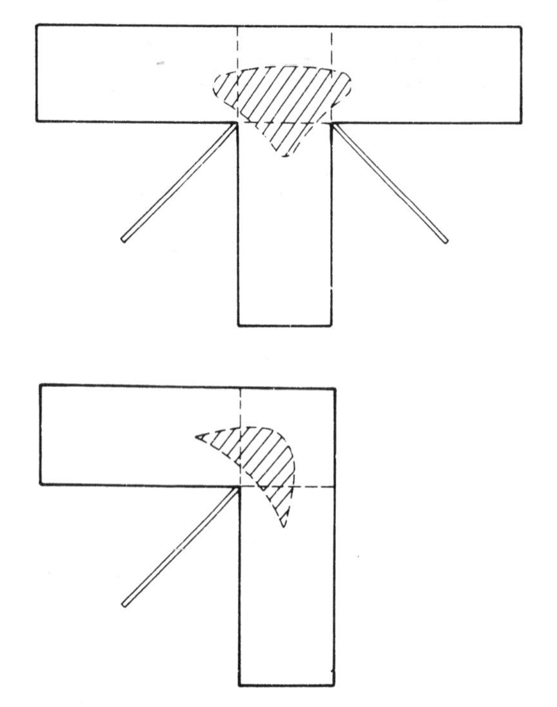

Figure 6.21 *T- and L-junctions in pure aluminium cast in oil-bonded greensand. The shape of porosity in these junctions is shown, and the region of the junction used to calculate the percentage porosity is shown by the broken lines. The position of fins added to eliminate the porosity is shown. Results are presented in Figure 6.22.*

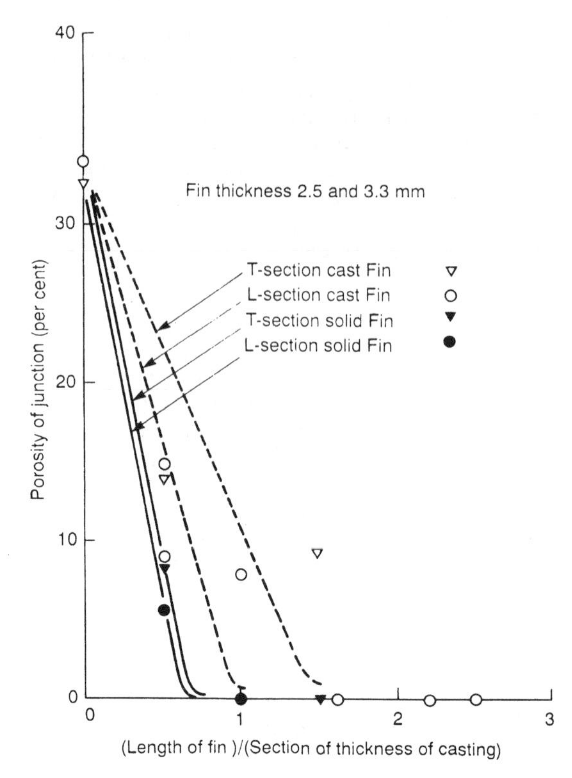

Figure 6.22 *Results from Creese and Sarfaraz (1987, 1988) showing the reduction in porosity as a result of increasing length of fins applied as in Figure 6.21.*

that were difficult to access in other ways. They cast on fins to T- and L-junctions as shown in Figure 6.21. The reduction in porosity achieved by this technique is shown in Figure 6.22. For these casting sections of 50 mm there was no apparent difference between fins of 2.5 and 3.3 mm thickness so these results are treated together in this figure. These fins at 5 and 6 per cent of the casting section happen to be close to optimum as is confirmed later below. The reason that they conduct away perhaps less effectively than might be expected is because of their unfavourable location at 45 degrees to two hot components of the junction.

Returning to the case where the upright of the T is sufficiently thin to act as a cooling fin, one further case that is not presented in Figure 2.34 is the case where the fin is so thin that it does not exist. This, you will say, is a trivial case. But think what it tells us. It proves that the fin can be too thin to be effective, since it will have insufficient area to carry away enough heat. Thus there is an optimum thickness of fin for a given casting section.

Similarly, an identical argument can be made about the fin length. A fin of zero length will have zero effect. As length increases, effectiveness will increase, but beyond a certain length, additional length will be of reducing value. Thus the length of fins will also have an optimum.

These questions have been addressed in a preliminary study by Wright and Campbell (1997) on a horizontal plate casting with a symmetrical fin (Figure 6.23). Symmetry was chosen so that thermocouple measurements could be taken along the centreline (otherwise the precise thermal centre was not known so that the true extension in freezing time may not have been measured accurately). In addition the horizontal orientation of the plate was selected to suppress any complicating effects of convection so far as possible. The thickness of the fin was $B.H$ and the length $L.H$ where B and L are dimensionless numbers to quantify the fin in terms of H, the thickness of the plate. From this study it was discovered that there was an optimum thickness of a fin, and this was less than one tenth of H. Figure 6.23a interpolates an optimum in the region of 5 per cent of the casting section thickness. The optimum length was $2H$, and longer lengths were not effective (Figure 6.23b). For these conditions the freezing

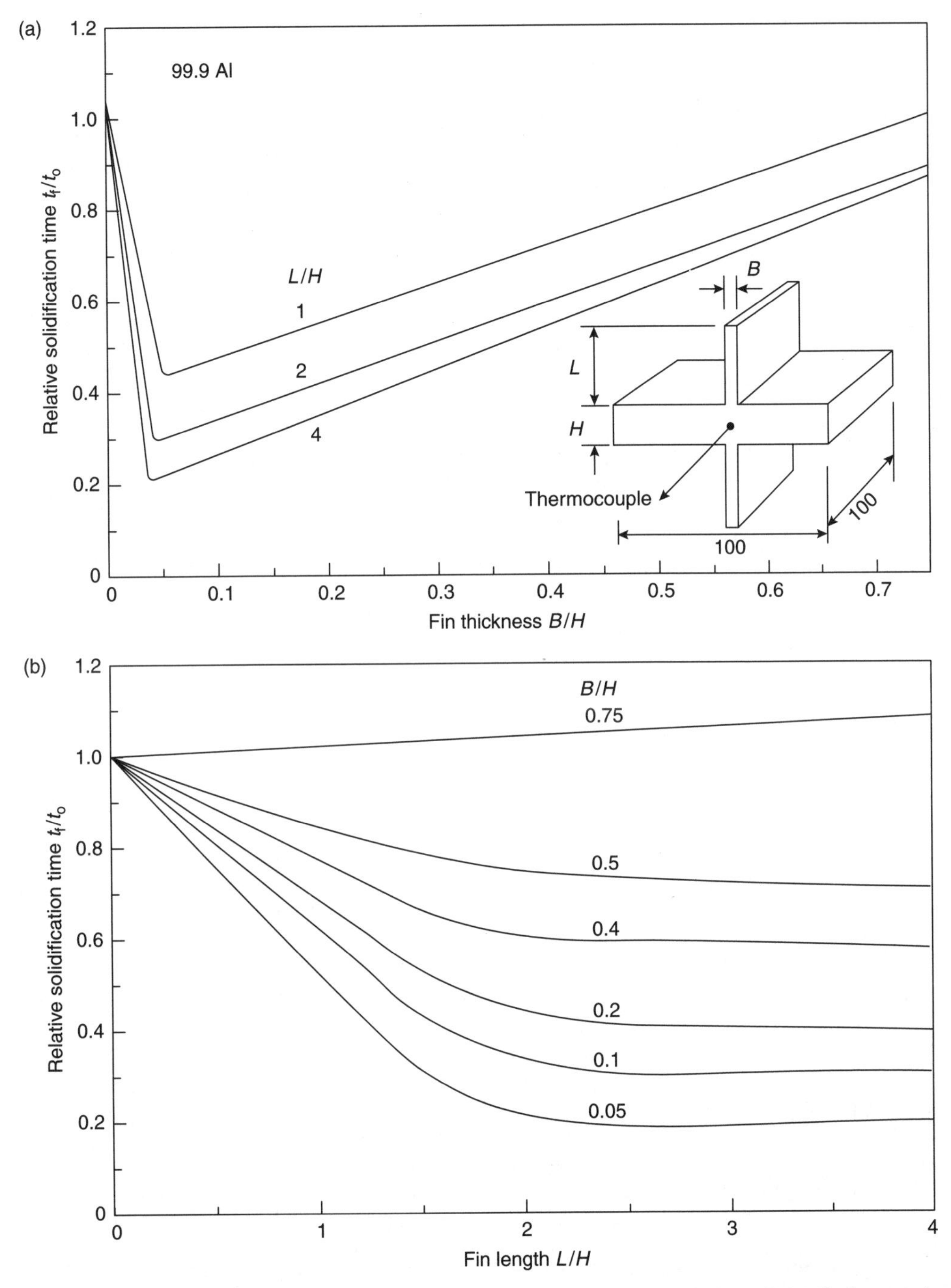

Figure 6.23 *The effect of a symmetrical fin on the freezing time at the centre of a cast plate of 99.9Al alloy as a function of the length and thickness of the fin (averaged results of simulation and experiment from Wright and Campbell 1997).*

time of the casting was increased by approximately ten times. Thus the effect is useful. However, the effect is also rather localized, so that it needs to be used with caution. Eventually, non-symmetrical results for a chill on one side of the plate would be welcome.

Even so, the practical benefits to the use of a fin as opposed to a chill are interesting, even

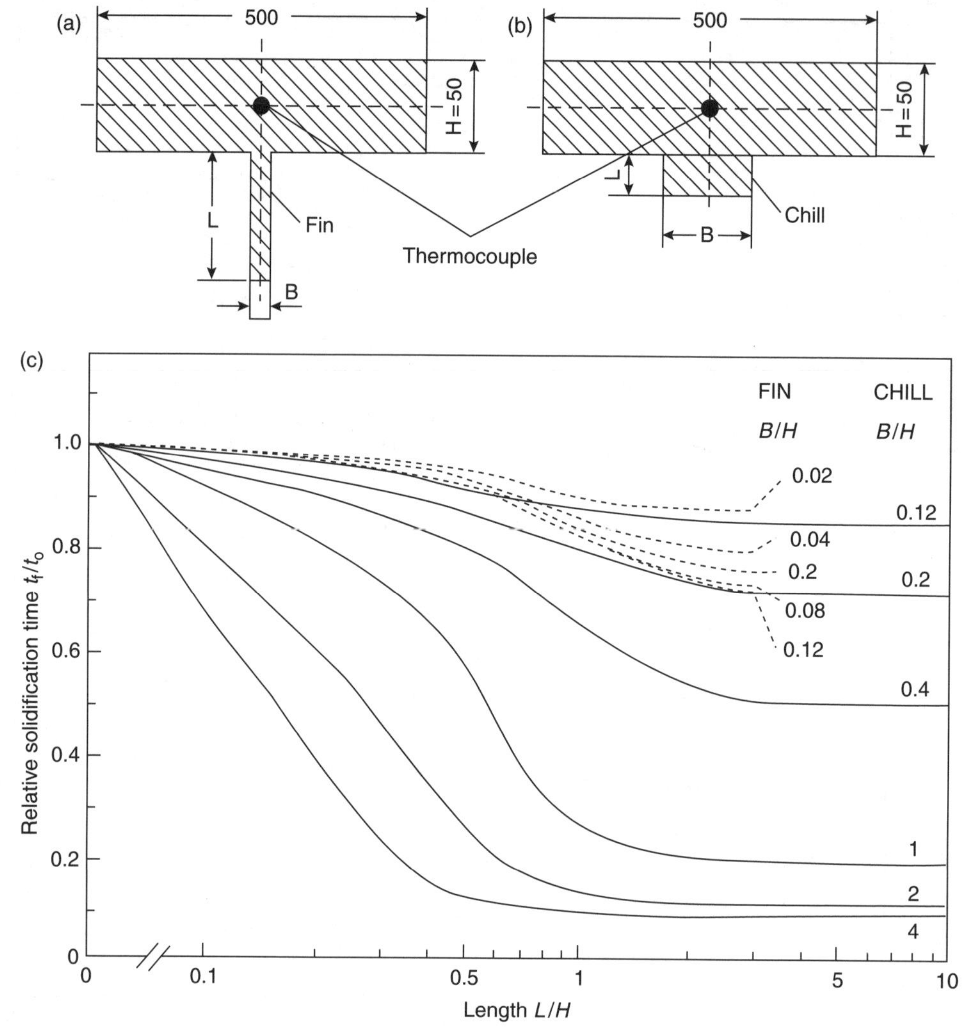

Figure 6.24 *A comparison of the action of chills and cooling fins in aluminium bronze alloy AB1 (Wen, Jolly and Campbell 1997).*

compelling. They are:

1. The fin is always provided on the casting, because it is an integral part of the tooling. Thus, unlike a chill, the placing of it cannot be forgotten.
2. It is always exactly in the correct place. It cannot be wrongly sited before the making of the mould. (The incorrect positioning of a chill is easily appreciated, because although the location of the chill is normally carefully painted on the pattern, the application of the first coat of mould release agent usually does an effective job in eliminating all traces of this.)
3. It cannot be displaced or lifted during the making of the mould. If the chill lifts slightly during the filling of the tooling with sand the resulting sand penetration under the edges of the chill, and the casting of additional metal into the roughly shaped gap, make an unsightly local mess of the casting surface. Displacement or complete falling out from the mould is a common danger, sometimes requiring studs to support the chill if awkwardly angled or on a vertical face. Displacement commonly results in sand inclusion defects around the chill or can add to defects elsewhere. All this is expensive to dress off.

4. An increase in productivity has been reported as a result of not having to find, place and carefully tuck in a block chill into a sand mould (Dimmick 2001).
5. It is easily cut off. In contrast, the witness from a chill also usually requires substantial dressing, especially if the chill was equipped with v-grooves, or if it became misplaced during moulding, as mentioned above.
6. The fin does not cause scrap castings because of condensation of moisture and other volatiles, with consequential blow defects, as is a real danger from chills.
7. The fin does not require to be retrieved from the sand system, cleaned by shot blasting, stored in special bins, re-located, counted, losses made up by re-ordering new chills, casting new chills (particularly if the chill is shaped) and finally ensuring that the correct number in good condition, re-coated, and dried, is delivered to the moulder on the required date.
8. The fin does not wear out. Old chills become rounded to the point that they are effectively worn out. In addition, in iron and steel foundries, grey iron chills are said to 'lose their nature' after some use. This seems to be the result of the oxidation of the graphite flakes in the iron, thus impairing the thermal conductivity of the chill.
9. Sometimes it is possible to solve a localized feeding problem (the typical example is the isolated boss in the centre of the plate) by chilling with a fin instead of providing a local supply of feed metal. In this case the fin is enormously cheaper than the feeder.

This lengthy list represents considerable costs attached to the use of chills that are not easily accounted for, so that the real cost of chills is often underestimated.

Even so, the chill may be the correct choice for technical reasons. Fins perform poorly for metals of low thermal conductivity such as zinc, Al-bronze, iron and steel. The computer simulation result in Figure 6.24 illustrates for the rather low thermal conductivity material, Al-bronze, that there are extensive conditions in which the chill is far more effective.

The kind of result shown in Figure 6.24 would be valuable if available for a variety of casting alloys varying from high to low thermal conductivity, so that an informed choice could be made whether a chill or fin was best in any particular case. These results have yet to be worked out and published.

Fins are most easily provided on a joint line of the mould, or around core prints. Sometimes, however, there is no alternative but to mould them at right angles to the joint. From a practical point of view, these upstanding fins on patternwork are of course vulnerable to damage. Dimmick (2001) records that fins made from flexible and tough vinyl plastic solved the damage problem in their foundry. They would carry out an initial trial with fins glued onto the pattern. If successful, the fins would then be permanently inserted into the pattern. In addition, only a few standard fins were found to be satisfactory for a wide range of patterns; a fairly wide deviation from the optimum ratios did not seem to be a problem in practice.

Sarfaraz and Creese (1989) investigated an interesting variant of the cast-on fin. They applied metal fins to the pattern, and rammed them up in the sand as though applying a normal external chill, in the manner shown in Figure 6.21. The results of these 'solid' or 'cold' fins (so called to distinguish them from the empty cavity that would, after filling with liquid metal, effectively constitute a 'cast' or 'hot' fin) are also presented in Figure 6.22. It is seen that the cold fins are more effective than the cast fins in reducing the porosity in the junction castings. This is the consequence of the heat capacity of the fin being used in addition to its conducting role. It is noteworthy that this effect clearly overrides the problem of heat transfer across the casting/chill interface.

The cold fin is, of course, really a chill of rather slim shape. It raises the interesting question, that as the geometry of the fin and the chill is varied, which can be the most effective. This question has been tackled in the author's laboratory (Wen and colleagues 1997) by computer simulation. The results are summarized in Figure 6.24. Clearly, if the cast fin is sufficiently thin, it is more effective than a thin chill. However, for normal chills that occupy a large area of the casting (effectively approaching an 'infinite' chill as shown in the figure), as opposed to a slim contact line, the chill is massively more effective in speeding the freezing of the casting.

Other interesting lessons to be learned from Figure 6.24 are that a chill has to be at least equal to the section thickness of the casting to be really effective. A chill of thickness up to twice the casting section is progressively more valuable. However, beyond twice the thickness, increasingly thick chills show progressively reducing benefit.

It is to be expected that in alloys of higher thermal conductivity than aluminium bronze, a figure such as Figure 6.24 would show a greater regime of importance for fins compared to chills. The exploration of these effects for a variety of materials would be instructive and remains as a task for the future.

The business of getting the heat away from the casting as quickly as possible is taken to a logical extreme by Czech workers (Kunes *et al.* 1990) who show that a heat pipe can be extremely effective for a steel casting. Canadian workers (Zhang *et al.* 2003) explore the benefits of heat pipes for aluminium alloys. The conditions for successful application of the principle are not easy, however, so I find myself reluctant at this stage to recommend the heat pipe as a general purpose technique in competition to fins or chills. In special circumstances, however, it could be ideal.

Rule 7

Avoid convection damage

7.1 Convection: the academic background

Convection is the flow phenomenon that arises as a result of density differences in a fluid.

In a solidifying casting the density differences in the residual liquid can be the result of differences in solute content as a consequence of segregation. This is a significant driving force for the development of channel defects known as the 'A' and 'V' segregates in steel ingots and as freckle trails in nickel- and cobalt-base investment castings. The name 'freckles' comes from the appearance of the etched components that shows randomly oriented grains in the channels that have been partly remelted in the convecting flow and detached from their original dendrites. These defects are discussed earlier in *Castings (2003)* and are not discussed further here. Although, for many reasons, channel defects are unwelcome, they are usually not life threatening to the product.

Convection can also arise as a result of density differences that result from temperature differences in the melt. There have been numerous theoretical studies of the solidification of low melting point materials in simple cubical moulds, of which one side is cooled and the other not. The resulting gentle drift of liquid around the cavity, down the cool face and up the non-cooled face, changes the form of the solidifying front. A schematic example is shown in Figure 7.1. These are interesting exercises, but give relatively little assistance to the understanding of the problems of convective flow in engineering systems.

The results due to Mampaey and Xu (1999) who studied the natural convection in an upright cylinder of solidifying cast iron showed that the thermal centre of the liquid mass was shifted upwards, and graphite nodules in spheroidal graphite irons were transported by the flow. Such studies reflect the gentle action of convection in small, simple shaped, closed systems; the kind of action one would expect to see in a cooling cup of tea. These facts have lulled us into a state of false security, assuming convection to be essentially harmless and irrelevant. We need to think again.

7.2 Convection: the engineering imperatives

Convection was practically unknown as an important factor in shaped castings until the early 1980s. Even now, it is not widely known nor understood. However, it can be life and death to a casting, and has been the death of a number of attempts to develop counter-gravity casting systems around the world. Most workers in this endeavour still do not know why they failed. The Cosworth Casting Process nearly foundered on this problem in its early days, only solving the problem by its famous (infamous?) roll-over system.

Thus convection is not merely a textbook curiosity. The casting engineer requires to come to terms with convection as a matter of urgency. The problem can be of awesome importance, and can lead to major difficulties, if not impossibilities, to achieve a sound and saleable casting.

Convection enhances the problems of uphill feeding in medium section castings, making them extremely resistant to solution. In fact increasing the amount of (uphill) feeding by increasing the diameter of the feeder neck, for

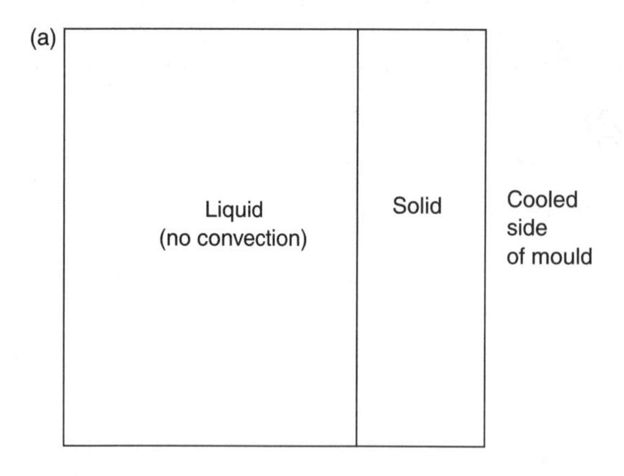

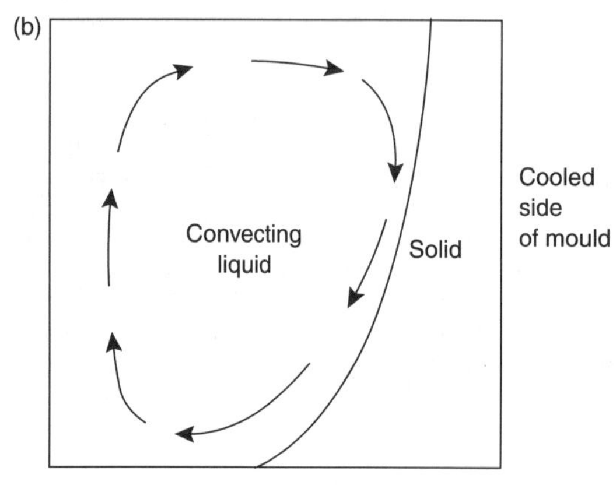

Figure 7.1 *Solidification in 2-D box, of which only the right-hand side is cooled. (a) planar front in the case of no convection; (b) the distortion caused by convective flow.*

instance, makes the feeding problem worse by increasing the opportunity for convection. Many of the current problems of low-pressure casting systems derive from this source.

In contrast, having feeders at the top of the casting, and feeding downwards under gravity is completely stable and predictable, and gives reliable results.

The instability of convective problems is worth emphasizing. Because the heavy, cool liquid overlays the hotter less dense liquid, the situation is metastable. If the stratified layers of liquid are not disturbed there is a chance that the heavy liquid will remain wobbling around on the top, and may solidify in place without incident. However, a small disturbance may upset the delicate balancing act. Once started, the cold melt will slip sideways, plunging downwards to the bottom, and the hot liquid surge upwards, so that a convective circulation will quickly establish. In practice therefore, a number of castings may be made successfully if the metastable equilibrium is not disturbed, but, inexplicably, the next may exhibit massive remelted and unfed regions.

Triggers to initiate the unstable flow could arise from many different kinds of uncontrolled events. A significant trigger could be an event such as the rising of a bubble from a core blow, as a result of an occasionally ill-fitting of a core print, leading to the chance sealing of the core vent by liquid metal.

Momchilov (1993) gives one of the very few accounts of the exasperating randomness of convection problems. He found that with the use of two riser tubes from a furnace containing liquid metal into one die cavity, successive castings could be observed to have completely different internal temperature histories. The first casting might be fine. However, the subsequent casting would suffer a die temperature inexplicably overheating by 120°C and the temperature in the furnace simultaneously dropping by 65°C. These are powerful and important exchanges of heat between the die and the crucible below. These changes caused the second casting to be partially remelted.

The use of twin riser tubes by Momchilov raises an important feature of convection. Convective flows require to be continuous, as in a circulation. Thus in the case of two riser tubes into one die cavity, the conditions for a circular flow, up one tube and down the other, are ideal. It is likely that Momchilov would have solved his problem, or at least greatly reduced it, simply by blocking off one of the tubes.

The elimination of ingates in this way to solve convection problems in counter-gravity fed castings should be considered as a standard first step. This was found to be a useful measure in the early days of the Cosworth Process when it operated merely as a static low-pressure casting process. (The later development of the roll-over concept represented a welcome total solution.)

The only other description of the problems of convection ever discovered by the author comes from a patent by Rogers and Heathcock (1990). They fall foul of convection during the attempt to make an aluminium alloy cylinder block casting in a counter-gravity filled permanent mould. They found that as the mould heated up the problem became worse, and the rate of flow of the convection currents increased. The microstructure of the casting was unacceptable in the area affected by convection. They dealt with the problem by providing strong cooling just above the ingates. This solution clearly threatened the provision of feed metal while the casting was solidifying, and so was a risky strategy. There is no record that the patent was ever implemented in production. Perhaps convection secured another victim.

(a)

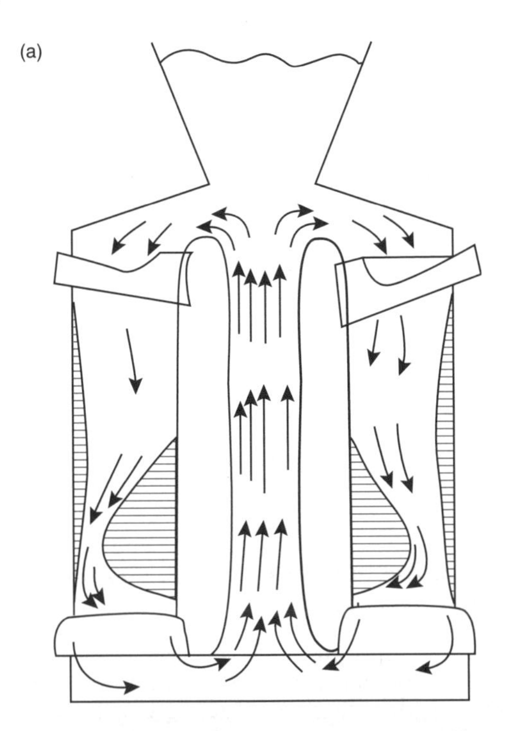

(b)

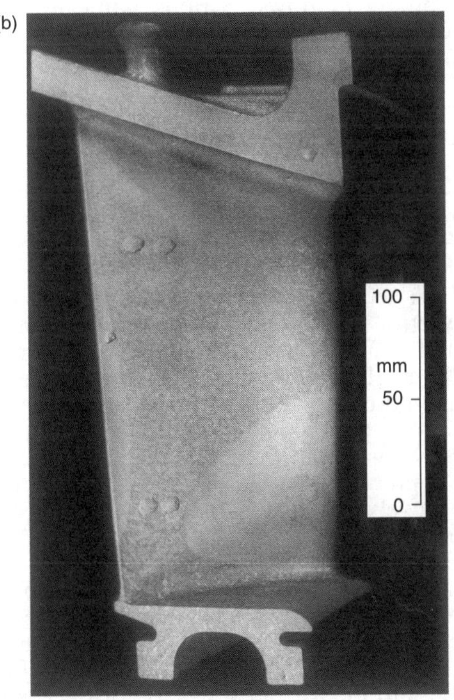

Figure 7.2 *(a) Lost wax assembly of six Ni-base turbine blades around a central feeder, showing the expected convective loops; (b) an etched blade, showing the remelting of the fine surface grains created by the cobalt-aluminate nucleant in the mould surface, and the subsequent growth of coarse grains that define the flow path.*

Castings that employ a third mould part to site the running system under the casting are at risk of convective effects causing the melt to circulate up some ingates and down others via the casting above and the runner underneath. This is especially dangerous if the runner is a heavy section. Pressurizing the runner with an adequate feeder is a way of maintaining the net upward movement of metal required for the feeding of the casting, thus reducing the deleterious effects of the convection to merely that of delaying freezing. In this case the worst that happens is the development of a locally coarser structure.

Investment casting often provides numerous convective loops in wax assemblies, as a result of attaching the wax patterns at more than one point to increase the strength of the complete wax assembly. A typical wax assembly for the casting of polycrystalline Ni-base turbine blades is illustrated in Figure 7.2a. The central upright is surrounded by six blades (only two are shown in the section), so that in addition to its heavy section designed to act as a feeder, it is kept even hotter by the presence of the surrounding blade castings that prevent loss of heat by radiation. Conversely, of course, the blades cool quickly because they can radiate heat freely to the cool surroundings. A convective loop is therefore set up, with hot metal rising up the central feeder, and falling through the cooling castings. The final grain structure seen on the etched component reveals the path of the flow (Figure 7.2b). The casting is designed to have fine surface grains nucleated by the cobalt aluminate addition to the primary coat of the mould. However, because additional hot metal enters the mould cavity from the top after the chill grains are formed, the original chill grains are remelted. The convective flow sweeps down through the casting, becoming a concentrated channel as it exits the base of the blade. The very narrow section of the trailing edge of the casting is not penetrated, and so escapes remelting, as does the large region in the bottom right that the flow has missed.

Very large blades for the massive land-based turbines for power generation are sometimes cast horizontally. In this case each end of the casting is subject to convective problems as is seen in Figure 7.3.

The cutting of convective links in wax assemblies is recommended, and cries out for wide attention in most current investment casting operations. The strengthening of wax assemblies by wax links inadvertently provides convective links and should be avoided. Ceramic rods can provide strengthening, or, if wax connections are used, they should be plugged with a

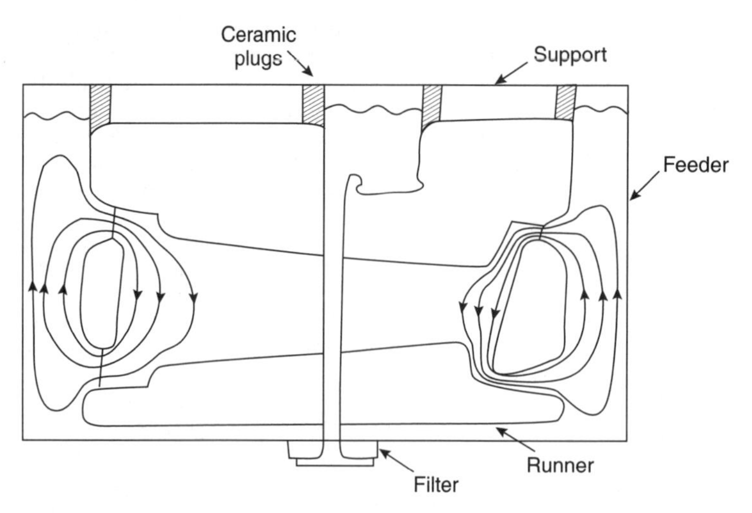

Figure 7.3 *Horizontal orientation of a large investment-cast turbine blade, illustrating convective loops in the root and shroud. The flows convey heat from the cylindrical feeders, remelting regions of the casting.*

ceramic disc to avoid metal flow. These simple modifications to the wax assembly will completely change the mode of solidification of the castings, allowing for the first time an accurate understanding of filling and feeding effects.

Other problems in sand castings are illustrated in Figure 7.4. Gravity die (permanent mould) castings are less prone to these problems because of their more rapid rate of heat extraction by the metal mould. For castings in metal moulds the sections have to be considerably larger before convection starts to be a threat. The aspect of the relative times for solidification and convection damage are dealt with in more detail in the section below on *casting section thickness*.

It is evident that many computer predictions of heat flow and the feeding of castings will be quite inadequate to deal with convection problems, since it is usual to consider the loss of heat from castings simply by conduction. Clearly, thicker sections in a loop will cool more quickly than the computer would predict, since convection allows them to export their heat. Conversely of course, thin sections in the same loop will suffer the arrival of additional heat that will greatly delay their solidification. In fact, if the hot section has an independent source of heating, such as the electrical heating provided in many counter-gravity systems, the sections in the loop can circulate for ever. The computer would have particular difficulty with this.

Even so, the greater speed and sophistication of computing will eventually provide the predictions containing the contribution of convection that are so badly needed. It is hoped that future writers and founders will not need to lament our poor abilities in this area.

7.3 Convection damage and casting section thickness

If the solidification time of the casting is similar to the time taken for convection to become established, extensive remelting can be caused by convective flows. Serious damage to the micro- and macro-structure of the casting can then occur. The time for convection to start appears to be in the region of 1 or 2 minutes. In 3 or more minutes convection can become well established, causing extensive remelting and a major redistribution of heat in castings.

Castings that freeze in a time either shorter than 1 minute, or longer than perhaps 10 minutes, are expected to be largely free from convection problems as indicated below.

Thin section castings are largely free from convection difficulties. They can therefore be fed uphill simply because (i) the viscous restraint of its nearby walls makes any convective tendency more difficult, and (ii) more rapid freezing allows convection less time to develop and wreak damage in the casting. Thus instability is (i) suppressed and (ii) given insufficient time, respectively, so that satisfactory castings can be made.

Conversely, thick section castings are also relatively free from convection problems, because the long time available before freezing allows the metal plenty of time to convect, re-organizing itself so that the hot metal floats gently into the feeders at the top of the casting, and the cold metal slips to the bottom. All this activity occurs and is complete before any significant amount of solidification has occurred. Thus the system reaches a stable condition before damage can be caused. Once again, castings are predictable.

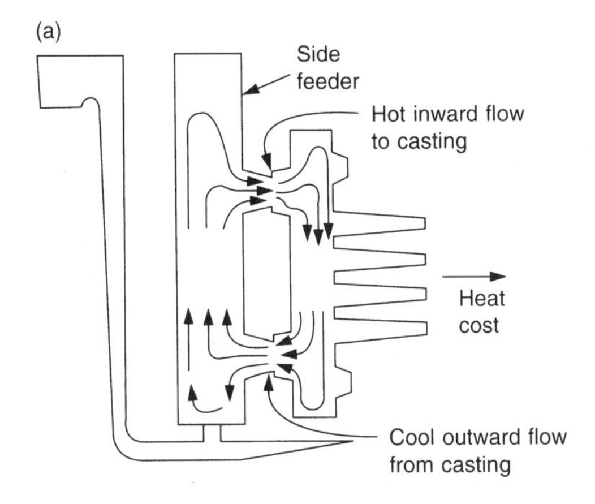

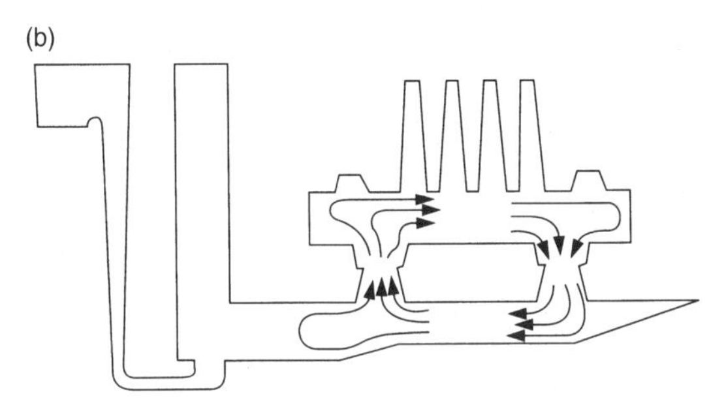

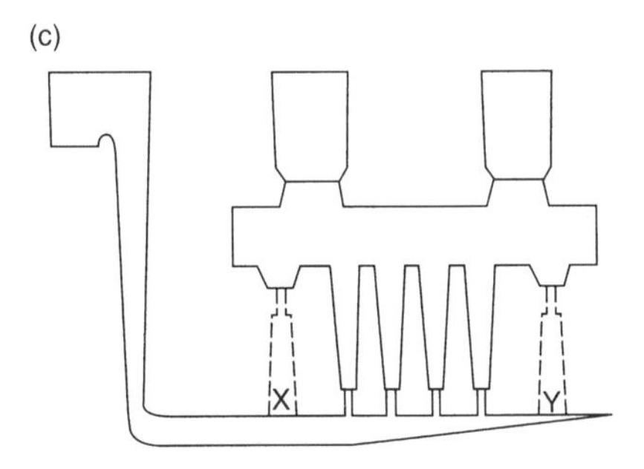

Figure 7.4 *Encouragement of thermal convection by (a) side feeding; (b) bottom feeding; (c) its elimination by top feeding.*

In what can only be described as a perverse act of fate, convection does its worst in the most common sizes of castings, the problem emerging in a serious way in the wide range of intermediate section castings. These include the important structural castings such as automotive cylinder heads and wheels, and the larger investment cast turbine blades in nickel-based alloys amongst many others. Convection can explain many of the current problems with difficult and apparently intractable feeding problems with such common products. The convective flow takes about one or two minutes to gather pace and organize itself into rapidly flowing plumes. This is occurring at the same time as the casting is attempting to solidify. The flows cut channels through the newly solidified material, remelting volumes of the casting.

The channels will contain a coarse microstructure because of their greatly delayed solidification, and in addition may contain shrinkage porosity if unconnected to feed metal. This situation is likely if the feeders solidify before the channels as undoubtedly happens on occasions, because the channels derive their energy for flow from some other heat source, such as a very heavy section low down on the casting, or the ingate attached to the riser tube of a counter-gravity system for instance.

For conventional gravity castings that require a lot of feed metal, such as cylinder heads and blocks, and which are bottom gated, but top fed, this will dictate large top feeders, because of their inefficiency as a result of being furthest from the ingates, and so containing cold metal. This is in contrast to the ingate sections at the base of the casting that will be nicely preheated. The unfavourable temperature regime is of course unstable because of the inverted density gradient in the liquid, and thus leads to convective flow, and consequent poor predictability of the final temperature distribution and effectiveness of feeding. It is the standard legacy of bottom filling: the favourable filling conditions leading to the worst feeding conditions. Life never was easy for the casting engineer.

The upwardly convecting liquid within the flow channels usually has a freezing time close to that of the preheated section beneath, which is providing the heat to drive the flow. In the case of many low-pressure systems, the metal supply system is artificially heated, leading to a constant heat input, so that the convecting streams rising out of these regions never solidify. This is what happened to the Cosworth system in the early days of its development. When the mould and casting (which should by now have been fully solidified) was hoisted from the casting unit, liquid poured from the base of the mould, emerging from such remelted channels to the amazement of onlookers who had assumed that after the appropriate length of time for solidification, no liquid could possibly still be present, and present in such quantity.

When removing a convecting casting from a counter-gravity filling system in this way, the draining of liquid from the interdendritic regions leaves regions in the casting that appear convincingly like shrinkage defects, and are usually confused as such.

The convection of hot metal up (and, of course, the simultaneous movement of cold metal down) the riser tube of a low-pressure

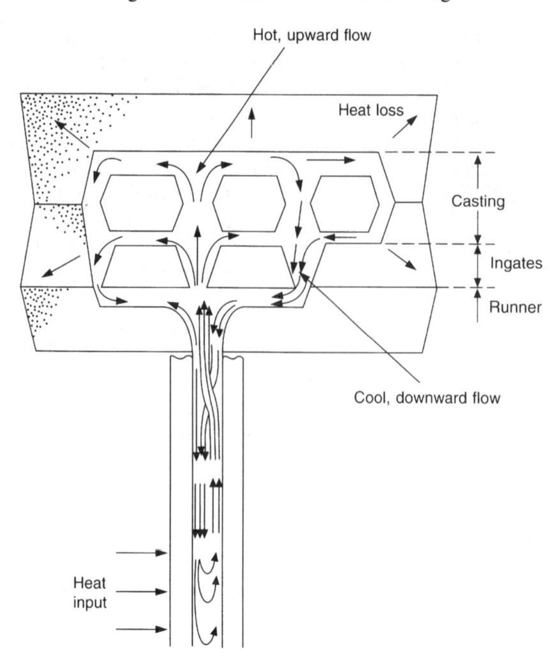

Figure 7.5 *Convection driven flow within a solidifying low pressure casting.*

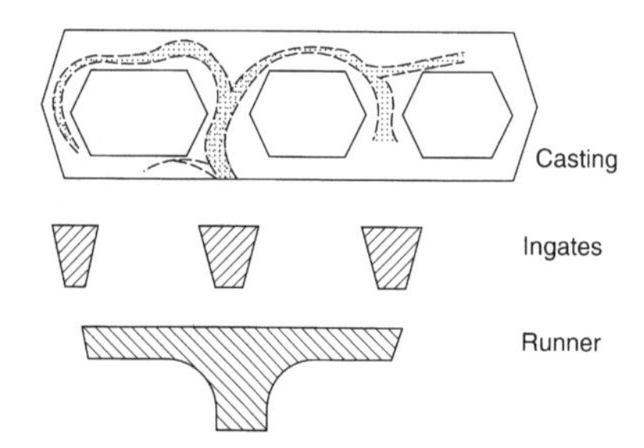

Figure 7.6 *Remnants of the convective plumes in a casting, defining regions of coarse structure and porosity.*

casting unit (Figures 7.5 and 7.6) delays the freezing of the casting in the mould above, and can lead to a significant reduction in productivity. The author is aware of a casting being made on a low-pressure machine whose freezing time kept increasing as the melt was subjected to increasingly thorough rotary degassing treatment. It seems that each rotary degassing treatment reduced the amount of bifilms in suspension. As the effective viscosity of the melt was progressively reduced in this way the convection increased, extending the time taken for the casting to solidify. Thus clean metal is free to convect, whereas melt with an internal semi-rigid lattice of bifilms will be more resistant to flow.

7.4 Countering convection

Solutions to the problems of convection are summarised as follows:

(i) The inversion of the mould after casting effectively converts the preheated bottom ingate filling system into a top-feeding system, thus gaining a really efficient feeding system.

Furthermore, of course, the massive technical benefit of the inversion of the system to take the hot metal to the top, and the cold at the bottom, confers stability on the thermal regime. Convection is eliminated. For the first time, castings can be made reliably without shrinkage porosity.

The massive productivity and economic benefit of this technique follows because the mould now contains its liquid metal all below the entry point, so that it can be detached from the casting station without waiting for the casting to freeze (which is of course the standard productivity delay suffered by most counter-gravity casting processes). In this way cycle times can be reduced from about 5 minutes to 1 minute. This is a powerful and reliable system used by such operations as the Cosworth Process and an increasing number of other processes at the present time. We can hope that techniques involving roll-over *immediately after* casting will become the norm for most castings in the future.

(ii) Tilt casting processes (where the roll-over is used *during* casting—actually to effect the filling process) can also satisfy the top-feeding requirement. However, in practice many geometries are accompanied by waterfall effects, if only by the action of the sliding of the metal in the form of a stable, narrow stream down the sloping side of the mould. Thus meniscus control is, unfortunately, often poor. Where the control of the meniscus can be improved to eliminate entrainment problems, tilt casting techniques are valuable. Ultimately, if the tilt is controlled to perfection, a kind of horizontal transfer of the melt can be achieved, as discussed in Section 2.3.3. This system does not seem difficult or costly to attain, and is to be recommended strongly.

(iii) Cut convective loops. Explore the elimination of ingates on counter-gravity feeding of castings. The widespread convective loops in investment castings wax assemblies should be cut by the wider use of ceramic supports and stops.

Rule 8

Reduce segregation damage

At regions in which the local cooling rate of the casting changes, such as at a change of section, or at a chill, or at a feeder, it is to be expected that a change in composition of the casting will occur. In many alloy systems such variations are so slight as to be negligible. Such problems are therefore normally neglected. However, there are alloy systems that are particularly prone to such severe segregation problems that the casting may be scrapped (if detected) or (if not corrected) can threaten its performance in service.

There are many good solidification texts that deal with the problem of segregation, so that it is hardly necessary to treat the subject in any length here. Thus the various types are simply listed as a reminder of the extent of the problem, and the many forms it can take.

Microsegregation is an unavoidable consequence of normal solidification in which solutes are concentrated (if the distribution coefficient is less than one, as is usual) in the residual liquid between dendrites. This inter-dendritic segregation can be cured, because it can be re-distributed by diffusion back into the depleted, rather pure, centres of the dendrites by homogenization heat treatment; the diffusion distances achievable during heat treatment are of the same order as the inter-dendritic spacing. Such heat treatments are usually carried out at temperatures close to the melting point of the alloy, and require up to several hours to achieve a reasonable re-distribution of solutes.

When macroscopic flow occurs during solidification, the contents of these microscopic regions may be dispersed or may be concentrated in distant regions of the casting depending on whether the pattern on flow diverges or converges respectively. *Macrosegregation* is the result. This re-organization of the pattern of chemical elements in the casting involves distances vastly greater than can be cured by subsequent heat treatment. Heat treatments times of perhaps the age of the earth (i.e. geological time scales) may cure it but is not recommended. Macrosegregation, if it occurs, is unfortunately therefore, for all practical purposes, a permanent feature. Some of the various types of macrosegregation are described below.

In the casting of steel ingots the segregation of impurities into the head of the ingot, generally known as *positive* or *normal segregation*, has been so bad that it has been necessary to cut-off and discard the top of the ingot. This has represented a massive loss to the efficiency of the steel industry over the years, and was the main driving force for the development of continuous casting in the 1960s, and now almost universal use for the casting of steel.

A well-understood type of segregation is so-called *inverse segregation*. Because this is the perfectly normal segregation to be expected in conditions of dendritic solidification the author prefers to call it simply *dendritic segregation*. In this case the partitioned solute is segregated preferentially to the face of the mould, especially if this is a chill mould (Figure 8.1a). A similar effect will occur, of course, at the junction with a thinner section that will act as a chill. The effect is shown in Figure 8.1b. The distribution of alloying elements in this figure is particularly disturbing, because in the case, for instance, of the high strength Al–4.5%Cu alloy, usually chosen for highly stressed applications, the deviations in chemistry can easily be outside the allowed specification of the alloy. The regions of the casting high in solute will normally be expected to be extra strong, possibly even brittle. The regions depleted in solute will,

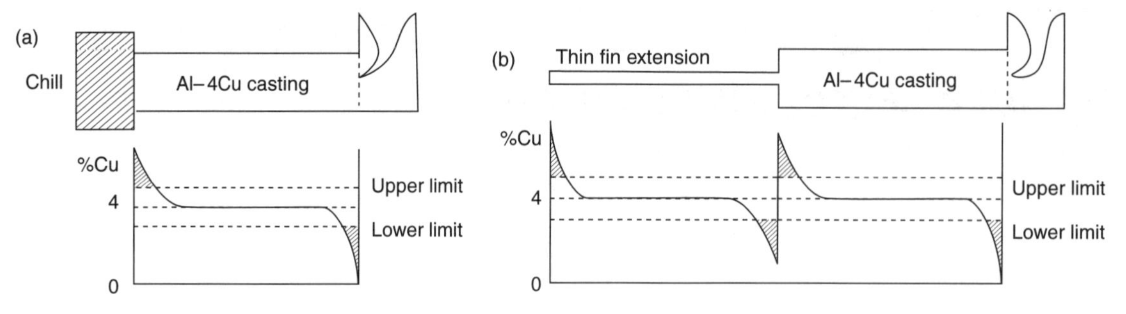

Figure 8.1 *(a) Dendritic segregation pattern, concentrating solute against a chilled face. (b) The analogous pattern produced by a reduction in section thickness acting as a cooling fin.*

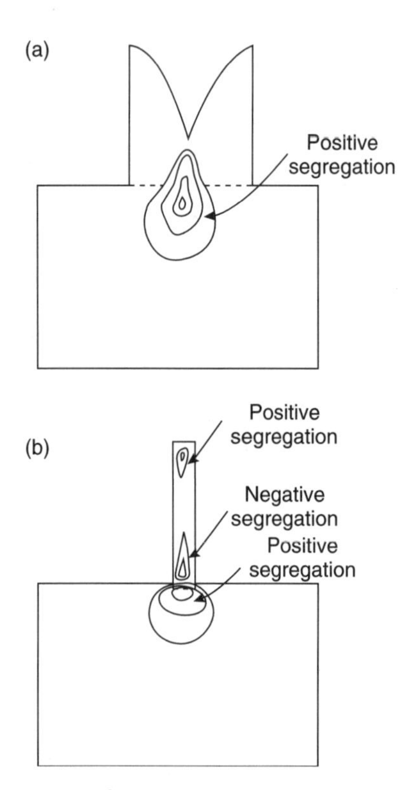

Figure 8.2 *(a) The positive segregation pattern under a feeder. (b) The positive segregation pattern under a cooling fin, but negative close by inside the fin. These extremes are both close to the vulnerable change in section.*

conversely, be weak. What gives even greater cause for concern is that these sharp differences in mechanical properties are sited so close to the change in section where any stress will normally be concentrated (Figure 8.1b).

Naturally, in a complex thermal field, and where the geometry of the casting is requiring a complex distribution of residual liquid to feed shrinkage, these chemical variations can be complex in distribution, and not always easily predicted, except perhaps by sophisticated computer simulation.

Another highly visible and severe type of inverse segregation occurs when the residual liquid, concentrated against the surface of the casting, actually penetrates the surface, to emerge as exuded segregated liquid on the surface of the casting. Such exudations are often low melting point eutectics. The historically famous example is 'tin sweat' on bronze castings and ingots, in which beads of tin-rich eutectic occur on the surface of the copper-based alloy. The driving force for such exudations is sometimes the general contraction of the casting as it cools, and sometimes the internal pressure generated by the precipitation of gas from solution. An example of an Al–Si eutectic exudation against a chill surface is common in Al–Si alloys. In this case the chill first causes the melt to freeze quickly. However, the surface of the casting then contracts away from the chill, whereupon it reheats and melts, allowing the segregated residual liquid to bleed into the air gap that has opened up between the casting and the chill (Figure 6.18).

In heavy steel castings and in steel ingots, the distribution of carbon is significantly increased under the feeder. This occurs simply because in the feeder the last liquid to solidify is high in carbon (and other elements such as phosphorus and sulphur) but this liquid is still being supplied to the casting. The liquid is finally sucked into the casting to compensate for the shrinkage accompanying the solidification of the casting (Figure 8.2). There has been much research into this problem, so that actions to reduce the problem are now reasonably well understood. An early description is given by Flemings (1971) in which he draws attention to the complicating fact that, depending on whether the flow from the feeder is converging or diverging, the segregation can be positive or negative respectively.

One of the easiest actions to reduce the problem is merely to make the feeder somewhat oversize so that the remaining liquid is less concentrated at the time it is demanded by the casting. Again, it is sufficiently complicated to require a computer solution for an ultimate quantitative description. It is perhaps sufficient for most of us to be aware that the problem exists, and check to see how important it may be.

Other *positive segregation* derives from the flow of the liquid, and is driven almost purely by gravity. A well-known example of *gravity segregation of the liquid* is the concentration of carbon, and other light elements such as sulphur and phosphorus, in the tops of large ferrous castings. In contrast, tool steel castings suffer from the concentration of heavy elements such as tungsten and molybdenum at the base of the casting.

Gravity segregation of the solid can occur by equiaxed crystals in the melt that sediment to the bottom of solidifying castings leading to these lower regions being generally more pure because they are composed of some of the first solid to solidify. Some additional purification during freezing may occur because of the divergence of flow of residual liquid through this zone. The overall effect is known as *negative segregation*.

Strong concentrations of segregated solutes and inclusions are found in *channel segregates*, that are once again a feature of larger, or slowly cooled castings. In steel ingots these are the once familiar *'A' and 'V' segregates*, nowadays much less prominent in continuously cast steels. In Ni-base superalloys they are commonly known as *freckle* defects.

As we have seen above, when extensive and/or intensive, all such changes in composition of the casting may cause the alloy of the casting to be locally out of specification. If this is a serious deviation, the coincidence of local brittleness in a highly stressed region of the casting might threaten the serviceability of the product. The possibility of such regions therefore needs to be assessed prior to casting if possible, and demonstrated to be within acceptable limits in the cast product. Otherwise, techniques to reduce the segregation may need to be implemented. This is probably easier said than done. One approach would be to attempt to cool the casting locally with chills or fins so as to achieve a more even temperature distribution throughout the casting.

Rule 9

Reduce residual stress (the 'no water quench' requirement)

9.1 Introduction

Action to reduce internal stress can be awesomely important. Unfortunately, it seems that in general, the engineering community has not been made aware of the central importance of this factor in the manufacturing of engineering components. All manufactured components contain internal stress, often high. The problem is that this very real danger is invisible.

The problem is widespread, and not confined to metal products. A common example we have all seen is the high stress revealed by the maze of cracks around the plug hole in some plastic wash basins. In this case the stress has been relieved by cracking, probably aided and abetted by the soaps and detergents that encourage crack growth—perhaps to be known as liquid surfactant embrittlement, analogous to liquid metal embrittlement or stress corrosion cracking in metals.

There are those metallurgists within the industry, some eminent, and whose opionions on other matters I respect, that have taken issue with me. They have argued that the presence of residual stresses, particularly those from quenching, are actually irrelevant since the whole component is in balance with its own stresses. The question of balance is certainly true. However, this argument overlooks the fact that the distribution of stress is usually far from uniform, and parts of the component may be near to their failure stress even prior to the application of any service stress. Usually, as we shall see, the major tensile stress is in the centre, and it is this part of the component that fails first under tensile load.

Admittedly, not all components are necessarily endangered by internal stress. Indeed, the stress can be beneficial in some cases (see somc examples in *Castings 2003*). However, the major risk is that the stress may not be beneficial. It may add to the service stress and so promote premature failure at only low service stress, to the bewilderment of the designer who imagines his component material to be inert. Because of the complexity of some castings, and the complexity of the state of stress, it is usually not easy to estimate the magnitude of either the internal residual stress or its precise action. Often, however, it is at least equal to or exceeds the yield stress. Thus it is not trivial. In fact at this level it will dominate all other designed loads in a fatigue condition, and certainly lead to early failure. It is ignored at our peril.

This section takes a look at the wide spectrum of stresses in castings, and attempts to clarify those that are important and which should be controlled, from those that can be safely neglected.

9.2 Residual stress from casting

In an aggregate mould, castings are cooled relatively slowly, so that the final internal stress in the product will normally be relatively low, and can often be neglected. It is true that the dimensions of the casting will often be changed by stress during cooling, but on shaking out from the mould the final, residual, stress will not normally be high. Some examples are given in *Castings* (2003). In addition, the distortions that have arisen during cooling in the mould are usually extremely reproducible. This is a consequence of the reproducible conditions of production, in which the mould is the same temperature each time, and the metal is the same temperature each time, so that the final shape is closely similar each time. This reproducibility is probably greater than for any other casting process.

This repeatable regime is not quite so well enjoyed by the various kinds of die-casting, particularly gravity die (permanent mould) casting, as a result of many factors, but in particular the variability of mould size and shape as a result of variation of mould temperature. The somewhat faster cooling, particularly because of the earlier extraction of the casting from the mould, is an additional factor that does not favour low final stress.

In general, internal stress remaining from the casting process is rarely high enough to be troublesome but we cannot always be complacent about this. The ability to predict stresses using computer simulation will be invaluable to maintain a cautious watch for such dangers.

Ultimately, however, particularly for aluminum alloys, the stresses from casting are usually eliminated by any subsequent high temperature solution heat treatment.

9.3 Residual stress from quenching

The final stresses in the component are dictated by the final stages of this treatment, which is normally a quench, and normally into water. Thus the major problems of internal stresses and distortion of the casting are usually created at this moment. Furthermore, the stresses are not significantly reduced by the subsequent ageing treatment. The temperatures for ageing treatments are too low to lead to stress relief.

It is unfortunate that many heat treatments require a quenching stage, intended to cool the casting sufficiently quickly to freeze solutes in to a solid solution, thereby preventing them from precipitating. If the quench is slow some solute may be lost by precipitation from solution, thus making it unavailable for subsequent hardening reactions, so that the final strength of the casting is reduced. This reasoning has driven the quest by metallurgists for quenching rates to be as fast as possible.

The problem has been that all such research by metallurgists to optimize heat treatments has been carried out on test bars of a few millimetres in diameter that represent no problem to cool quickly. The outside and inside of the bars is in excellent thermal communication, and the high thermal conductivity of most metals ensures that the cooling throughout the section is essentially uniform. Thus the world's standards on heat treatment often dictate water quenching to obtain the highest material properties.

Quite clearly, the problem of larger components, or certain components of special geometrical complexity in which uniform cooling is an impossibility, has been overlooked. This is a most serious oversight. The performance of the whole component may therefore be undermined by the application of these techniques that have been optimized by work on small test bars, and which therefore are inappropriate, if not actually dangerous, for many large and complex components.

This is such a common problem, that when a troubled casting user telephones me to say words to the effect 'My aluminium alloy casting has broken. What is wrong with it?' this is such a regular question that my standard, and rather tired, reply now is 'Do not bring the casting to me. I will tell you now over the telephone why it has failed. It has failed because it has been poured badly and therefore contains bifilms that reduce its strength. However, in addition, you have carried out a solution heat treatment accompanied by a water quench.' The caller is usually stunned, incredulous that I know that he has water quenched his casting, and asks how I know. My experience is this: in all my life investigating the causes of failure of perhaps hundreds of Al alloy castings, only one failed because of serious embrittlement caused as a result of the alloy being outside chemical specification. All the rest failed for only two reasons; (i) weakening by bifilms, together with (ii) massive internal stresses that have loaded the already weakened casting close to its failure stress even before any service stress was applied. I have to record, with some sadness, that all the standard and costly investigations by metallurgists into the chemical specification, the metallurgical structure, the mechanical properties and other standard metallurgical tests, are nearly always irrelevant. It underlines the importance of understanding the new metallurgy of cast metals in which the residual stresses and bifilms together play the dominating roles in the performance of engineering components, particularly cast engineering components.

The key role of internal stress in the failure of castings (and other components such as forgings) is explained in Figure 9.1. The stress ε is given by

$$\varepsilon = \alpha \cdot \Delta T \qquad \textbf{(9.1)}$$

where α is the coefficient of thermal expansion, and ΔT is the temperature change experienced by the part.

Equation 9.1 explains why not all shapes and sizes of castings necessarily suffer a problem. Compact or small castings, and those for which

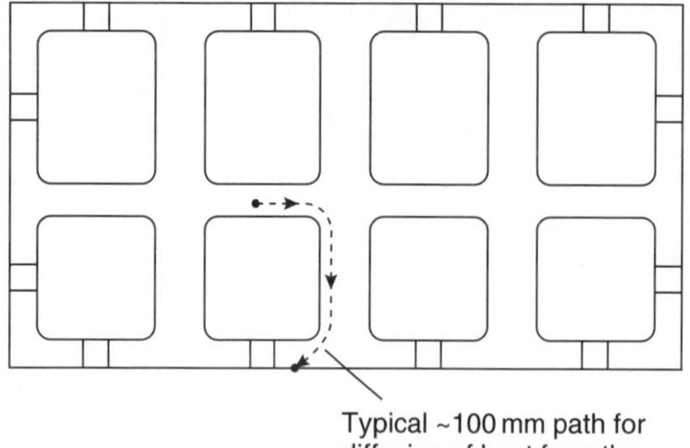

Figure 9.1 *Schematic representation of a hollow casting with small ports to the outside world and internal walls, such as a cylinder head, illustrating the long diffusion path for heat during a quench, together with the high internal tensile stress that might result in failure.*

the quenchant can easily reach all parts, are often not seriously affected, because ΔT is necessarily small.

The actual magnitude of the strain ε is salutary to estimate. For aluminium α is about $20 \times 10^{-6}\ K^{-1}$ and the temperature fall during a quench is approximately 500 K. The strain works out to be therefore approximately 1 per cent. For steels α is approximately 14×10^{-6}, and the temperature change for quench from many heat treatments in the region of 900 K, again giving a strain close to 1 per cent. Since the yield (or proof) strain is only approximately 0.1 per cent, these imposed quench strains are about ten times the yield strain, and can therefore be seen to take the component well into the plastic deformation range.

Parts that are particularly susceptible include large, thick section castings, where the heat of the interior takes time to reach the outside of the casting, giving high ΔT. Ingots or other block-type products can be seriously stressed for this reason. Direct chill (continuously cast) ingots aluminium alloys are severely cooled by water, but are often over 300 mm diameter. While sitting on the shop floor awaiting further processing the strong 7000 series alloy ingots have sometimes been seen to explode like bombs. (As an aside, the length of time taken before the ingot decides to fail is curious and interesting. It seems likely that the failure under the high internal stresses is initiated from one of the large bifilms that is expected to be entrained during the turbulent start of casting. The gradual precipitation of hydrogen into the bifilm will gradually increase the pressure in the bifilm crack, encouraging it to extend as a stress crack. The hydrogen may be already in solution in the metal, or may be gradually accrued by reaction with water vapour in the atmosphere during storage, especially if the bifilm is connected to the exterior of the ingot surface, allowing direct penetration of water vapour. Other penetrating contaminants may include air to cause additional internal oxidation, or fluxes, or traces of chlorine gas, or sulphides from greases, to act as surface active additions to reduce the surface energy of the metal and so further encourage crack growth. Research to clarify these possibilities would be valuable.)

Other varieties of castings that are susceptible to damaging levels of residual stress include those that are hollow, with limited access for the quenchant into the interior parts of the casting, and which also have interior geometrical features such as dividing walls and strengthening ribs (Figure 9.1). This latter series of geometrical requirements might seem to eliminate most castings. Perhaps surprisingly therefore, the list of castings that fulfils these requirements is rather long, and includes such excellent examples as automotive cylinder heads and blocks, and housings for components such as compressors and pumps. When immersed in the water quench, the water attempts to penetrate the entrances into the hollow interior of the casting. However, because the casting is originally above 500°C, any water that succeeds in entering will convert almost instantaneously to steam, blowing out any additional water that is attempting to enter. The result is that the outside of the casting cools rapidly, whereas the interior can cool only at the rate that thermal conduction will conduct the heat along the tortuous path via interior walls of the casting to the outer surfaces.

The rate of conduction of heat from the interior to the exterior of the casting can be estimated from the order of magnitude relation

$$x = (Dt)^{1/2} \tag{9.2}$$

where x is the average diffusion distance, D is the thermal diffusivity of the alloy, and t is the time taken. The thermal diffusivity is defined as

$$D = K/\rho C \tag{9.3}$$

where K is the thermal conductivity, about $200\ W\,m^{-1}\,K^{-1}$ for aluminium, the density ρ is about $2700\ kg\,m^{-3}$ and the specific heat C is approximately $1000\ J\,kg^{-1}\,K^{-1}$. These values yield a value for the thermal diffusivity D close to $10^{-4}\ m^2\,s^{-1}$. (The corresponding value for steel is approximately $10^{-5}\ m^2 s^{-1}$.) Equation 9.1 is used to generate Figure 9.2 in which the

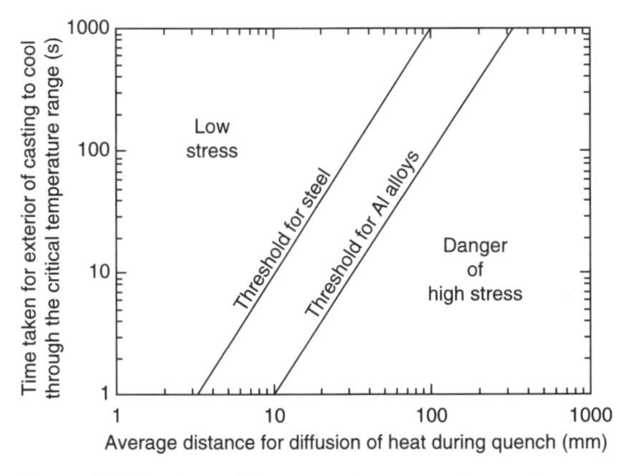

Figure 9.2 *Regime of low stress in terms of quench rate and distance for heat flow.*

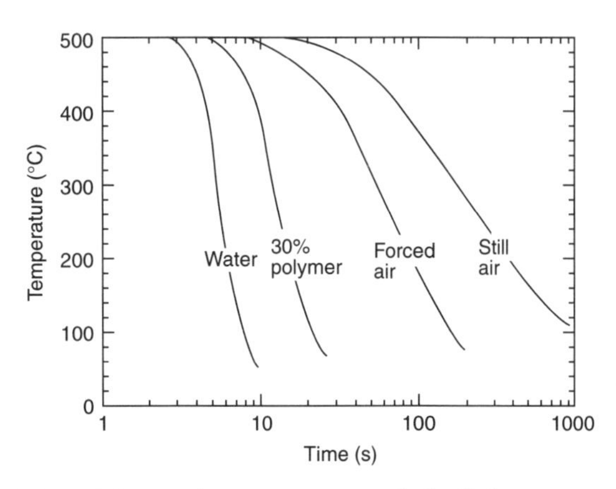

Figure 9.3 *Quench rates in a 10 mm thick Al plate casting in a variety of quench media.*

distance for diffusion of heat out of a product indicates the approximate boundaries of safe regimes constituting conditions in which sufficient time is available for the diffusion of heat from the interior during the quench. The time of cooling in different quenchants over the critical range of approximately 500°C down to 250°C is provided by results such as that shown in Figure 9.3. These results were obtained by siting a thermocouple in the centre of a 10 mm wall of an Al–7Si–0.4Mg alloy casting. Similar results would be valuable for ferrous materials.

For a solid aluminium bar of 20 mm diameter, or a solid plate of 10 mm thickness, Figure 9.3 indicates that quenching in water will reduce the temperature from 500 to 250°C in about 5 seconds. Substituting 5 s in Equation 9.2 shows that on average heat will have travelled 20 mm in this time. The 20 mm bar or 10 mm plate will both therefore enjoy a reasonably uniform temperature so that minimal stress will be generated.

If, when quenching castings following high temperature heat treatment, the time for cooling the outer sections of castings is shorter than the time required for heat to diffuse out from interior sections, the outer parts of the casting cool to form a rigid frame. However, the inner sections will cool and contract later, but by that time unfortunately experiencing the restraint of the outside rigid sections. Thus the interior sections go into tension, and the outer parts into compression.

As stated above, this situation is common in such castings as automotive cylinder heads, whose links between the internal sections and the outside world are via tortuous routes around the water jackets. The total distance that heat now has to diffuse is of the order of 100 mm. However, the walls of the casting are 10 mm or less, so that the cooling of the exterior of the casting will again occur in a time of the order of 5 seconds. However, Figure 9.3 indicates that approximately 100 seconds is required for the heat to diffuse the 100 mm distance from the inside to the outside, so the interior of the casting will be expected to experience high tensile stress.

If the cylinder head casting had been subjected to quenching in a blast of air, Figure 9.3 indicates that cooling will now take a leisurely 100 seconds or so. Thus sufficient time is available for the internal sections to lose their heat to the outside so that the casting maintains a reasonably uniform temperature during the quench. The generation of high internal stress is avoided.

The author has personal experience of quenching complex cylinder heads into water, and has suffered the consequences of banana-shaped castings that required to be straightened with a 50 000 kg press specially bought-in to rectify the damage. Those were the castings that did not crack in the quench itself (the internal cracks inside the water jackets were often difficult to locate). In addition, castings failed by fatigue in service after only short lives. The introduction of *polymer quenching* eliminated the problem. As explained in *Castings* (2003) there are a number of polymers that can be used. One commonly in use is a solution of polyalkylene glycol in water. The polymer precipitates out of solution at 73°C, and so deposits over the surface of the hot castings, forming a sticky, viscous layer. The layer is resistant to boiling, so that a vapour blanket is avoided, and a steady, uniform flow of heat from the casting into the water is achieved. When the casting finally cools below 73°C the polymer goes back into solution.

However, the polymer quench was not a long-term practical solution for production for an automotive product. It had to be cleaned out of internal cavities where it could lodge, becoming concentrated and carbonized during the subsequent ageing treatment. In addition, any residual core sand in such locations would be effectively cemented into place to cause damage later in the life of the engine when it finally became dislodged.

In contrast to its use with automotive castings, polymer was excellent for aerospace castings where the extra trouble to clean each casting individually did not outweigh the benefit of superb heat treatment response and reduced internal stress.

Air quenching was, however, a complete solution for automotive castings. It was low cost and quickly and easily implemented in a series production environment. The castings retained their accuracy, and quench failures and fatigue failures disappeared. We were able to restore productivity and profitability (and get our money back for the press).

Figure 9.4 illustrates how the overall strength of a casting can be reduced by a heat treatment designed to increase the strength of its material. Figure (a) shows the stress–strain curve for the alloy, and the imposition of 1 per cent; tensile strain on the inner parts of the casting as a result of a water quench. This quench strain results in a quench stress close to the failure stress of the material. If no ageing treatment is carried out this stress is locked into the component for the rest of its life. Naturally, it has little residual strength left, and is likely to fail on the first application of a stress in service.

However, after an ageing treatment in which the yield strength of the alloy was intended to be doubled, the situation is shown in Figure 9.4(b). Assuming the benefit of a small amount of stress relief (the amount indicated in the figure may be rather generous), the residual quench stress is only slightly lower; substantially unchanged. If additional service stress in tension is applied to the central parts of the casting, the residual tensile stress in these parts is effectively a starting point for the additional loading. Thus, effectively, the new stress–strain curve for the component is shown in (c). It is clear that the new overall stress–strain response has been reduced compared to the original unheat-treated material; as a result of our lengthy, complex and expensive heat treatment the component is effectively weakened.

In summary, the residual stress in aluminium alloy castings quenched into water in this way is well above the yield point of the alloy. Even after the strengthening during the ageing treatment, the stress remains at between 30 and 70 per cent of the yield stress, with a useful working approximation being 50 per cent. Thus the useful strength of the alloy is reduced from its unstressed state of 100 per cent, down to around 50 per cent. This massive loss of effective strength makes it inevitable that residual tensile

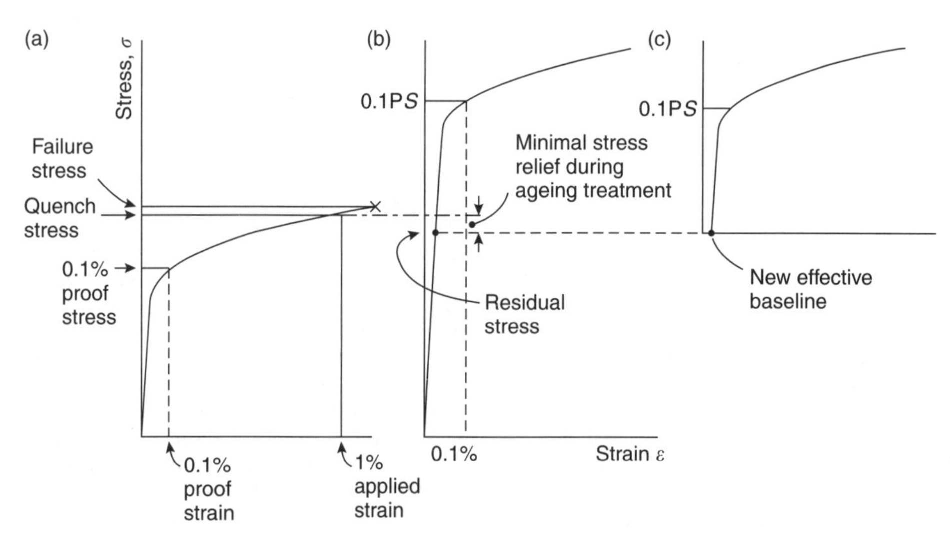

Figure 9.4 *Evolution of the stress/strain curve of an Al alloys as heat treatment progresses (a) after quench; (b) after ageing to double the 0.1 per cent proof stress; and (c) the final effective stress/strain curve showing properties effectively less than the as quenched properties.*

stresses are a significant cause of casting failure in service, particularly fatigue failure, since the residual stress is always generously above the fatigue limit of the alloy.

For many castings, the use of a boiling water quench has been demonstrated to be of negligible help in reducing the stresses introduced by water quenching (*Castings 1991*). Thus although the rate of quench is certainly reduced by the use of hot or boiling water the results are not always reliable. This is almost certainly the consequence of the variability of the vapour blanket that forms around the hot casting. The blanket forms and disappears irregularly, depending on many factors including the precise geometry of the part, its inclination during the quench, and the proximity of other hot castings, etc. In addition, from a practical point of view, a hot water quench is not cheap to install, run or maintain.

Turning now to steels; in contrast to the behaviour of Al alloys, the thermal diffusivity D is approximately ten times lower, of the order of only 10^{-5}. The reader can quickly show that the corresponding distances to which heat can flow are 7 mm in 5 seconds but only 30 mm in 100 seconds. For a given rate of quench therefore, steels will suffer a higher residual stress (Figure 9.2). Nevertheless, they are much more able to withstand such disadvantages, having higher strength, but more particularly, higher elongation. Thus although the final internal stress is high, the steel product is nowhere near the failure condition experienced by the aluminium alloy casting. The aluminium alloy casting experiences about 1 per cent imposed elongation but has only a few per cent, perhaps even less than 1 per cent elongation prior to failure. Thus it can fail actually in the quench, or early in service. In contrast, the steel casting has ten or twenty times greater elongation (as a result primarily of its reduced bifilm content). Thus although the 1 per cent or so of imposed quench strain resulting from unequal cooling may result in 1 per cent or so of distortion of the product, its condition is far from any dangerous condition that might result in complete failure, since enormously greater strain has to be imposed to reach the failure condition (Figure 9.5).

The above statements are so important they are worth repeating in different words for additional clarity. The rapid quenching of steels for metallurgical purposes (such as the stabilization of austenite for Hadfield Manganese steel) is not usually a problem. The reason is that most steels are particularly clean, because of the rapidity with which entrainment defects are deactivated and/or detrained after pouring. The result is that steels typically have an elongation to failure of 40 or 50 per cent. In contrast, most Al alloys (and probably most Mg alloys) do not enjoy this benefit; suffering from a high

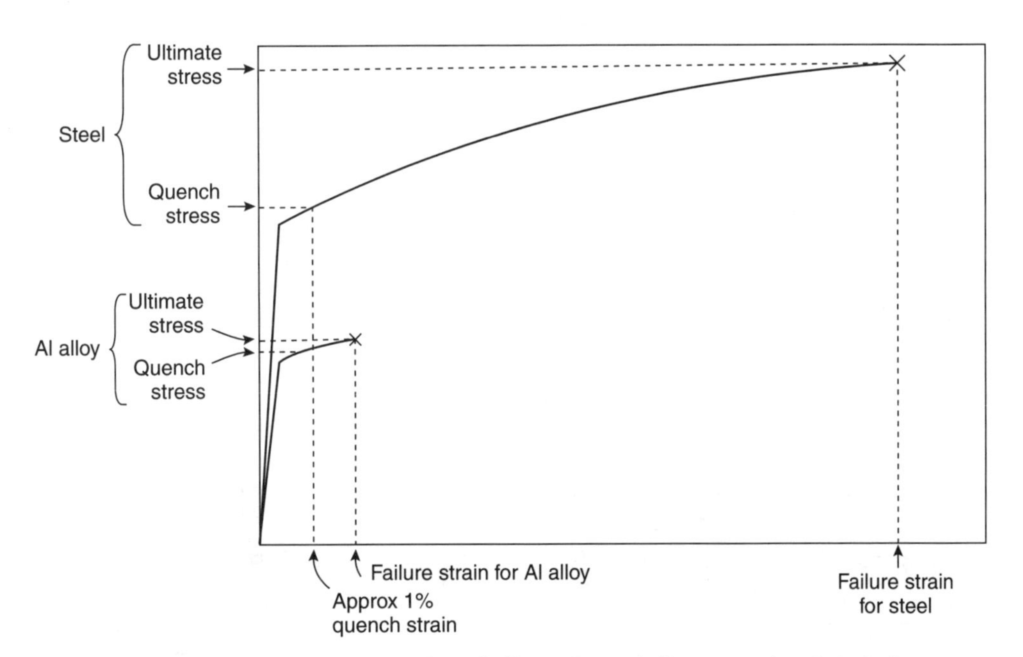

Figure 9.5 *A comparison of stress/strain curves of an Al alloy and a steel, illustrating the relatively dangerous condition of the Al alloy after a quench.*

density of bifilms they typically achieve less than a tenth of this ductility. Thus the application of 1 per cent strain takes the aluminium alloy close to, or even sometimes in excess of its failure strain. For steels, even though the 1 per cent strain applied by the quench will take the part into the plastic region, causing huge stresses, the steel remains safe; its greater freedom from bifilms permits it to endure enormously greater extension before it will fail (Figure 9.5).

For the future, the production of Al alloys with low bifilm concentration promises to offer ductilities in the range of that of steels. Already, good foundries know that high strengths together with elongations of 10 to 20 per cent are achievable, if good care is taken.

Slower quenching techniques are safer, although, of course, the strength attained by the heat treatment is somewhat reduced. Even so, the reduced mechanical strength when using slower and more controlled quenches such as a polymer or a forced air quench is more than compensated by the benefit of increased reliability from putting unstressed (or more accurately, low-stressed) castings into service. Thus the casting designer and/or customer needs to specify somewhat reduced mechanical strength and hardness requirements in order to gain a superior performance from the product. The reductions of strength and hardness are expected to be in the range 5 to 10 per cent, but the improvement in casting performance can be expected to be approximately 100 per cent. These are huge benefits to be gained at no extra cost.

9.4 Distortion

Residual stresses in castings are not only serious for parts that require to withstand stress in service. They are also of considerable inconvenience for parts that are required to retain a high degree of dimensional stability. This problem was understood many years ago, being first described as early as 1914 in a model capable of quantitative development by Heyn. The model of a three-bar casting is shown in Figure 9.6. The internal stresses are represented by two outer springs in compression, each carrying half of the total load of internal compressive stress, and an inner spring in tension carrying all of the internal tensile stress. If one of the surfaces of the casting is machined away, one of the external stresses is removed. It is predictable therefore that the casting will deform to give a concave curvature on the machined side as illustrated in the figure.

The distortion of castings both before and after machining is a common fault, and typical of castings that have suffered a water quench.

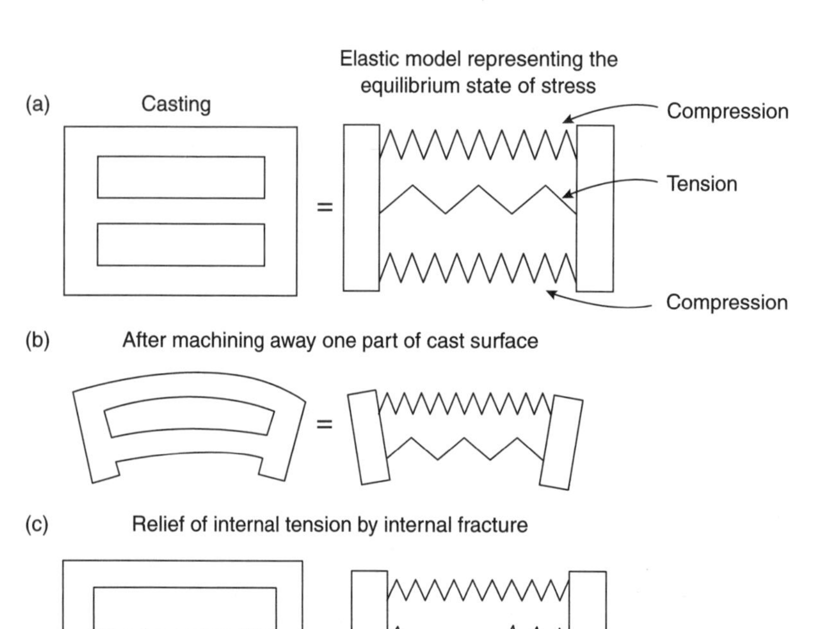

Figure 9.6 *Heyn's (1914) model of the balance of internal stresses after rapid cooling: (a) the quenched casting showing high internal tensile stress and relatively low external compressive stress; (b) the distortion of the casting after one side is machined away; and (c) the condition of internal tensile failure.*

Once again, it is a problem so frequently encountered that I have, I regret, wearied of answering the telephone to these enquiries too. After all, it is difficult to understand how a casting could avoid distortion if parts of it are stressed up to or above its yield point.

For light alloy castings in particular a more gentle quench, avoiding water (either hot or cold), and choosing polymer or air will usually solve the problem instantly. As mentioned briefly above, such polymer performs well for aerospace castings but is expensive and messy, whereas air is recommended as being clean, economical and practical for high volume automotive work. Otherwise, stress relieving castings by heat treatment prior to machining is strongly recommended (*Castings 2003*). In either case, of course, some fraction of the *apparent* strength of the product has to be sacrificed.

9.5 Heat treatment developments

Although not strictly relevant to the question of reducing residual stresses, it is worth emphasizing the newer developments in heat treatments that give approximately 90 per cent or more of total attainable strength, but with much reduced stress and greatly reduced cost. The reduced cost is always an attention-grabbing topic, and materially helps the introduction of technology that can deliver an improved product.

Figure 9.7 illustrates the progression of recent developments in heat treatment of Al alloys where the problem of stress is central.

The traditional full heat treatment of a precipitation-hardened alloy, that constitute the bulk of cast structural components at this time, consists of a solution treatment, water quench and age as illustrated in (a). The treatment results in excellent apparent strength for the material, but is energy intensive in view of the long total times.

Illustration (b) shows how the traditional treatment can be reduced significantly in modern furnaces that enjoy accurate control over temperature, thereby reducing the risk of overheating the charge because of random thermal excursions. An increase in temperature by 10°C will allow, to a close approximation, an increase in the rate of treatment by a factor of 2. Thus times at temperature can be halved. These benefits are cumulative, such that a rise of 20°C will allow a reduction in time by a factor of $2 \times 2 = 4$, or a rise of 30°C a reduction in time of a factor $2 \times 2 \times 2 = 8$, etc. Both (a) and (b) require separate furnaces for solution and ageing treatments (if long delays waiting for the

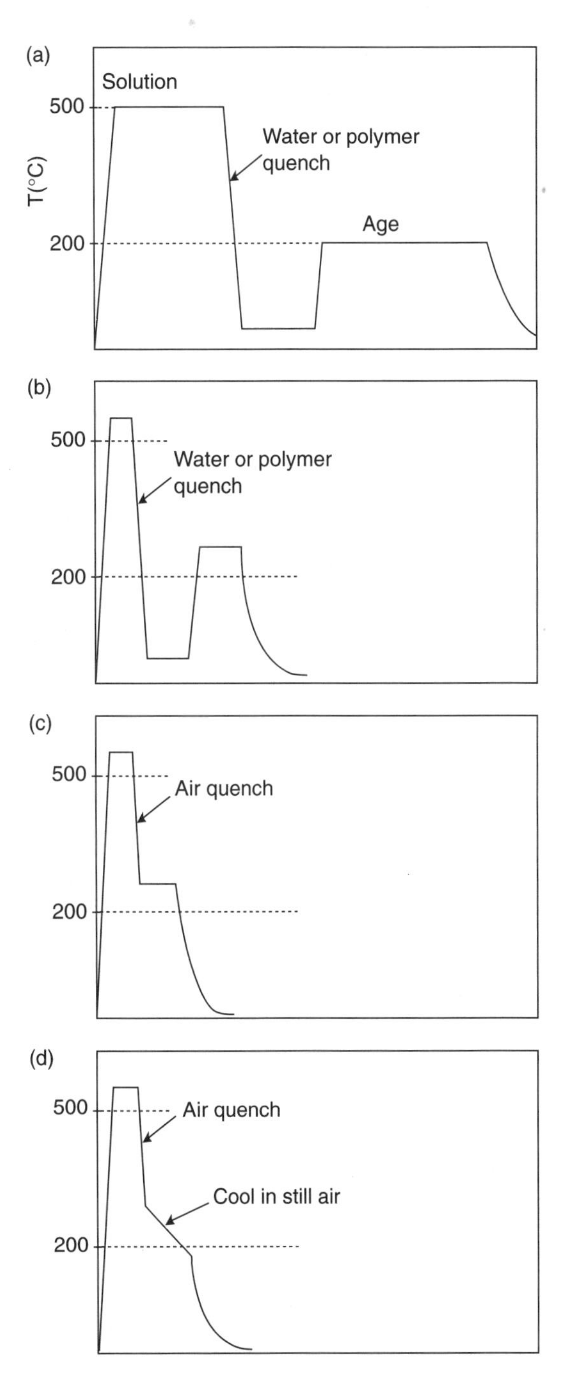

Figure 9.7 *A progression of precipitation heat treatment developments for Al alloys. (a) A traditional full treatment, giving excellent apparent properties but taking between 12 and 24 hours; (b) shortened treatment giving nearly equivalent result; (c) the use of air quench to reduce time, energy, and residual stress; (d) an ultimate short and simple cycle.*

solution furnace to cool to the ageing temperature are to be avoided). Thus floor space requirement is high. Floor space requirement is increased further by the quench station, and, if a polymer quench is used, by a rinsing tank station.

The reader will appreciate that the tiny additional energy required by the higher temperature is of course completely swallowed by the huge savings in overall time at temperature.

If an air quench is used to gain the benefits of reduced residual stress, the additional benefits to the overall cycle time are seen in Figure (9.7c) because the quench can now be interrupted and the product transferred to the ageing furnace already at the correct temperature for ageing, saving time and reheat energy. Additional benefits include the fact that the air quench is environmentally friendly; the castings are not stained by the less-than-clean water; the conveyor is straightforward to build and maintain; there is no mechanism required for lowering into water that normally results in complex and rusting plant. As we have repeatedly emphasized, the products from this type of furnace have somewhat lower apparent strengths and hardness, but greatly improved performance in service.

Figure 9.7d shows an ultimate system that might be acceptable for some products. The ageing treatment is simply carried out by interrupting the air quench slightly above the normal ageing temperature, and allowing the part to cool in air (prior to final rapid cooling by fans if necessary). This represents a kind of natural ageing process in which no ageing furnace is required. Strengths will suffer somewhat, but the lower costs and simplicity of the process may be attractive, making the process suitable for some applications.

9.6 Epilogue

Although the *strength of the material* will be lowered by a slower quench, the *strength of the component* (i.e. the failure resistance of the complete casting acting as a load bearing part) in service will be increased.

If water quench is avoided with a view to avoiding the dangers of internal residual stress, it is common for the customer to complain about the 5 per cent or so loss of apparent properties. In answer to such understandable questions, an appropriate reply to focus attention on the real issue might be 'Mr Customer, with respect, do you wish to lose 5 per cent or 50 per cent of your properties?'

In the experience of the author, a number of examples of castings that have been slowly quenched, losing 5 or 10 per cent of their strength, are demonstrated to double their performance in service (*Castings* 2003).

Finally therefore, it remains deeply regrettable, actually a scandal, that many national standards for heat treatment continue to specify water quenching. This disgraceful situation requires to be remedied. In the meantime the author deeply regrets having to recommend that such national standards be set aside. It is easy for the casting supplier to take refuge in the fact that our international and national standards on heat treatment often demand quenching into water, and thereby avoid the issue that such a production practice is risky for many components, and in any case provides the user with a casting of inferior performance. However, the ethics of the situation are clear. We are not doing our duty as responsible engineers and as members of society if we continue to ignore these crucial questions. We threaten the performance of the whole component merely to fulfil a piece of metallurgical technology that from the first has been woefully misguided.

The fact is that our inappropriate heat treatments have been costly to carry out, and have resulted in costly failures. It has to be admitted that this has been nothing short of a catastrophe for the engineering world for the past half century, and particularly for the reputation of light alloy castings, not to mention the misfortune of users. As a result of the unsuspected presence of bifilms they have suffered poor reliability so far, but as a result of the unsuspected presence of residual stress this has been made considerably worse by an unthinking quest for material strength that has in fact reduced component performance.

Rule 10

Provide location points

This Rule, *provide location points*, is added simply because the foundry can accomplish all the other 9 Rules successfully, and so produce beautiful castings, only to have them scrapped by the machinist. This can create real-life drama if the castings have been promised in a just-in-time delivery system. This Rule is added to help to avoid such misfortunes, and allow all parties to sleep more soundly in their beds.

Before describing location points, their logical precursors are datum planes. We need to decide on our datums first.

10.1 Datums

A datum is simply a plane defining the zero from which all dimensions are measured. For a casting design it is normal to choose three datum planes at right angles to each other. In this way all dimensions in all three orthogonal directions can be uniquely defined without ambiguity.

In practice, it is not uncommon to find a casting design devoid of any datum, there being simply a sprinkling of dimensions over the drawing, none of the dimensions being necessarily related to each other. On other designs the dimensions relate with great rigour to each other and to all machined features such as drilled holes, etc., but not to the casting. In yet other instances that the author has suffered, datums on one face have not been related to datums on other views of the same casting. Thus the raft of features on one face of the casting shifts and rotates independently of the raft of features on the opposite face.

Figure 10.1a shows a sump (oil pan) for a diesel engine designed for gravity die-casting in an aluminium alloy. The variations in die temperature and ejection time result in variability of the length of the casting that are well known with this process, and not easily controlled. The figure shows how the dimensioning of this part has made the part nearly unmanufacturable by this method. Three fundamental criticisms can be made:

1. The datum is at one end of the product. If the datum had been defined somewhere near the centre of the part, then the variability produced by the length changes of the casting would have been approximately halved.
2. There is only one feature on the component whose location is critical; this is the dipstick boss. If the boss is slightly misplaced then it fouls other components on the engine. It will be noticed that the dipstick boss is at the far end of the casting from the datum. Thus variability in length of the casting will ensure that a large proportion of castings will be deemed to have a misplaced boss. If the datum had been located at the other end of the casting, near to the boss, the problem would have been reduced to negligible proportions. If the datum had been chosen as the boss itself, the problem would have disappeared altogether, as in Figure 10.1b.
3. The datum is not defined with respect to the casting. It is centred on a row of machined holes, which clearly do not exist at the time the casting is first made and when it is first required to be checked. Depending on whether the machinist decides to fix the holes in the centre of the flange, or relate them by measurement to the more distant dipstick boss, or to the centre of the casting averaged from its two ends, or any number of alternative strategies, the drilled holes could be almost anywhere in or even partly off the flange!

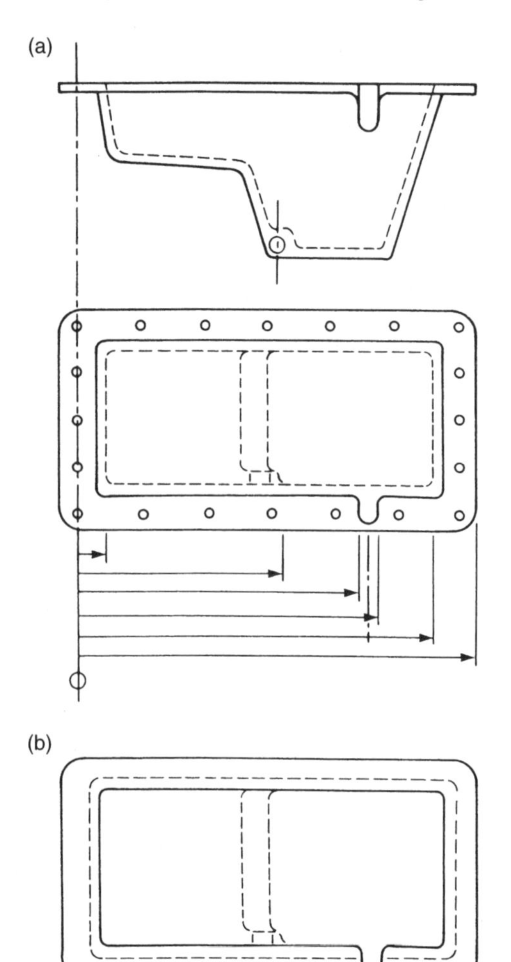

Figure 10.1 *(a) A badly dimensioned sump, resulting in a casting that can only be manufactured with difficulty and with high scrap losses. (b) A sympathetically dimensioned casting, datumed on the most critical casting feature, that can be easily and efficiently manufactured.*

Figure 10.1b shows how these difficulties are easily resolved. The datum is located against the side of the dipstick boss, and hence is fixed in its relation to the casting and goes some way to halving errors in the two directions from this plane. It also means that the dipstick boss itself is now impossible to misplace, no matter how the casting size varies; all the other dimensions are allowed to float somewhat because movement of other points on the casting is not a problem in service. This part can then be produced easily and efficiently, without trauma to either producer or customer!

In summary, the rules for the use of datums (partly from Swing 1962) are:

1. Choose three orthogonal datum planes.
2. Ensure the planes are parallel to the axes of motion of the machine tools that will be used to machine the part (otherwise unnecessary computation and opportunity for error is introduced).
3. Fix the planes on real casting features, such as the edges of a boss, or the face of a wall (i.e. not on a centreline or other abstract constructional feature). Choose casting features that are:
   (a) critical in terms of their location, and
   (b) as near the centre of the part as possible.

10.2 Location points

Location points are those tiny patches on the casting that are used to locate the casting precisely and unambiguously in three dimensions. They are required by the toolmaker, since he can construct the tooling with reference to them; the founder, to check the casting once it has been made; and the machinist, who uses them to locate the casting prior to the first machining operations. These features therefore integrate the manufacture of the product, ensuring its smooth transmission as it progresses from toolmaker to founder to machinist.

Whereas the casting datums are invisible planes, defining the concept of a zero in the dimensional space in and around a casting, the tooling points are real bits of the casting. The datums are the software, whereas the tooling points are the hardware, of the dimensioning system. It is useful, although not essential, for the datums to be defined coincident with the tooling points.

Location points are known by several different names, such as machining locations (which is a rather limiting name) or pick-up points. On drawings, TP for tooling point is used as the common abbreviation for the drawing symbol. Although in practice I tend to use all the names interchangeably it is proposed that 'location points' describes their function most accurately and will be used here.

Before the day of the introduction of location points in the Cosworth engine-building operation, I was accustomed to a complex cylinder head casting taking a skilled man at least 2 hours to measure to assess how to pick up the casting for machining. The casting was repeatedly re-checked, re-orienting it slightly with shims to test whether wall thicknesses were adequate, and whether all the surfaces required to be machined would in fact clean up on machining. After the 2 hours, it was common to see it, our most expensive casting, dumped on the scrap heap; no orientation could be found to ensure that it was dimensionally satisfactory and could be completely cleaned up on machining. All this changed on the day when the new foundry came on stream. A formal system of location points to define the position of the casting was introduced. After this date, no casting was subjected to dimensional checking. All castings were received from the foundry and on entering the machine shop were immediately thrown on to a machine tool, pushed up against their location points, clamped, and machined. No tedious measurement time was subsequently lost, and no casting was ever again scrapped for machining pick-up problems.

It is essential that every casting has defined locations that will be agreed with the machinist and all other parties who require to pick up the casting accurately.

For instance, it is common for an accurate casting to be picked up by the machinist using what appear to be useful features, but which may be formed by a difficult-to-place core, or a part of the casting that requires some dressing by hand. Thus although the whole casting has excellent accuracy, this particular local feature is somewhat variable in location. The result is a casting that is picked up inaccurately, and does not therefore clean up on machining. As a result it is, perhaps rather unjustly, declared to be dimensionally inaccurate.

The author suffered precisely this fate after the production of a complex pump body casting for an aerospace application that achieved excellent accuracy in all respects, except for a small region of the body that was the site where three cores met. The small amount of flash at this junction required dressing with a hand grinder, and so, naturally, was locally ground to a flat, but at various slightly different depths beneath the curved surface of the pump body. This hand-ground location was the very site that that machinist chose to locate the casting. The result was disaster. Furthermore, it was not easily solved because of the loss of face to the machinist who then claimed that the location options suggested by the foundry were inconveniently awkward. The fault was not his of course. The fundamental error lay in not obtaining agreement between all parties before the part was made. If the location point used by the machinist really was the only sensible option for him, the casting engineer and toolmaker needed to ensure that the design of the core package would allow this.

Ultimately, this Rule is designed to ensure that all castings are picked up accurately, and conveniently if possible, so that unnecessary scrap is avoided.

Different arrangements of location points are required for different geometries of casting. Some of the most important systems are listed below.

10.2.1 Rectilinear systems

1. Six points are required to define the position of a component with orthogonal datum planes that is designed for essentially rectilinear machining, as for an automotive cylinder head or block. (Any fewer points than six are insufficient to define the position of the casting, and any more than six will ensure that one or more points are potentially in conflict.)

On questioning a student on how to use a six-point system to locate a brick-shaped casting, the reply was 'Oh easy! Use four points around the outside faces and one top and one bottom.' This shows how easy it is to get such concepts wildly wrong!

In fact, the six points are used in a 3, 2, 1 arrangement as shown in Figure 10.2. The system works as follows: three points define plane A, two define the orthogonal plane B, and one defines the remaining mutually orthogonal plane C (Figure 10.2). The casting is then picked up on a jig or machine tool that locates against these six points. Example (a) shows the basic use of the system: points 1, 2 and 3 locate plane A; points 4 and 5 define plane B; and point 6 defines plane C. Planes A, B and C may be the datum planes. Alternatively, it is often just as convenient for them to be parallel to the datum planes, but at accurately specified distances away.

Clearly, to maximize accuracy, points 1, 2 and 3 need to define a widely based triangle, and points 4 and 5 similarly need to be as widely spaced as possible. A close grouping of the locations will result in poor reproducibility of the pickup of the casting; tiny errors in the position or surface roughness of the tooling points would be magnified if they were not widely spaced.

Example (b) shows an improved arrangement whereby the use of a tooling lug on the longitudinal centreline of the casting allows the dimensions along the length of the casting to be halved. The largest dimension of the casting is usually subject to the largest variability, so halving its effect is a useful action.

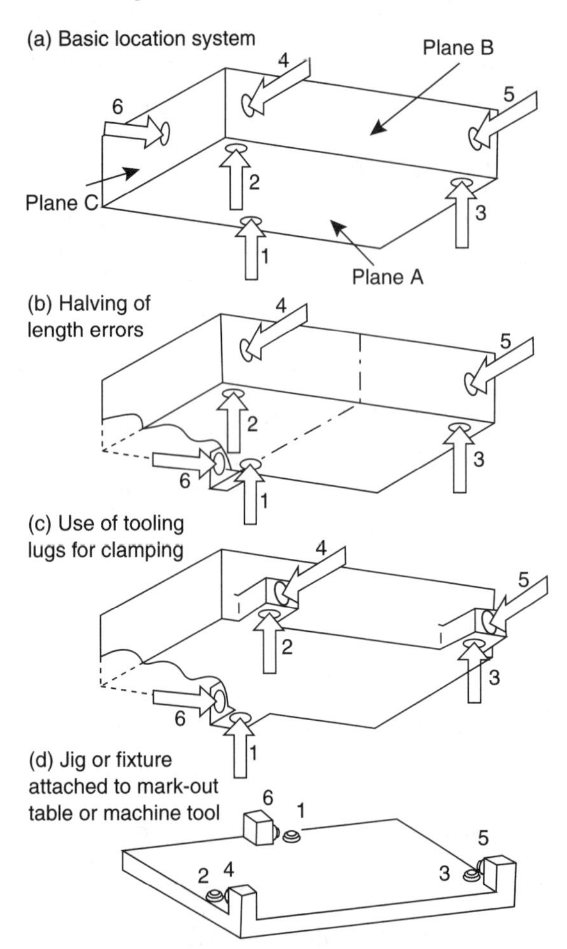

Figure 10.2 *(a, b) Increasingly improved ways in which the six-point location method can be used on a casting. (c) Further recommended technique using lugs to facilitate clamping. (d) The jig used as a cradle for the six-point location system (clamps are omitted for clarity).*

Example (c) is a further development of this idea, creating lugs that serve the additional useful purpose of allowing the part to be clamped immediately over the points of support, and off the faces that require to be machined. In practice the lugs may be existing features of the casting, or they may be additions for the purpose of allowing the casting to be picked up for inspection or machining. This concept is capable of further development, using lugs arranged on all the centrelines of the casting so as to halve the errors in all directions.

10.2.2 Cylindrical systems

Most cylindrical parts do not fall nicely into the classical six-point location systems as described above for rectilinear components. The errors of

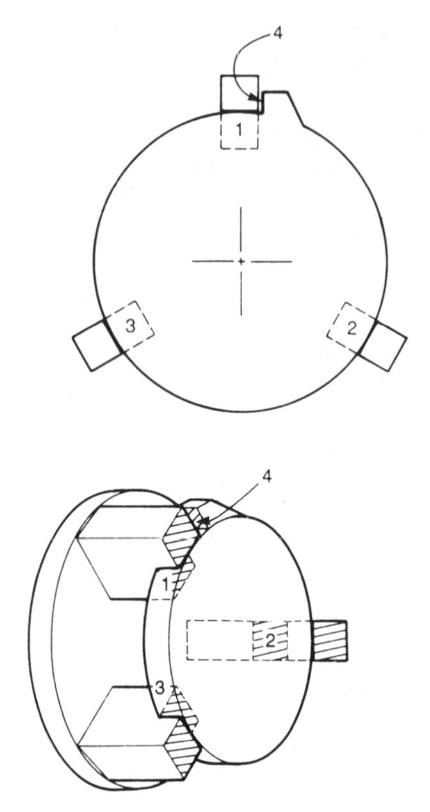

Figure 10.3 *The use of a three-jaw self-centring chuck for casting location and clamping.*

eccentricity and diameter both contribute to a rather poor location of the centre using this approach. The unsuitability of the orthogonal pick-up system is analysed nicely by Swing (1962).

In fact, the obvious way to pick up a cylinder is in a three-jaw chuck. The self-centring action of the chuck gives a useful averaging effect on any out of roundness and surface roughness, and is of course insensitive to any error of diameter.

In classical terms, the three-jaw chuck is equivalent to a two-point pickup, since it defines an axis. We therefore need four more points to define the location of the part absolutely. Three points abutting the jaws will define the plane at right angles to the centre axis, and one final point will provide a 'clock' location. Figure 10.3 shows the general scheme.

Another location method that is occasionally useful is the use of a V block. This is a way of ensuring that a cylindrical part, or the round edge of a boss, is picked up centrally, averaging errors in the size and, to some extent, the shape of the part. The method has the disadvantage that errors in diameter of the part will cause the whole part to be shifted either nearer to or away from the block, depending on whether the diameter is smaller or larger. (The reader can quickly confirm these shifts in position with rough sketches.)

A widely used but poor location technique is the use of conical plugs to find the centre of a cored hole. Even if the hole is formed by the mould, and so relatively accurately located, any imperfection in its internal surface is difficult to dress out, and will therefore result in mislocation. If a separate core forms the hole then the core-positioning error will add to the overall inaccuracy of location. Location from holes is not recommended.

It is far better to use external features such as the sides of bosses or walls, as previously discussed. These can be more easily cast and maintained clean.

10.2.3 Trigonal systems

For some suitable parts of triangular form, such as a steering gear housing, a useful and fundamentally accurate system is the cone, groove and plane method (Figure 10.4).

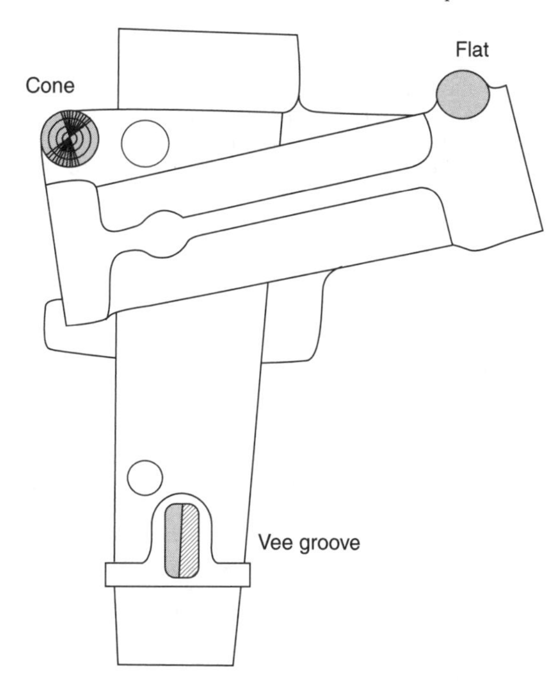

Figure 10.4 *A plan view of a steering housing for a car, showing a flat, groove and cone location system.*

10.2.4 Thin-walled boxes

For prismatic shapes, comprising hollow, box-like parts such as sumps (oil pans), the pickup may be made by averaging locations defined on opposite internal or external walls. This is a more lengthy and expensive system of location often tackled by a sensitive probe on the machine tool, that then calculates the averaged datum planes of the component, and orients the cutter paths accordingly. This technique is especially useful where an average location is definitely desirable, as a result of the casting suffering different degrees of distortion of its relatively thin walls.

The tooling points should be defined on the drawing of the part, and should be agreed by (i) the manufacturer of the tooling, (ii) the caster, and (iii) the machinist. It is essential that all parties work from ONLY these points when checking dimensions and when picking the part up for machining.

For maximum internal consistency between the tooling points, all six should be arranged to be in one half of the mould, usually the fixed or lower half, although sometimes all in the cope. The separation of points between mould halves, or having some defined from the mould and some from cores, will compromise accuracy. However, it is sometimes convenient and correct to have all tooling points in one internal core, or even one half of an internal core (defined from one half of the core box) if the machining of the part requires to be defined in terms of its internal features.

Clearly therefore, the location points are required to be *actual cast-on features* of the casting. This point cannot be over-emphasized. It is not helpful, for instance, to define a location feature as a centreline of a bore. This invisible feature only exists in space (perhaps we should say 'free air'). Virtual features such as centrelines have to be found by locating several (at least three) points on the internal as-cast solid surface of the bore, and its centre thereby calculated. Clearly, these 'virtual' or 'free-air' so-called location points necessarily rely on their definition from other nearby as-cast surfaces. These ambiguities are avoided by the direct choice of as-cast location features.

These features need to be cast nicely, without obscuring flash, or burned-on sand, and definitely should not be attacked by enthusiastic finishers wielding an abrasive wheel.

It is essential that the location points are *not* machined. If they are machined, the circumstance poses the infinitely circular question 'What prior datums were used to locate and position the casting accurately to ensure that the machining of the machining locations (from which the casting would be picked up for machining) were correctly machined?' Unfortunately, such indefensible nonsense has its committed devotees.

In general it is useful if the tooling locations on the casting can remain in position for the life of the part. It is reassuring to have the tooling points always in place, if only to resolve disputes between the foundry and machine shop concerning failure of the part to clean up on machining. It is therefore good practice to try to avoid placing them where they will be eventually machined off. Using existing casting features wherever possible avoids the cost of additional lugs, and possibly even the cost of subsequent removal if their presence on the final product is not allowed.

The definition of the six-point locations, preferably on the drawing, prior to the manufacture of the casting, is the only method of guaranteeing the manufacturability of the part. The method allows an integrated approach right from the start of the creation of the tooling, because the patternmaker can use the tooling points as the critical features of the tooling in relation to which all measurements will be defined. The foundry engineer will know how to pick up the casting to check dimensions after the production of the first sample castings. The machinist will use the same points to pick up the casting for machining. They all work from the same reference points. It is a common language and understanding between design, manufacture, and inspection of products. Disputes about dimensions then rarely occur, or if they do occur, are easily settled. Casting scrap apparently due to dimensioning faults, or faulty pickup for machining, usually disappears.

This integrated manufacturing approach is relatively easily managed within a single integrated manufacturing operation. However, where the pattern shop, foundry and machinist are all separate businesses, all appointed separately by the customer, then integration can be difficult to achieve. It is sad to see a well-designed six-point pick-up system ignored because of apparent cussedness by one member of the production chain. The industry and its customers very much need purchasing and manufacturing policies based on the adoption of integrated and fundamentally correct systems.

10.3 Location jigs

Figure 10.2d illustrates a basic jig that is designed to accept a casting with a six-point location system. The jig is simply a steel plate with a series of small pegs and blocks. It contrasts with many casting jigs, which are a nightmare of constructional and operational complexity.

Our simple jig is also simple to operate. When placing the casting on the jig, the casting can be slid about on locations 1, 2 and 3 to define plane A, then pushed up against locators 4 and 5 to define plane B, and finally slid along locators 4 and 5 until locator 6 is contacted. The casting is then fixed uniquely in space in relation to the steel jig plate. It can then be clamped, and the casting measured or machined. The six locations can, of course, be set up and fixed in the machine tool that will carry out the first machining operation.

After the first machining operations it is normal to remove the casting from the as-cast locations and proceed with subsequent machining using the freshly machined surfaces as the new location surfaces. McKim and Livingstone (1977) go on to define the use of functional datums which may become useful at this stage. They are machined surfaces that normally relate to features locating the part in its intended final application.

Other jigs can easily be envisaged for cylindrical and other shaped parts.

10.4 Clamping points

During machining the forces on the casting can be high, requiring large clamping loads to reduce the risk of movement of the casting. Clamping points require to be thought about and designed in to the casting at the same time as the location points. This is because the application of high clamping loads to the casting involves the risk of distortion of the casting, and of spring-back after release of the clamps at the end of machining. Flat machined surfaces are apt to become curved after machining because of this effect.

The great benefit of using tooling lugs as shown in Figure 10.2c can therefore be appreciated. The location point and the clamping point are exactly opposed on either side of the lug. In this way the clamping loads can be high, without introducing the risk of the overall distortion of the casting.

Further essential details of the design of the clamping action include the requirement for the action to move the part on to, and hold it against, the location point.

For softer alloys that are easily indented, the clamp face needs to be 5–10 mm in diameter, similar to the working area of the tooling point. Even so, a high clamping load will typically produce an indentation of 0.5 mm in a soft Al alloy, decreasing to 0.2 mm in an Al alloy hardened by heat treatment, and correspondingly less still in irons and steels.

10.5 Mould design: the practical issues

The problem for the casting engineer is to achieve a successful design of the mould. This problem is not to be underestimated, since it requires the simultaneous solving of a list of issues including

(i) The design of the mould and core assembly can be a problem in itself. It is not uncommon to find that it is impossible to assemble the cores because some shape feature of neighbouring cores has been overlooked. It is all too easy to stumble into such pitfalls in a complex core assembly. When the first set of cores are made from the new patternwork, with its shining new varnish and paintwork, the discovery of such 'passing problems', where one core will not pass another and so fit into the assembly, are greeted with embarrassment and dismay.

The other common problem for the casting engineer and toolmaker is the design of the assembly so that cores fit, in logical order, only into the drag if possible (Figure 10.5). Cores in the cope are not usually an option for horizontally parted greensand moulds, since, if the sand strength is not high, they are in danger of falling out of their prints when the cope is turned over and closed onto the drag prior to casting. Gluing cores into the cope is possible in the case of strong chemically bonded sand moulds. However, gluing takes time and is therefore costly, and introduces the danger that any excess glue may cause a blow hole defect in the casting if it contacts the metal. In addition, glue applied to a core print may prevent the core from venting, leading to a rather different form of blow defect from the core itself. The use of glues should therefore be avoided if at all possible.

It is common for complex core assemblies to be assembled at a separate station sited off the mould assembly line. Core assembly can then be accurate since the assembly is built up in a jig. The cores are designed to be lifted by the jig, transferring from the assembly station and lowered into the mould as a complete package. Castings that require lengthy core assembly times are not thereby allowed to slow the cycle time of the moulding line.

(ii) The filling system. The provision of a good filling system, and its integration with the rest of the mould and core system is sometimes not easy, and in some cases the additional trouble or expense to provide a good filling system is by-passed. (The minefield of poor castings and high scrap rates is always entered for apparently good reasons.) The filling system design forms the major part of this book. It is mandatory reading. Its rules are recommended to be followed in all cases.

(iii) The feeding system. Naturally, following the first rule for feeding, it is clearly best if feeders can be completely avoided. However, if they are considered to be necessary, it is usually not a problem to place feeders high on a casting. Thus the provision of feeders rarely involves difficulties of mould design. One of the key issues is to place the feeders so that they are easy to cut off or machine away subsequently.

(iv) The avoidance of infringement of any of the 10 Rules. For instance, convection considerations might force the issue of rotating the mould through 180 degrees after filling. This action usually confers other benefits and makes integration of the filling and feeding systems powerfully effective and economic. It is a strategy to be recommended.

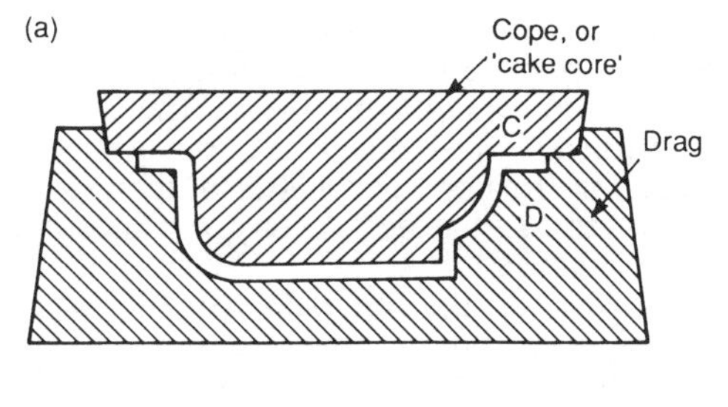

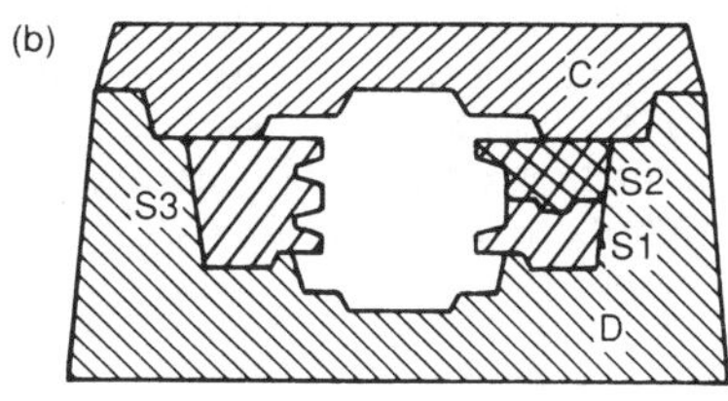

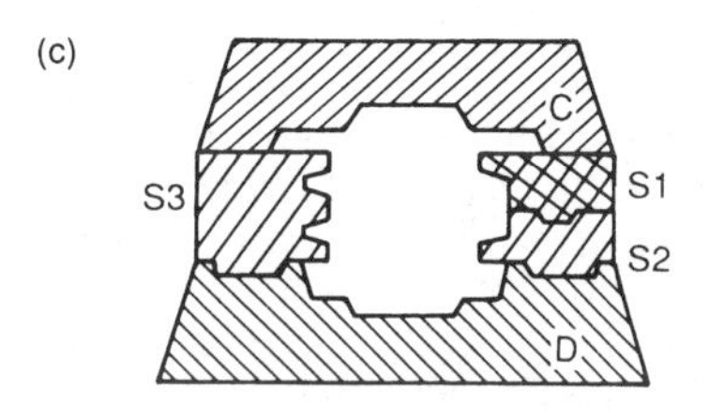

Figure 10.5 *(a) Simple cake core and drag assembly; and (b) a cope and drag with side cores, all located in the drag; (c) an apparently lower-cost alternative to (b), but resulting in possible loss of dimensional control.*

However, sometimes the solution to all these issues is not straightforward. For instance much time may be spent attempting to solve the issues with the casting oriented in one direction, only to realize that such an orientation involves insoluble problems. The casting is then turned upside down and the exercise is repeated in the hope of a better outcome. Such experiences are the day to day routine of the casting engineer.

Furthermore, the complexity of the issues is not easily solved at this time by computer. There have been many such attempts, but it is fair to say at this stage such efforts have not been developed to such a degree that the professional casting expert is offered any significant help. However, of course, we may look forward to the day when the computer can provide useful solutions.

10.6 Casting accuracy

There is, of course, very little reason to go to great lengths regarding the provision of casting location points if the casting is hopelessly inaccurate in other ways. This section draws attention to the general problems associated with casting accuracy.

Any casting that we make is, in common with all other manufactured products, never quite perfect in terms of size and shape. To allow for this, tolerances are quoted on engineering drawings. So long as the casting is within tolerance, it will be acceptable.

Some reasons for the casting being out of tolerance include elementary mistakes like the patternmaker planting the boss in the wrong place. This leads to an obvious systematic error in the casting, and is easily recognized and dealt with by correcting the pattern. It is an example of those errors that can be put right after the first sample batch of castings is made and checked.

Another common systematic error in castings is the wrong choice of patternmaker's contraction allowance. The contraction of the casting during cooling in the mould is often of the order of 1 or 2 per cent. However, it depends strongly on the strength of the mould. For instance, in an extreme case, a perfectly rigid mould will fix the casting size; in such a situation the casting simply would have to stretch during cooling since it would be prevented from taking its natural course of contracting. To summarize, the choice of contraction allowance prior to the making of the first casting is often not easy, and is often not exactly right. This point is taken up at length in *Castings 2003*, with recommendations on how to live with the problem.

Other errors are less easily dealt with. These are random errors. No two nominally identical castings are precisely alike. The same is true for any product, including precision-machined parts. The ISO Standard (1984) for casting tolerances indicates that although different casting processes have different capabilities for precision, in general the inaccuracies of castings grow with increasing casting size, and the standard therefore specifies increasing linear tolerances as linear dimensions increase. (Nevertheless it is worth pointing out that the corresponding percentage tolerance actually falls as casting size increases.) Other work on the tolerancing of castings suggests that the ISO standard is still in its early days, and has considerable potential for further improvement (Reddy *et al.* 1988).

Because of the effects of random errors being superimposed on systematic errors, it is of course risky to attempt to correct the patternmaker's error by moving the boss into an apparently correct location simply after the production of the first trial casting. Figure 10.6 illustrates that the random scatter in positions might mean that the boss appeared to be in the

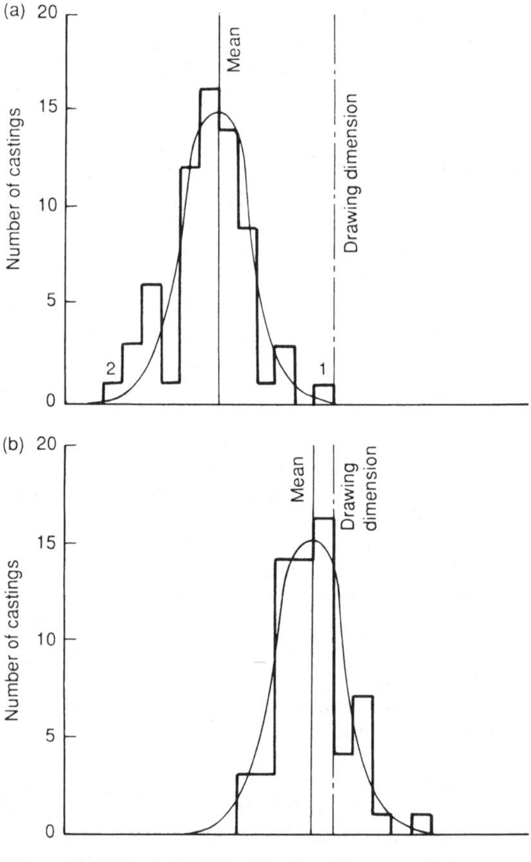

Figure 10.6 *Statistical distribution of casting dimensions (a) before, and (b) after pattern development. Based on Osborn (1979).*

correct place first time if the casting happened to be number 1 in Figure 10.6, or might have been over twice as far out of place, compared to its average position, if the first casting had been number 2. A sample of at least two or three castings is really needed, and preferably ten or a hundred. The mean boss location and its standard deviation from the mean position can then be known and the appropriate actions taken.

At the present time it will be of little surprise to note that such exemplary action is not common in the industry. This is because companies are not generally equipped with a sufficient number of fast, automated three-coordinate measuring machines. As such standards of measurement become more common, so the attainments in terms of accuracy of castings will increase.

Finally, as we have seen, the ISO Standard gives the general trend of increase in the size of random errors as casting size increases. However, the casting designer and engineer require much more detailed knowledge of the sources of individual contributions to the final total error. The remainder of this chapter is an examination of these contributions. The reduction of these errors allows the production of castings that are considerably more accurate than the minimum accuracy requirements of the International Standard Organization document.

10.7 Tooling accuracy

Tooling is taken to include the pattern and its coreboxes, or the die, and any measuring or checking jigs and gauges.

The problem of constructing the pattern, allowing correctly for the contraction and distortion of the casting, has already been discussed and will not be dealt with here.

Patterns used in sand casting, and dies used in die-casting, are subject to wear, so that the casting gradually becomes oversize. Conversely, the tooling of many processes is also subject to build-up problems associated with the deposition of small amounts of mould aggregate and binder on the surface of the tooling, and the gradual accumulation of release agents that may be used, causing parts of the casting to become undersize.

Distortion is another problem. Wood is a useful and pleasant material for pattern construction. It is easily worked, light to handle, and easily and quickly repaired or modified as necessary. Even so, it is not a contender for really accurate work because of its tendency to warp. A good patternmaker will attempt to reduce such movement to a minimum by the careful use of ply and the alignment of the grain of the wood, together with strengthening battens. The use of various stabilized woods and synthetic wood-like materials has also helped considerably (Barrett 1967). Nevertheless the ultimate stability in tooling is only achieved with the use of metal, or cast resin that is properly supported in metal frames. Cast-resin patterns, especially when cast into aluminium alloy frames for strength and rigidity, are usually extremely reliable. However, some resin systems such as polyurethanes tend to suffer from the absorption of solvents from the chemical binders in the sand, and so suffer swelling and degradation (Gouwens 1967). Cast-resin patterns that are backed with wood frames are not reliable; the warpage of the wood distorts the internal resin shape, usually within a month or so. After a year the tooling is seriously inaccurate, so that cores produced from such equipment will not assemble properly.

The working temperature of tooling affects the casting size directly; a warm pattern will give a slightly larger casting. If we consider an epoxy resin corebox cast into an aluminium alloy frame, the box will largely take its size from the temperature of the metal frame (i.e. not the internal lining of epoxy resin). If the temperature at the start of the Monday morning shift is 10°C, and if the returning sand creeps up to 30°C by the end of the morning, then for a 500 mm long casting the 20°C temperature rise will cause the castings to grow by $20 \times 20 \times 10^{-6} \times 500 = 0.2$ mm. This is not large in itself, but when it is added to other random variables the uncertainty in the final casting length becomes increasingly out of control.

Anderson (1987) emphasized the important requirement that for the most accurate work the pattern or die should be utilized as an adaptive control element in production. Thus it needs to be built in such a way that it can be modified to produce the required size and shape of the casting. The use of patterns split transversely across their major length is common. The prior insertion of a spacer in this split allows the spacer to be removed and replaced by something thinner or thicker as necessary. Such simple techniques involve only modest extra expense during the construction of the pattern but are a reassurance against the possibility of major expensive rebuilds later.

10.8 Mould accuracy

Mould accuracy depends strongly on the mould assembly method. Usually, a mould assembly simply involves two parts; a cope and a drag. This makes for maximum accuracy. On other occasions the mould assembly can be complicated, requiring many parts, and requiring much discussion between the pattern shop, foundry, and casting designer to find an appropriate solution. Accuracy can now become illusive and troublesome.

As a general rule, it is useful to ensure that even in the most complicated of mould assemblies, the design of the assembly consists principally of a drag and a cope. In addition, all parts should be interrelated via a single mould part, which will normally be the drag. In this case the assembly of cores will be in the drag, with each core located separately to prints in the drag, and the final operation will simply consist of closing with the cope. The cope also should have features that contact and locate directly on the drag as the key mould part.

Such simple rules are easily forgotten. One can see many ambitious castings that have failed to achieve dimensional acceptability because the mould assembly has resembled a random heap of assorted blocks: one layer of cores tottering upon another, with finally the cope perched precariously on top. Clearly, the details of the casting formed in the cope bear no constant relation to the details formed in the drag.

The simplest form of construction of moulds that was widely popular consisted of a drag with a cope in the form of a 'cake core' as shown in Figure 10.5a. Accuracy was excellent. However, this simple construction did not allow for the placement of any useful bottom-gated running system, and the top pouring that had to be accepted as a consequence might have been marginally acceptable for some rough and ready grey iron castings in greensand moulds but did not give good results for the majority of casting alloys.

Figure 10.5b shows a simple type of cope–drag arrangement with side cores, all located in the drag, apart from side core S2, which rests on S1. It was judged that the small accumulation of errors in the positioning of S2 would be acceptable in this case. If the positioning of S2 had been critical, it could still have been located in the drag by stepping the contour of the drag appropriately to form a convenient location.

Figure 10.5c shows how it would have been easy to have saved some sand by abandoning the deep drag construction, and having a core assembly that consisted simply of a pile of cores, rather than a proper cope and drag. The overall accuracy of the casting now suffers from the accumulation of errors introduced by the intermediate side cores S1, S2 and S3. There will, for instance, be a poor match between the top and bottom features of the casting.

The dimensional problems that arise in setting cores are examined by Skarbinski (1971). In general it needs to be said about the accurate printing of cores that the core print should be designed assuming that the core will be produced with errors in its size and shape, and will have to fit into a mould which will also have suffered some distortion during its manufacture. The print also sometimes has to restrict the movement of the core during mould filling because of buoyancy, and yet may also have to allow the relative movement of the core in its print to permit the thermal expansion of the core. All this is a seemingly impossible task to achieve accurately. However, it is usually solvable by applying the following simple rules:

1. The print requires tolerance where it needs to fit (i.e. must not be made size-for-size, that would have produced an interference fit. The only exception to this is the heights of cores where cores are stacked one on top of the other, since in this case the accumulation of errors requires to be kept to a minimum).
2. The print requires clearance where it is not required to fit, and where expansion clearance is required.

Rules often appear pedantic or even pedestrian when they are spelled out! However, the application of the rules involves much work that is seldom expended on the design of patternwork. Although some prints are easily and quickly designed, others require lengthy agonized consideration resulting in compromises that have to be carefully assessed. Every print requires such detailed design. It is attention to details such as these that makes the difference between the inadequate and the excellent casting.

Nevertheless, these problems are eliminated if the use of the core can be avoided altogether. (The situation is reminiscent of Feeding Rule 1: Avoid using a feeder if possible. The useful equivalent rule for cores, that makes an additional rule for mould assembly, is:

3. The number of cores should be reduced to a minimum, moulding as much as possible directly in the cope and drag.

Not only does the application of this principle reduce dimensional errors, but also the addition of each core involves considerable extra tooling cost, and an additional cost in the production of the casting, sometimes approaching the cost of the production of a cope or drag. The addition of a core between a cope and drag represents the third piece of sand to be added to the two original mould halves; thus the costs at this stage may increase by 50 per cent for the addition of the first core. At other times a small core can save money by avoiding extra complexity of the tooling. Each case needs separate evaluation.

A high hidden addition to total casting costs results from the use of cores. These difficult to assess costs arise from the accumulation of a number of minor operations, most of which are usually overlooked. For instance, the core needs to be scheduled, made (perhaps on a capital-intensive core-blowing machine), deflashed, stored on special racks (taking up valuable floor space), retrieved from storage, transported to the moulding line, and then correctly assembled into the mould. Errors arise as a result of the incorrect core being made or transferred, or sufficient are broken in storage or transit to cause the whole process to be repeated. Alternatively, its assembly into the mould gets forgotten at the last moment! Cores are therefore almost certainly more expensive than most foundry accounting systems are aware of. (The costs of chills, and of scrapped castings, are similarly illusive and not widely understood.)

A further use of cores, in addition to their obvious purpose in providing detail that cannot be moulded directly, is that the running system can often be integrated behind and underneath them, the main runner and gates being located beneath a base, side or end core. This is a valuable facility offered by the use of a core and should not be overlooked. In a number of castings the addition of a core may be for the sole purpose of providing a good running system. Such a core is often money well spent!

The problem with the automation of core-assembly systems is finding the core again after it has been put down on, for instance, a conveyor or a storage rack. This is a difficult job for a robot, since extreme accuracy is required, and the cores are often of extreme delicacy. Clearly, one method of solving this problem is never to put the cores down in the first instance.

Schilling (1987) succeeded in developing this concept with a unique system of making and assembling cores in which the cores are not released from one half of the opened corebox until the other half of the core has already been located in the core-assembly package. In this way the cores are assembled completely automatically and with unbelievable precision. Cores are located to better than 0.03 mm, allowing them to be assembled with clearances which are so small that the cores could not be assembled by hand. In fact the cores are sprung into place with interference fits. The rigorous application of this technique means that castings need to be designed for the process, since the assembly of each core is by vertical placement over the previous core. For instance, any threading of cores in through holes in the sides of other cores, such as often occurs with port cores through the water jacket core of a cylinder head casting, is not possible. This disadvantage will limit the technique to partial application, loading some but not all cores of a cylinder head, for instance. Even this would be an important advance.

A final note in this section relates to cope-to-drag location. This is, of course, of primary importance. Failure to achieve good location results in a mis-match defect. Mis-match is a lateral location error, and not to be confused with the vertical precision with which cope and drag meet, which is normally of the order of ±0.05 to 0.10 mm.

In foundries using moulds contained in moulding boxes, however, mis-match is unfortunately all too common and is usually the result of the use of worn pins and bushes that are used to locate the boxes. Southam (1987) analyses the effect of the errors involved in the pin and bush location system. These are numerous and serious. The pin-to-bush clearance is typically 0.25 mm, and given an apparently acceptable additional wear of 0.35 mm, he finds that the total possible mis-match between cope and drag moulds is as much as 1.5 mm.

He proposes, therefore, a completely different system, in which pins and bushes are eliminated. The cope and drag boxes are simply guided by wear blocks fixed to the outside edges of the box. These slide against two guides on the long side of the box, and one guide against the narrow side of the box during moulding and closing operations. The boxes are held against the guides by light spring pressure, or by pneumatic cylinders. The system appears deceptively simple, but actually requires a certain amount of good engineering to ensure that it operates correctly on mould closure, as Southam describes. Although Southam calls his method the three-point registration system, it is in reality a classical six-point location system, since he uses a further three points to locate the drag in a parallel plane to the cope during closure.

The ability to locate cope to drag with negligible error has a number of benefits that Southam lists. The maintenance and replacement of worn pins and bushings is a foundry chore and expense that is eliminated. (In fact anyone who has not experienced the problem of carrying out such an operation in a jobbing foundry will have a problem to comprehend the awesome scale of the task, because of the hundreds of pins and bushes, and the relentless wear problem, requiring the operation to be repeated at regular intervals despite the multitude of pressing problems elsewhere in the foundry environment.) Instead, only three guides on the closer and three each on the cope and drag pattern need to be checked, and the effect of wear of these parts on mis-match is minimal because the resultant displacement is largely self-compensating.

For the case of precision core packages, the sand mould is not contained in a box, and thus has to be located directly with a sand-to-sand location. Since this is defined from the patternwork, the location relates perfectly to the casting details, and mismatch is therefore not possible.

10.9 Summary of factors affecting accuracy

Some of the many factors that control the accuracy of the final casting have been dealt with above, and some are planned for inclusion in *Castings processes* to follow this publication. We shall therefore content ourselves here with a brief summary:

1. Pattern inaccuracy.
2. Mould inaccuracy.
3. Mould expansion and/or contraction because of temperature and pressure.
4. Casting expansion because of precipitation of less dense phases such as graphite or gases.
5. Casting contraction on freezing causing local sinks.
6. Casting contraction on cooling leading to (a) different overall casting size, depending on the constraint by the mould, and (b) distortion if unevenly constrained or unevenly cooled.
7. Casting overall change of size on heat treatment or on slow ageing at room temperature.
8. Casting distortion if unevenly cooled by an inappropriate quenchant or too rapid quench from heat-treatment temperature.
9. Casting distortion caused by shot blasting. This effect has not been dealt with previously.

The compressive stresses introduced into the surface by a peening effect can lead to the distortion of the casting as reported by Kasch and Mikelonis (1969). The effect is widely used in the sheet metal industry to induce the controlled forming of curved surfaces; aircraft wing panels are formed from flat sheets in this way; the flat product gradually curves away, becoming convex towards the direction of the impingement of the shot. Controlled shot peening is also used to increase the fatigue resistance of castings as discussed by Lawrence (1990) and O'Hara (1990).

10.10 Metrology

Even if it were possible to produce an absolutely accurate casting, it would not be possible to prove it! This apparently curious statement is the consequence of errors that occur during measurement. Inexact measuring of the casting will cause apparent random deviations in the dimensions of the casting. Svensson and Villner (1974) point out this problem, and work out the influence of measuring accuracy on the apparent dimensional accuracy of the casting. Table 10.1 is based on their work.

It is clear that even if the casting has dimensions that are quite correct, even careful measurement will introduce a certain amount of apparent error, and careless measurement will, of course, introduce even more. These errors have been a traditional problem within the industry but the introduction of large-size three-dimensional coordinate measuring machines has significantly helped.

Even so, problems still remain. For instance, Swedish workers point out that for small dimensions, and where high accuracy is required, the surface roughness will influence the apparent accuracy of the casting. Thus a change in the surface finish from 75 μm to 200 μm will give an increase of one tolerance grade in the ISO system.

Table 10.1 Limits of accuracy of measurements

Measuring equipment and range (mm)	*Accuracy of measurement* (±mm)	*Mean dimension* (mm)	*Accuracy* (%)	*ISO tolerance grade IT*
Steel tape >1000	1	1000	0.10	13
		2000	0.05	12
		5000	0.02	11
Steel rule 500–1000	0.5	500	0.10	12
		1000	0.05	11
Vernier calliper 0–500	0.1	50	0.20	10
		100	0.10	10
		200	0.05	9
		500	0.02	9

The surface finish influences the measurement and location processes in other ways. For instance, the modern touch probes, which locate dimensions on the casting with the most delicate of contact pressure, effectively only measure to high spots, thus biasing the measurements in one direction: exterior dimensions on the casting are measured oversize, and cored holes appear undersize.

Results from mark-out equipment using a scribing line tend to give more averaged results, since minor surface irregularities are cut through.

Similarly, when castings are clamped on to their location points, the small area of the contact points, typically 5 mm diameter, and the high loads which can be exerted by the clamps, ensure that the locating jig point actually indents the surface of the casting by up to 0.25 mm for some aluminium alloy sand castings. Harder materials such as cast irons will, of course, indent less. All surface irregularities are effectively locally smoothed and averaged in this operation. The indentation effect sets an upper limit to the accuracy and repeatability with which castings can be picked up for measurement or machining.

A traditional method of checking the profile of a casting is by the use of template gauges. These are typically sheets of metal that have been cut to the correct contour. On applying them to the casting, the contour on the casting can be seen to be correct or not, depending on the clearance that can be seen between the two. This is an analogue technique that can no longer be recommended in these modern times. The gauges are expensive to make. They are also subject to wear, and thus need to be checked regularly and occasionally replaced. However, what is much more serious, they are difficult to use in any effective way. This is because in practice the contours never match exactly. The problem for the user then is how inaccurate can the contour of the casting be allowed to become before remedial action must be taken?

The use of 'go/no-go' gauges removes the matter of judgement. However, the gauges are again subject to wear, and thus require the cost and complexity of a calibration system. More fundamentally, their use is similarly not helpful in terms of providing useful data to assist process control.

All these difficulties can be removed by the use of a much simpler technique: the use of simple goalpost fixtures that straddle the casting and are equipped with one or more spring-contact probes, such as dial gauges. The readings from the gauges are read and recorded. The operation becomes even simpler with the use of digital read-out devices (Figure 10.7). Linear transducers are easily fitted and operated, and give an immediate numerical signal of the degree of inaccuracy.

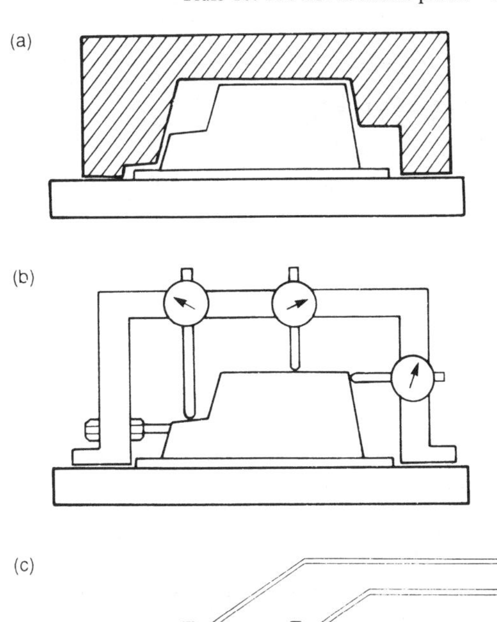

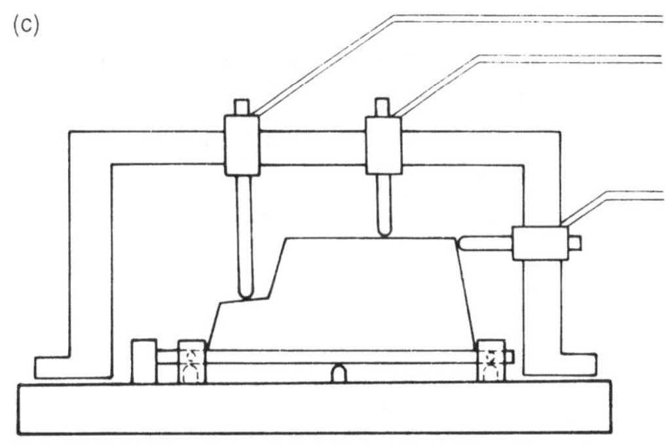

Figure 10.7 *Comparison of checking techniques for the monitoring of the size and shape of castings by: (a) template, with the casting and template sat on a baseplate; (b) an equivalent analogue measurement using spring dial gauges; and (c) digital measurement using linear displacement transducers, with the casting located on a six-point jig. The six-point locations for the goalpost frame on the jig are omitted for clarity.*

The goalpost would be calibrated and stored on a standard casting, and thereby always be seen to be in calibration by being set to zero in this position. (For calibration away from the zero, other readings can be obtained by the insertion of slip gauges under the probe.)

The use of digital electronic read-out in this way allows its incorporation into data-logging and quality-monitoring systems, such as statistical process control. By watching the trends on a daily or weekly basis, the gradual drifts in casting dimensions can be used to predict, for instance, that tooling wear will reach a level that will require the tooling to be replaced in three weeks' time. Such prior warning allows the appropriate action to be planned well in advance.

Appendix

The 1.5 factor

Experimental results for side-gated 99.8Al plate castings plotted in Figure A1 show that casting time t_c may be estimated for the plates and other castings from an equation;

$$t_c = 1.5 \times \text{casting volume/initial casting rate} \quad \textbf{(1)}$$

This is equivalent to

$$\text{initial fill rate/average fill rate} = 1.5 \quad \textbf{(2)}$$

These experimental results gives support to the value of 1.5 chosen by previous authors, particularly those of the British Non-ferrous Research Association (now no longer with us) researching for the UK Admiralty (Ship Department 1975).

Exploring the 1.5 factor further by a theoretical approach is not quite so straightforward, but an attempt is outlined below.

Consider Figure A2, The velocity at the base of the sprue is given by

$$V_2 = (2gH)^{0.5}$$

If the area of the base of the sprue is A_2 and the mould cavity is of uniform area A_C the initial velocity of rise in the mould will be given by

$$V_i = (A_C/A_2) \cdot (2gH)^{0.5}$$

Similarly, at some later instant, when the melt has reached height h, the net head driving the filling is now reduced to $(H - h)$ so that the rate of rise is now

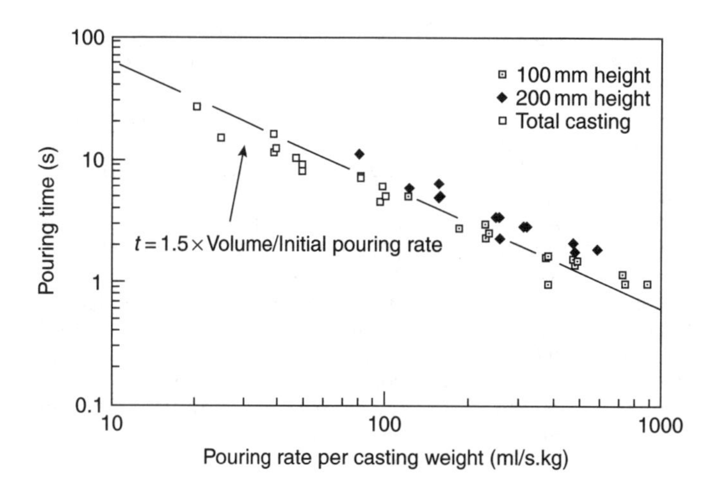

Figure A1 *Experimental demonstration of the relation between initial and average filling rates (Data from Runyoro and Campbell 1992).*

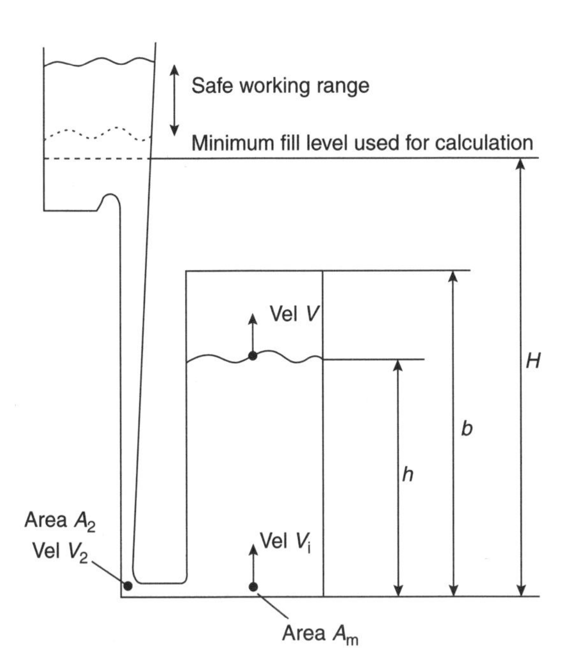

Figure A2 *Schematic view of the filling of a uniform casting.*

$$V = (A_C/A_2) \cdot (2g(H-h))^{0.5}$$

Substituting dh/dt for the rate of rise V, rearranging and integrating between the limits of time $t=0$ at $h=0$, and $t=t_c$ at $h=b$, we find the casting time (the time to fill the mould) t_c is given by

$$t_c = (A_m/A_2) \cdot (2/g)^{0.5} [H^{0.5} - (H-b)^{0.5}]$$

Now writing simple definitions of the initial rate of casting Q_i and the average rate of casting Q_{av} in such units as volume of liquid per second, defined by the appropriate velocity times the area, we have

$$Q_i = A_2 \cdot (2gH)^{0.5}$$

$$Q_{av} = A_m \cdot b/t_c$$

It follows that

$$Q_i/Q_{av} = 2H^{0.5}[H^{0.5} - (H-b)^{0.5}]/b$$

This solution to the filling problem is interesting. There are various combinations of H and b that can fulfil the conditions defined by the equation. For instance, if $H=b$, then $Q_i/Q_{av}=2$, which is actually an obvious result meaning simply that the average is half of the start and finishing rates.

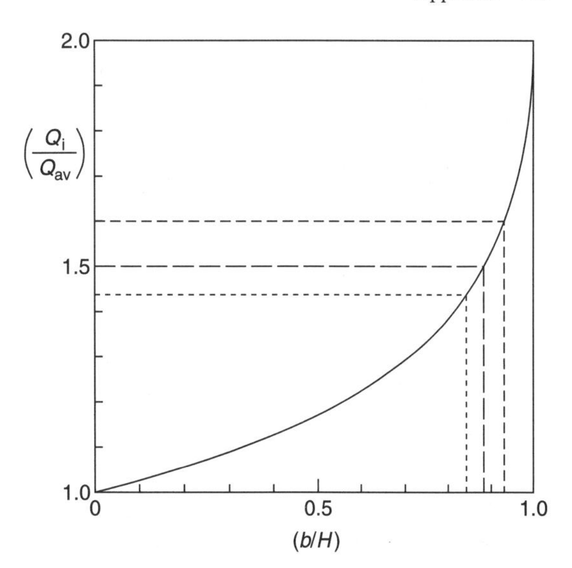

Figure A3 *The relation between the initial and average fill rates for a uniform casting as a function of the relative heights of the casting and the pouring basin.*

On the other hand, $Q_i/Q_{av}=1.5$ only when $b=0.89H$. This represents an intriguing result, indicating that for most castings the top of the pouring basin is on average only about 10 per cent higher than the height of the casting. Thus it seems the factor 1.5 is quite fortuitous, and results simply from the geometry we happen to select for most of the castings we make. If, in general, we were to raise (or lower) our pouring basins in relation to the tops of our castings, the factor would have to be revised.

However, all is not so bad as it seems. Figure A3 shows that the factor 1.5 does not change rapidly with changes in relative height of basin, varying over reasonable changes in basin height of b/H from 85% to 95% from roughly 1.55 to 1.60. These changes are of the same order as errors arising from other factors such as frictional losses, etc. and so can be neglected for most practical purposes.

The Bernoulli equation

Daniel Bernoulli represents the revered name in flow. He published his equation in 1738 in one of the first books on fluid flow. This magnificent result is the one used for all descriptions of flow in pipes and channels. Whole books are devoted to its application.

There are of course, excellent examples of the power of Bernoulli's equation. Sutton (2002) made good use of the equation to describe the

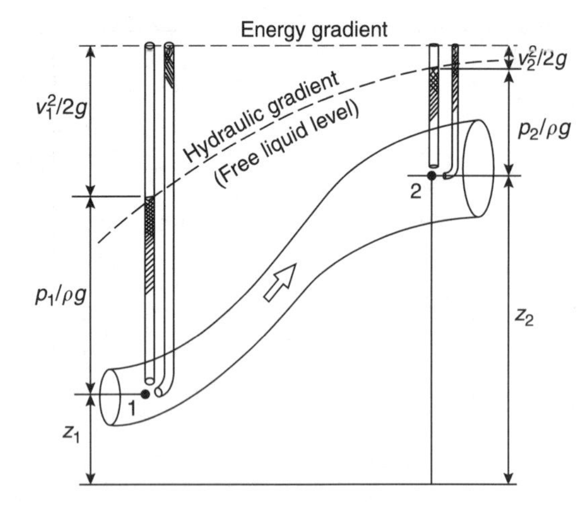

Figure A4 *A pictorial representation of the terms of the Bernoulli equation.*

pressures along a long runner, explaining the early partial filling of gates at different positions along the runner, and thus resulting in part-filled castings. He used the equation in its simplest form, derived from a statement of conservation of energy along a flow tube as illustrated schematically in Figure A4:

$$p_1/\rho g + v_1^2/2g + z_1 = p_2/\rho g + v_2^2/2g + z_2 = \text{constant}$$

where all the component terms of this equation have units of length, conveniently metres. For this reason each term can be regarded as a 'head'. Thus $p/\rho g$ = pressure head, $v^2/2g$ = kinetic or velocity head, and z = potential or elevation head.

In application to the running system used by Sutton (Figure A5) at location 1 the height above the centreline of the runner is 0.5 m, the kinetic head at this point is zero because the melt has zero downwards velocity, and the elevation head z is considered zero because the runner is horizontal. At point 2, the pressure head requires to be known, since this is the pressure raising the melt level in the vertical ingate. The elevation is zero once again, and the velocity head is close to 0.25 m, easily deduced from the total fall height and allowing for a small loss factor of 0.70 (probably overestimated, since I think this should be more like 0.80 or even 0.85) as a result of the turn at the base of the sprue. Thus the Bernoulli equation becomes

$$0.5 + 0 + 0 = p_2/\rho g + 0.25 + 0$$

Thus $\qquad p_2/\rho g = 0.25$ metre

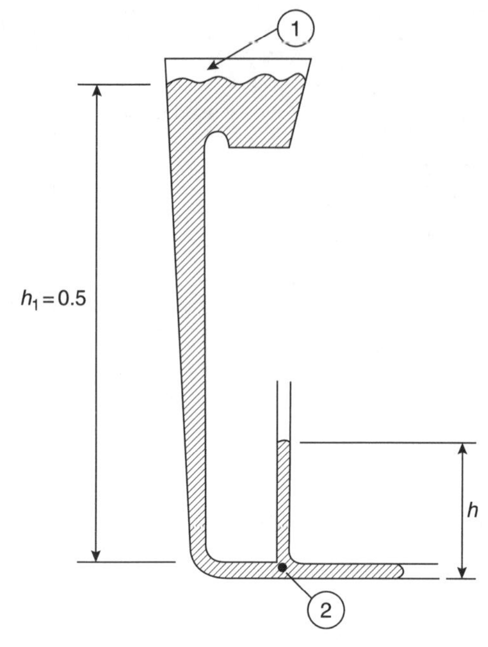

Figure A5 *An example of the use of the Bernoulli equation by Sutton (2002) to calculate the rise of metal in a vertical gate.*

It is not necessary to find p_2 alone; the whole term is the height distance. Thus this answer would be the same for aluminium or iron.

Sutton found that because of this kinetic head, ingates were filling before the runner was fully filled. The first impression in his multi-impression mould was only about 200 mm above the runner so that metal entered the mould cavity under only about 50 mm net head. The result was a premature dribble into the cavity that quickly froze. The arrival of melt at the intended full flow rate a few seconds later was too late to remelt and thus assimilate the frozen droplets. An apparently mis-run casting was the result.

In general however, the application of the Bernoulli equation to filling systems is not quite so straightforward as has sometimes been assumed. There are various reasons for this.

1. In general, Bernoulli's equation relates to steady state flow. However, of course, in filling systems most of the interest necessarily lies in the priming of the flow channels. In this situation the surface tension of the advancing meniscus can be important, as enshrined in the Weber number. If the priming is not carried out well, the casting is likely to suffer severely.

2. The surface tension of liquid metals is over ten times higher than that of water, and even higher still compared to most organic liquids. Thus pressures due to surface tension have been neglected and are neglectable for such common room temperature liquids on which most flow research has been conducted. The additional pressure generated because of the curvature of the meniscus at the flow front, and the curvature at the sides of a flow stream affect the behaviour of metals in many examples involved in the filling of moulds. For instance at the critical velocity that is targeted in mould filling, the effects of surface tension and flow forces are equal. At velocities lower than this, surface tension dominates.
3. The presence of the oxide (or other thin, solid film) on the surface of an advancing liquid is a further complication, and is not easily allowed for. The flow adopts a stick-slip motion as the film breaks and re-forms. The advance of the unzipping wave is a classic instance that could not be predicted by a purely liquid model such as that described by Bernoulli.
4. The frictional losses during flow, that can be explicitly cited in Bernoulli's equation, are known to be important. However, in general, although they are assumed to be known, they have been little researched in the case of the flow of liquid metals. Furthermore, it is unfortunate that most of the research to date in this field has used such poor designs of filling systems that the existing figures are almost certainly misleading. The losses need to be confirmed by new, careful, accurate studies, supplemented by accurate computer simulation together with video X-ray radiography of real flows.
5. The presence of oxide films floating about in suspension is another uncertainty that can cause problems. The density of such defects can easily reach levels at which the effective viscosity of the mixture can be very much increased (although it is to be noted that viscosity does not appear explicitly in the Bernoulli equation). The suppression of convection in such contaminated liquids is common. Flow out of thick sections and into very thin sections can be prevented completely by blockage of the entrance into the thin section.

From the above list it is clear that the application of Bernoulli is more accurate for thicker section flows where surface effects and internal defects in the liquid are less dominant. As filling systems are progressively slimmed, and casting sections are thinned, Bernoulli's equation has to be used with greater caution.

As a result of the problems of the application of Bernoulli to the priming of the filling system, it has been relatively little used in this book because the concentration of effort has focused on the control of the priming of the system. The subsequent flow of the system when completely filled, as nicely described by Bernoulli, is, with the greatest respect to the Great Man and his magnificent equation, much less important.

Rate of pour of steel castings from a bottom-pour ladle (Figures A6, A7, A8)

All three parts of the nomogram have to be used in conjunction to obtain the time of pour of a casting. The reduced scale illustration is merely to show the intended arrangement of the three components of the nomogram. The following is an example of how the nomogram is used.

A ladle contains 5000 kg of steel, from which we wish to pour a casting of total weight 1250 kg. Thus we follow the arrows from these start points to the junction A. From here a horizontal line connects to the next figure, where we select a pouring nozzle for the ladle of 60 mm diameter. At this junction B we drop a vertical line down to intersect with the line denoting that our ladle is about 1.5 m internal diameter. From this junction C we continue with a parallel line to the family of sloping lines, to find that our casting will pour in approximately 23 seconds.

Interestingly, the reader can check that the next 1250 kg casting in line (now starting with a ladle of 5000 − 1250 = 3750 kg will be found to pour in about 29 seconds, and the next in 34 seconds, and the next in 77 seconds, as the ladle progressively empties.

Running system calculation record

A typical work sheet record (Table A1) for the calculation of a filling system. Successive iterations may be required after initial trials because of unforeseen reasons that can be noted and used for the tuning of subsequent filling system designs.

The system below can, of course, be used most effectively in simple spread-sheet format.

Table A1 Running system record

		Casting name		
		Part number		
		Customer		Alloy
		Design 1 signed Date	Design 2 signed Date	Design 3 signed Date
Casting weight $= M_c$	kg			
Rigging system weight $= M_r$	kg			
Total poured weight	kg			
$M_c + M_r = M$				
Select fill time $= t$	s			
Average mass flow rate $= M/t$	$kg\,s^{-1}$			
Liquid metal density $= \rho$	$kg\,m^{-3}$			
Average volume flow rate $= M/t\rho$	$m^3\,s^{-1}$			
Initial volume flow rate $Q = 1.5M/t\rho$	$m^3\,s^{-1}$			
Height of liquid in basin $= h$	m			
Velocity into sprue entrance $V_1 = (2gh)^{1/2}$	$m\,s^{-1}$			
Area of sprue entrance $= Q/V_1$	m^2			
Total casting height $= H$	m			
Velocity at base of sprue $V_2 = (2gH)^{1/2}$	$m\,s^{-1}$			
Area of sprue base $A_2 = Q/V_2$	m^2			
Radius of turn at sprue base $R = (A_2)^{1/2}$	m			
Area of runner $= A_2$ or possibly up to $1.2A_2$	m^2			
Select critical velocity $V_c = 0.5$ to $1.0\ m\,s^{-1}$	$m\,s^{-1}$			
Total area of gates $A_3 = A_2 \cdot V_2/V_c$	m^2			
Min fill depth of basin $= h$	m			
Basin depth $= 2h$ to $4h$ for safety	m			
Basin volume for 1 s response $= Q$	m^3			
Basin sides, if square, for 1 s response $= (Q/h)^{1/2}$	m			

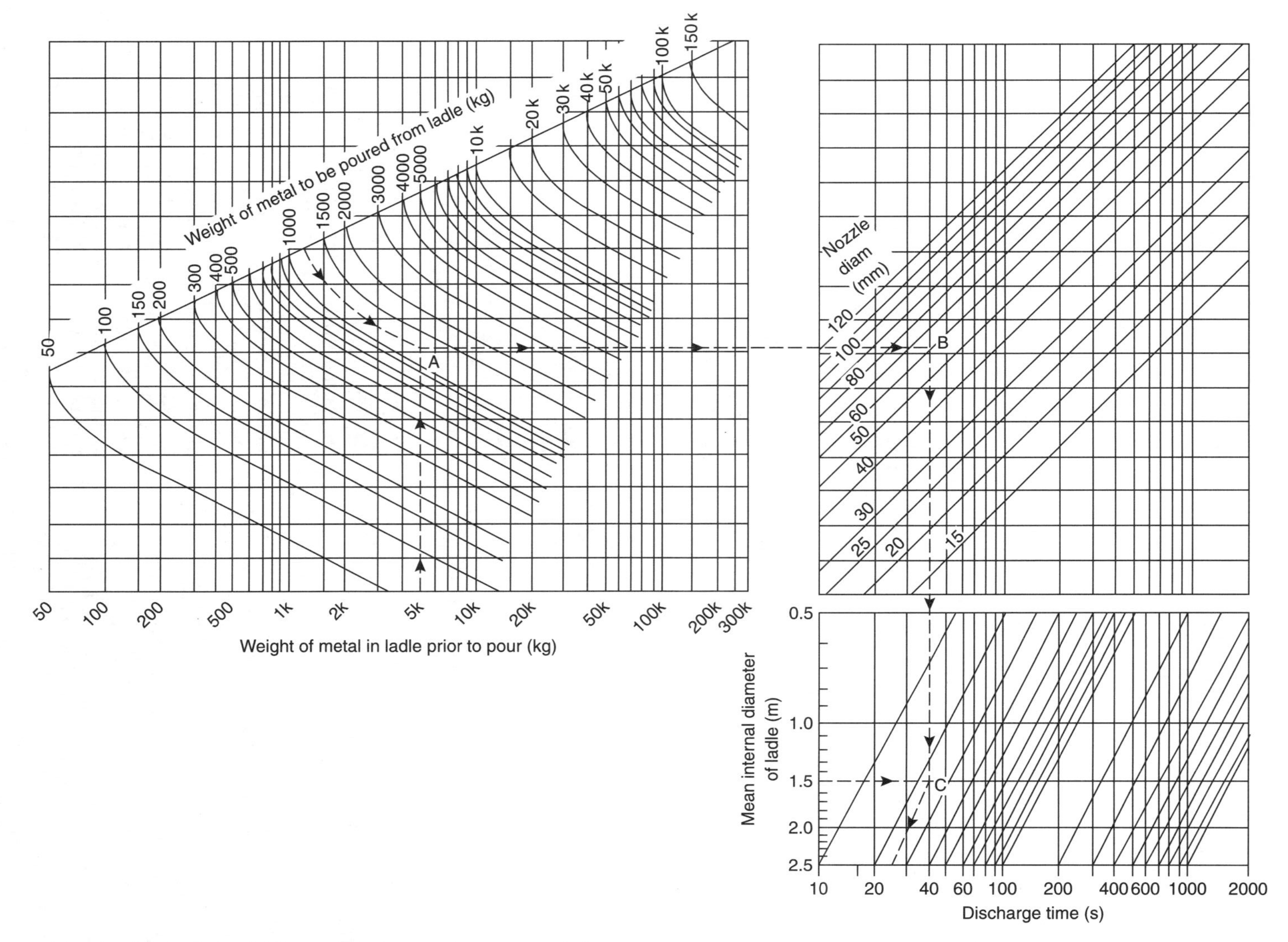

Figures A6, A7 and A8 *Rate of delivery of steel from a bottom-pour ladle.*

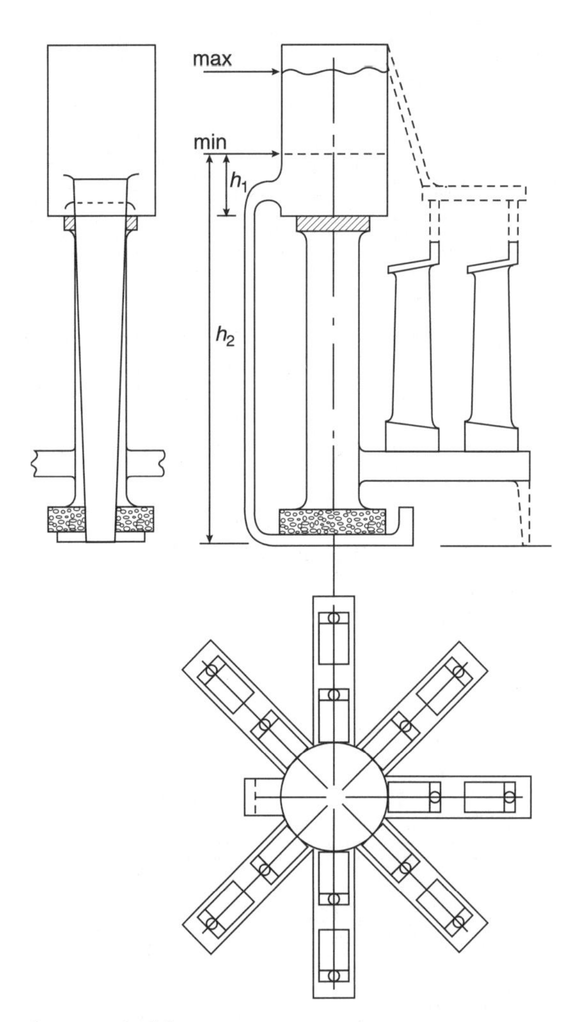

Figure A9 *Investment casting, design methodology.*

Appendix A9

Design methodology for investment castings (Figure A9)

The wax assembly is required to be strong to withstand the shell-making process, so that good bottom-filling systems are not easily applied to investment castings. Here the main support column forms a central feeder, filled upwards because of the ceramic disc at the base of the pouring basin. The basin is sharp edged and offset from the sprue. The sprue is cut from sheet wax, is only a few millimetres thick and so is relatively flimsy, and not expected to confer any strength on the whole assembly. The sprue turns and expands under the filter to maximize coverage. The filter is tangentially placed, with some scope for the sideways displacement of bubbles or other buoyant material into a trap. Small 'feet' at the ends of the radial runners help the assembly to stand upright during processing. Other small extensions from the tops of the castings can form vents for dewaxing and/or escape of entrapped gas in the mould during pouring. The breaking open of the shell into the top of the vents will allow most effective venting. Connection of the vents to the top of the pouring basin can be used to strengthen the assembly if necessary, and plugged with fireclay to prevent the accidental ingress of metal. Most of the Rules are fulfilled by this design. Convection is controlled by the avoidance of circular flow paths and rapid cooling of the casting because of their thin section and their placement around the outside of the assembly where they can radiate away heat effectively.

References

Adams A; 2001 *Modern Castings* **91** (3) 34–36.

Adams C M and Taylor H F; 1954 *TAFS* **61** 686–693.

Ali S, Mutharasan R and Apelian D; 1985 *Met Trans* **16B** 725–742.

Anderson G R; 1985 Patt-Tech 85 Conference, Oxford, UK.

Anderson G R; 1987 *TAFS*, **95** 203–210.

Ashton M C and Buhr R C; 1974 Phys. Met. Div. Internal Report PM-1-74-22 Canada Dept Energy Mines and Resources.

Askeland D R and Holt M L; 1975 *TAFS* **83** 99–106.

Barrett L G; 1967 *TAFS*, **75** 326–329.

Bates C and Wallace J F; 1966 *TAFS*, **74** 174–185.

Beck A, Schmidt W, and Schreiber O; 1928 US Patent 1, 788, 185.

Beeley P R; 1972 *Foundry Technology* Butterworths, London.

Berry J T and Taylor R P; 1999 *TAFS* **107** 203–206.

Bex T; 1991 *Modern Casting* November p 56 only.

Bird P G; 1989 Foseco Technical Service Report MMP1.89 'Examination of the factors controlling the flow rate of aluminium through DYPUR units'. 1 March 1989.

Biswas P K, Rohatgi P K and Dwarakadasa E S; 1985 *Br Foundryman* **78** 511–516.

Bishop H F, Myskowski E T and Pellini W S; 1955 *TAFS* **63** 271–281.

Bossing E; 1982 *TAFS* **90** 33–38.

Bromfield G; 1991 *The Foundryman* July 261–265.

Brown N and Rastall D; 1986 European Patent Application No 87111549.9 filed 10.08.87. pp 9.

Campbell J; 1991 *Castings* Butterworth Heinemann.

Campbell J and Caton P D; 1977 Institute of Metals Conference on Solidification Sheffield, UK, 208–217.

Campbell J and Isawa T; 1994 UK Patent GB 2,284,168 B. (Filed 04–02–1994).

Chvorinov N; 1940 *Giesserei*, **27** 177–186, 201–208, 222–225.

Clegg A L and Das A A; 1987 *Br Foundryman*, **80** 137–144.

Cox B M, Doutre D, Enright P and Provencher R; 1994 *TAFS* **102** 687–692.

Creese R C and Safaraz A; 1987 *TAFS* **95** 689–692.

Creese R C and Safaraz A; 1988 *TAFS* **96** 705–714.

Creese R C and Xia Y; 1991 *TAFS* **99** 717–727.

Cunliffe E; 1994 PhD Thesis. University of Birmingham, UK.

Datta N and Sandford P; 1995 3rd AFS Internat Permanent Mold Casting of Aluminum Conference paper 3 pp 19.

Davis K G; 1977 *AFS Int Cast Metals J*, March, 23–27.

Davis K G and Magny J -G; 1977 *TAFS*, **85** 227–236.

Daybell E; 1953 *Proc Inst Br Foundryman* **46** B46-B54.

Din T and Campbell J; 1994 University of Birmingham UK unpublished work.

Din T, Kendrick R and Campbell J; *TAFS* 2003 **111** paper 03-017.

Durville P H G; 1913 British Patent 23,719.

Eastwood L W; 1951 AFS Symposium on Principles of Gating 25–30.

Emamy M, Taghiabadi R, Mahmudi M and Campbell J; 2002 Statistical study of tensile properties of A356 aluminum alloy, using a new casting design. ASM 2nd Int Al Casting Technology Symposium 7–10 October 2002 Columbus Ohio.

Evans J, Runyoro J and Campbell J; 1997 SP97 4th Decennial International Conference on Solidification Processing, Sheffield, 74–78.

Flemings M C; 1974 *Solidification Processing* McGraw Hill.

Forslund S H C; 1954 21st Int Foundry Congress, Florence, paper 15.

Gebelin J -C and Jolly M R; 2003 Modeling of Casting, Welding and Solidification Processing Conference X, Destin USA.

Gebelin J -C Jolly M R and Jones S; *INCAST* **13** (11) (Dec) 22–27.

Geskin F S, Ling E and Weinstein M I; 1986. *TAFS* **94** 155–158.

Gouwens P R; 1967 *TAFS* **75** 401–407.

Grote R E; 1982 *TAFS* **90** 93–102.

Groteke D; 2002 *Internat J Cast Metals Research* **14**(6) 341–354.
Halvaee A and Campbell J; 1997 *Trans AFS* **105** 35–46.
Hansen P N and Rasmussen N W; 1994 BCIRA International Conference, University of Warwick, UK.
Hansen P N and Sahm P R; 1988 Modelling of casting, welding and advanced solidification processes IV (Giamei A F; Abbaschian G J, editors) The Mineral, Metals and Materials Society.
Hayes K D, Barlow J O, Stefanescu D M and Piwonka T S; 1998 *TAFS* **106** 769–776.
Hedjazi D, Bennett G H J and Kondic V; 1975. *Br Foundryman* **68** 305–309.
Heine R W; 1982 *TAFS* **90** 147–158.
Heine R W and Uicker J J; 1983 *TAFS* **91** 127–136.
Hess K; 1974 *AFS Cast Metals Research J* March 6–14.
Heyn E; 1914 *J Inst Metals* **12** 3.
Hiratsuka S, Niyama E, Anzai K, Horie H and Kowata T; 1966 4th Asian Foundry Congress 525–531.
Hirt C W; 2003 www.flow3d.com
Hodjat Y and Mobley C E; 1984 *TAFS* **92** 319–321.
Huang H, Lodhia A V and Berry J T; 1990 *TAFS* **98** 547–552.
IBF Technical Subcommittee T571; 1969 *Br Foundryman* **62** 179–196.
IBF Technical Subcommittee TS71; 1971 *Br Foundryman* **64** 364–379.
IBF Technical Subcommittee T571; 1976 *Br Foundryman* **69** 53–60.
IBF Technical Subcommittee T571; 1979 *Br Foundryman* **72** 46–52.
Isawa T; 1993 PhD Thesis University of Birmingham Department of Metallurgy and Materials Science 'The control of the initial fall of liquid metal in gravity filled casting systems'.
Isawa T and Campbell J; 1994 Trans *Japan Foundrymen's Soc* **13** (Nov) 38–49.
ISO Standard 8062; 1984 *Castings—System of Dimensional Tolerances.*
Itamura M, Murakami K, Harada T, Tanaka M and Yamamoto N; 2002 *Int J Cast Met Res* **15** 167–172.
Itamura M, Yamamoto N, Niyama E and Anzai K; 1995 Proc 3rd Asian Foundry Congress, eds Lee Z H, Hong C P, Kim M H. 371–378.
Jacob S and Drouzy M; 1974 Int Foundry Congress 41 Liege, Belgium. Paper 6.
Jeancolas M and Devaux H; 1969 *Fonderie* **285** 487–499.
Jeancolas M, Devaux H and Graham G; 1971 *Br Foundryman* **64** 141–154.
Jirsa J; 1982 *Foundry Trade J* Oct 7, 520–527.
Johnson S B and Loper C R; 1969 *TAFS* **77** 360–367.
Johnson W H and Baker W O; 1948 *TAFS* **56** 389–397.
Jolly M J, Lo H S H, Turan M, Campbell J and Yang X; 2000 Modeling of Casting, Welding and Advanced Solidification Processing IX Aachen, editors Sahm P R, Hansen P N, Conley J G.
Kahn P R, Su W M, Kim H S, Kang J W and Wallace J F; 1987 *TAFS* **95** 105–116.
Karsay S I; *Ductile Iron Production 1; state of the art 1992*. Published by QIT Fer et Titane Inc 1992 and '*Ductile Iron; the essentials of gating and risering system design*' revised 2000 published by Rio Tinto Iron & Titanium Inc.
Kasch F E and Mikelonis P L; 1969 *TAFS* **77** 77–89.
Kim M H, Moon J T, Kang C S and Loper C R; 1993 *TAFS* **101** 991–998.
Kim M H, Loper C R and Kang C S; 1985 *TAFS* **93** 463–474.
Kim S B and Hong C P; 1995 Modeling of casting, welding and advanced solidification processes VII TMS 155–162.
Kono R and Miura T; 1975 *Br Foundryman* **69** 70–78.
Kotschi T P and Kleist O F; 1979 *AFS Int Cast Metals J* **4** (3) 29–38.
Kotschi R M and Loper C R; 1974 *TAFS* **82** 535–542.
Kotzin E L; 1981 *Metalcasting and Molding Processes.* American Foundrymen's Soc., Des Plaines, Illinois, USA.
Kubo K and Pehlke R D; 1985 *Met Trans* **16B** 359–366.
Kunes J, Chaloupka L, Trkovsky V, Schneller J and Zuzanak A; 1990 *TAFS* **98** 559–563.
Kuyucak S; 2002 *TAFS* 2002 **110**.
Lai N-W and Griffiths W D; 2003 University of Birmingham, UK. To be published.
Laid E; 1978 US Patent Application 16 Feb 1978 Number 878,309.
Latimer K G and Read P J; 1976 *Br Foundryman* **69** 44–52.
Lawrence M; 1990 *Modern Castings* February 51–53.
Lerner Y and Aubrey L S; 2000 *TAFS* **108** 219–226.
Lewis R W and Ransing R S; 1998 *Met and Mat Trans B* **29B**(2) 437–448.
Lewis R W, Ransing M R and Ransing R S; 2002 *Internat J Cast Metals Research* **15**(1) 41–53.
Lin H J and Hwang W-S; 1988 *TAFS* **96** 447–458.
Lo H and Campbell J; 2000 *Modeling of Casting, Welding and Advanced Solidification Processes IX* Eds Sahm P R, Hansen P N, Conley J G. pp 373–380.
Loper C R; 1981 *TAFS* **89** 405–408.
Loper C R, Javaid A and Hitchings J R; 1996 *TAFS* **104** 57–65.
Maeda Y, Nomura H, Otsuka Y, Tomishige H and Mori Y; 2002 *Int J Cast Met Res* **15** 441–444.
Mampaey F and Xu Z A; 1999 *TAFS* **107** 529–536.
Manzari M T, Lewis R W and Gethin D T; 2000 'Optimum design of chills in sand casting process' *Proc IMECE 2000 Int Mech Eng Congr* Florida, USA DETC98/DAC-1234 ASME pp 8.
McDavid R M and Dantzig J; 1998 *Met Mater Trans* **29B** 679–690.
McKim P E and Livingstone K E; 1977 *TAFS* **85** 491–498.
Mertz J M and Heine R W; 1973 *TAFS* **81** 493–495.
Mi J, Harding R A and Campbell J; *Int J Cast Met Res* 2002 **14**(6) 325–334.
Midea A C; 2001 *TAFS* **109** 41–50 and *Foundryman* 2003 (March) 60–63.
Mikkola P H and Heine R W; 1970 *TAFS* **78** 265–268.
Miles G W; 1956 *Proc Inst Br Foundryman* 49 A201–210.
Minakawa S, Samarasekera I V and Weinberg F; 1985 *Met Trans* **16**B 823–829.
Mollard F R and Davidson N; 1978 *TAFS* **86** 479–486.

Momchilov E; 1993 *J Mater Sci Technol*, Institute for Metal Science, Bulgarian Academy of Sciences, Sofia. **1**(1) 5–12.
Morthland T E, Byrne P E, Tortorelli D A and Dantzig J A; 1995 *Metall Mater Trans B* **26B** 871–885.
Mutharasan R, Apelian D and Romanowski C; 1981 *J Metals* **83**(12) 12–18.
Mutharasan R, Apelian D and Ali S; 1985 *Met Trans* **16B** 725–742.
Nguyen T and Carrig J F; 1986 *TAFS* **94** 519–528.
Nikolai M F; 1996 *TAFS* **104** 1017–1029.
Noguchi T, Kano J, Noguchi K, Horikawa N and Nakamura T; 2001 *Int J Cast Met Res* **13** 363–371.
Nyamekye K, An Y-K, Bain R, Cunningham M, Askeland D and Ramsay C; 1994 *TAFS* **102** 127–131.
O'Hara P; 1990 *Engineering*, February, 41–42.
Osborn D A; 1979 *Br Foundryman* **72** 157–161.
Pellini W S; 1953 *TAFS* **61** 61–80 and 302–308.
Pillai R M, Mallya V D and Panchanathan V; 1976 *TAFS* **84** 615–620.
Prodham A, Carpenter M and Campbell J; 1999 CIATF Technical Forum.
Rabinovich A; 1969 *AFS Cast Met Res J* March 19–24.
Rao G V K and Panchanathan V; 1973 *Cast Met Res J* **19**(3) 135–138.
Rao T S V and Roshan H Md.; 1988 *TAFS* **96** 37–46.
Rao G V K, Srinivasan M N and Seshadri M R; 1975 *TAFS* **83** 525–530.
Rashid A K M B and Campbell J; 2004 Oxide defects in a vacuum-investment-cast Ni-base turbine blade. Accepted for publication in *Met and Mat Trans* 2004.
Rasmussen N W, Aagaard R and Hansen P N; 2002 *Internat J Cast Metals Research* **14**(6) 325–383.
Reddy D C, Murty S S N and Chakravorty P N; 1988 *TAFS* **96** 839–844.
Rezvani M, Yang X and Campbell J; 1999 *TAFS* **107** 181–188.
Richins D S and Wetmore W O; 1951 AFS Symposium on Principles of Gating 1–24.
Roedter H; 1986 *Foundry Trade J Int* Sept 1986 6.
Rogers K P and Heathcock C J; 1990 US Patent 5,316,070 date 31 May 1994.
Romero J M, Smith R W and Sahoo M; 1991 *TAFS* **99**, 465–468.
Ruddle R W; 1956 'The Running and Gating of Sand Castings' Monograph and Report Series No 19, Institute of Metals, London.
Ruddle R W and Cibula A; 1956–57 *J Inst Metals* **85** 265–292.
Runyoro J; 1992 PhD Thesis 'Design of the Gating System', University of Birmingham.
Runyoro J, Boutorabi S M A and Campbell J; 1992 *TAFS* **100** 225–234.
Runyoro J, Boutorabi S M A and Campbell J 1992 *Trans AFS* **100** 225–234.
Runyoro J and Campbell J; 1992 *The Foundryman* **85**(April) 117–124.
Safaraz A R and Creese R C; 1989 *TAFS* **97** 863–870.
Sandford P; 1988 *The Foundryman* (March) 110–118.
Sandford P; 1993 *TAFS* **101** 817–824.
Sciama G; 1974 *TAFS* **82** 39–44.
Sciama G; 1975 *TAFS* **83** 127–140.
Sciama G; 1993 *TAFS* **101** 643–651.
Schilling H; 1987 Patent PCT WO 87/07543.
Schmidt D G and Jacobson A E; 1970 *TAFS* **78** 332–337.
Sexton A H and Primrose J S G; 1911 *The Principles of Ironfounding* The Technical Publishing Co., London.
Ship Department Publication 18; 1975 *Design & Manufacture of Nickel-Aluminium-Bronze Sand Castings* Ministry of Defence (Procurement Executive) Foxhill, Bath, UK.
Sirrell B and Campbell J; 1997 *TAFS* **105** 645–654.
Skarbinski M; 1971 *Br Foundryman* **44** 126–140.
Southam D L; 1987 *Foundry Manage Technol* **7** 34–38.
Stahl G W; 1961 *TAFS* **69** 476–469.
Stahl G W; 1963 *TAFS* **71** 216–220.
Stahl G W; 1986 *TAFS* **94** 793–796.
Stahl G W; 1989 'The gravity tilt pour process'; *Proc AFS Int Conf: Permanent Mold Castings*, Miami. Paper 2.
Steel Founders Society of America (Anon) 2000 *Foundry Trade J* **May** 40–41.
Stefanescu D M, Giese S R, Piwonka T S, Lane A M, Barlow J and Pattabhi R; 1996 *TAFS* **104** 1233–1264.
Sullivan E J, Adams C M and Taylor H F; 1957 *TAFS* **65** 394–401.
Sun L and Campbell J; 2003 *TAFS* paper 03-018 in press.
Sutton T; *The Foundryman* 2002 **95** (June) (6) 223–231.
Suzuki S; 1989 *Modern Castings*, October, 38–40.
Svensson I L and Dioszegi A; 2000 *Modelling of Casting, Welding and Advanced Solidification Processes IX* Eds Sahm P, Hansen P and Conley 102–109.
Svensson I and Villner L; 1974 *Br Foundryman* **67** 277–287.
Swift R E, Jackson J H and Eastwood L W; 1949 TAFS **57** 76–88.
Swing E; 1962 *TAFS* **70** 364–373.
Taylor K C and Baier A; 2003 *Casting Plant + Technol Int* **123** (2) 36–46.
Taghiabadi R, Mahmoudi M, Emamy Ghomy M and Campbell J; 2003 *Materials Science and Technology* **19** 497–502.
Tiryakioğlu E; A Study of the Dimensioning of Feeders for Sand Castings, PhD Thesis, University of Birmingham, UK, 1964.
Tiryakioğlu M, Tiryakioğlu E and Askeland D A; 1997a *Int J Cast Met Res* **9** 259–267.
Tiryakioğlu M, Tiryakioğlu E and Askeland D A; 1997b *TAFS* 907–915.
Tiryakioglu M, Tiryakioglu E and Campbell J; 2002 *Internat J Cast Metals Research* **14**(6) 371–375.
Trojan P K, Guichelaar P J and Flinn R A; 1966 *TAFS* **74** 462–469.
Turner G L; 1965 *Br Foundryman* **58** 504–505.
Turner A and Owen F; 1964. *Br Foundryman*, **57** 55–61, and 355–356.
Villner L; 1969 *Br Foundryman* **62** 458–468. Also published in *Giesserei* 1970 **57**(27), 837–844 and *Cast Met Res J* 1970 **6** (3) 137–142.
Wall A J and Cocks D L; 1980 *Br Foundryman* **73** 292–300.
Ward C W and Jacobs I C; 1962 *TAFS* **70** 332–337.
Webster P D; 1967 *Br Foundryman* **60** 314–319.

Wen S W, Jolly M R and Campbell J; 1997 Proc 4th Decennial Int Conf on Solidification Processing, Sheffield. Eds Beech J and Jones H. 66–69.

Weins M J, de Botton J L S and Flinn R A; 1964 *TAFS* **72** 832–839.

Wieser P F and Dutta I; 1986 *TAFS* **94** 85–92.

Wright T C and Campbell J; 1997 *TAFS* **105** 639–644 and *Modern Castings* June 1997 (see also T Dimmick 2001 *Modern Castings* March 91 (3) 31–33).

Xu Z A and Mampaey F; *TAFS* **105** 853–860.

Yang X and Campbell J; 1998 *Int J Cast Met Res* **10** 239–253.

Yang X, Dai X, Campbell J and Wood J; 2003 *Materials Science and Engineering* **A354**(1–2) 315–325.

Yang X, Din T and Campbell J; 1998 *Int J Cast Met Res* **11** 1–12.

Yang X, Jolly M R and Campbell J; 2000 *Aluminum Trans* **2**(1) 67–80 and *Modelling of Casting, Welding and Advanced Solidification Processing IX* 2000 Eds Sahm P, Hansen P and Conley 420–427.

Zadeh A H and Campbell J; 2002 *TAFS* paper 02–020.

Zhang C, Mucciardi F, Gruzleski J, Burke P and Hart M; 2003 AFS Trans paper 010-03.

Index